Positive
Youth
Development
&
Spirituality

To Mrinalini,
A bright spiritual light
in the firmament of my life

Warmly,

Positive
Youth
Development
&
Spirituality

From Theory to Research

Edited by Richard M. Lerner,
Robert W. Roeser, & Erin Phelps

TEMPLETON FOUNDATION PRESS
West Conshohocken, Pennsylvania

Templeton Foundation Press
300 Conshohocken State Road, Suite 670
West Conshohocken, PA 19428
www.templetonpress.org

*Templeton Foundation Press helps intellectual leaders and others learn about
science research on aspects of realities, invisible and intangible. Spiritual realities include unlimited love, accelerating creativity, worship, and the benefits of
purpose in persons and in the cosmos.*

Designed and typeset by Kachergis Book Design

LIBRARY OF CONGRESS CATALOGING-IN-PUBLICATION DATA
Positive youth development and spirituality : from theory to research /
edited by Richard M. Lerner, Robert W. Roeser, and Erin Phelps.
     p. cm.
Includes bibliographical references and index.
ISBN-13: 978-1-59947-143-3 (alk. paper)
ISBN-10: 1-59947-143-4 (alk. paper)
1. Youth—Religious life.   2. Youth development.   I. Lerner, Richard M.
II. Roeser, Robert W.   III. Phelps, Erin.
BL625.47.P67 2008
204.0835—dc22
                              2008012692

Printed in the United States of America

08   09   10   11   12   13   10   9   8   7   6   5   4   3   2   1

Chapter 2: Figure 1 is reprinted with permission from R. Nozick, *The
Examined Life* (New York: Simon and Schuster, 1989).
    Chapter 7: Figure 1 is reprinted with permission from M. H. Grosbras,
K. Osswald, M. Jansen et al., "Neural Mechanisms of Resistance to Peer
Influence in Early Adolescence," *Journal of Neuroscience* 27 (2007): 8040–96.
    Figure 2 is reprinted with permission from T. Paus, R. Toro, G. Leonard
et al., "Morphological Properties of the Action-observation Cortical
Network in Adolescents with Low and High Resistance to Peer Influence,"
Social Neuroscience DOI: 10.1080/17470910701563558; First published Sep-
tember 20, 2007.
    Chapter 8: Figure 1 is reprinted with permission from D. H. Feldman,
Beyond Universals in Cognitive Development, 2nd ed. (Norwood, NJ:
ABLEX, 1994).

Contents

Foreword vii
Peter L. Benson

Preface xi

1. Positive Development, Spirituality, and Generosity in Youth:
   An Introduction to the Issues 3
   Richard M. Lerner, Robert W. Roeser, & Erin Phelps

**PART 1 Conceptual Issues in Operationalizing Spirituality
in Positive Youth Development**

2. The Spirit of Spiritual Development 25
   Carl N. Johnson

3. Spirituality and Positive Youth Development: The Problem
   of Transcendence 42
   W. George Scarlett

4. Spirituality as Fertile Ground for Positive Youth
   Development 55
   Pamela Ebstyne King

5. Self and Identity Processes in Spirituality and
   Positive Youth Development 74
   *Robert W. Roeser, Sonia S. Issac, Mona Abo-Zena, Aerika Brittian,
   & Stephen C. Peck*

**PART 2 Biological Contributions to the Spirituality–
PYD Relation**

6. Genetics of Faith: An Oxymoron Worth Examining 109
   Elena L. Grigorenko

7. Cooperative Behavior in Adolescence: Economic Antecedents
   and Neural Underpinnings 128
   Tomáš Paus, Simon Gächter, Chris Starmer, & Richard Wilkinson

8. How Religious/Spiritual Practices Contribute to Well-Being:
The Role of Emotion Regulation   145
Heather L. Urry & Alan P. Poey

PART 3: Individual Contributions to the Spirituality–PYD Relation

9. The Role of Developmental Change in Spiritual Development   167
David Henry Feldman

10. Spirituality, "Expanding Circle Morality," and Positive Youth
Development   197
Janice L. Templeton & Jacquelynne S. Eccles

11. The Role of Spirituality and Religious Faith in Supporting Purpose
in Adolescence   210
Jennifer Menon Mariano and William Damon

12. From "Worm Food" to "Infinite Bliss": Emerging Adults' Views of
Life after Death   231
Jeffrey Jensen Arnett

PART 4: Social and Cultural Contexts of the Spirituality–PYD Relation

13. Immigrant Civic Engagement and Religion: The Paradoxical Roles
of Religious Motives and Organizations   247
Lene Arnett Jensen

14. Ethnic Identity and Spirituality   262
Linda Juang & Moin Syed

15. Considering Context, Culture, and Development in the Relationship
between Spirituality and Positive Youth Development   285
Na'ilah Suad Nasir

16. Application of the Ecological Model: Spirituality Research with
Ethnically Diverse Youths   305
Guerda Nicolas & Angela M. DeSilva

17. Possible Interrelationships between Civic Engagement, Positive Youth
Development, and Spirituality/Religiosity   322
Lonnie R. Sherrod & Gabriel S. Spiewak

18. A Palace in Time: Supporting Children's Spiritual Development
through New Technologies   339
Marina Umaschi Bers

List of Contributors   359

Index   367

Foreword

Some of the most exquisite phenomena of human life are hardly visible in the behavioral sciences. Take, for example, these issues:

How persons explore the mystery of the self and the mystery of the universe;

How the many facets of development—emotional, social, cognitive, moral—are integrated into whole cloth;

How and why we create a narrative about who we are in the context of space and time. As Robert Coles has so elegantly put it, we are the creatures who, to keep from feeling alone, weave a story "to gain for ourselves a sense of where we came from and where we are and where we're going" (1990, 8);

How art, poetry, and even religion emerge out of human imagination;

The experiences of awe and wonder and joy;

The inclination to covet community, interdependence, and connectedness;

The marshalling of personal energy and the investment of it in growth and thriving; and

The process of embedding the self in something larger than the self.

So what is this territory of human life? Certainly, there are cognitive, emotional, social, and biological factors at play. But there is something more, something that requires new thinking, new theory, and new research. We've got terms for this territory. Among them are *spirit*, *spirituality*, and *spiritual development*. The terms are alive and well in self-help sections of bookstores, in the mass media, and in daily conversation. Fields of practice, including medicine, nursing, and social work, are building a viable body of literature. But the slowest and most reticent participant in this exploration is the mainstream academy.

My field, psychology, has been particularly negligent. Recently, a team of us at the Search Institute tried to quantify this oversight. We searched the major social science databases to determine the extent to which spirituality and spiritual development in childhood and adolescence were being addressed by the research community. We found that less than 1 percent of the published articles (from 1990 to 2002) addressed these topics (Benson, Roehlkepartain, and Rude 2003). The oversight was even more graphic when we looked at references to spiritual development in six leading human development journals (e.g., *Child Development* and *Developmental Psychology*). Across a thirteen-year period, exactly zero articles discussed spiritual development as compared to many hundreds, of course, for cognitive, psychosocial, and moral development.

Even the field of positive youth development has kept its distance from spirit. There is literature that speaks to religion and religious community as protective factors and/or as developmental assets that prevent risk behaviors and promote forms of thriving such as generosity and service. Researchers and practitioners in the youth development field tend to know this, and perhaps even tend to affirm it, but lack vocabularies and tools to explore the underlying terrain of the spiritual life. There is some movement forward, however. This volume, happily and significantly, is one of the first efforts to explore the youth development and spirituality intersection. Richard Lerner, in particular, has been an early and persuasive voice in making this link.

A variety of explanations have been given for the persistent marginalization of spirit. The prime one is the academy's bias toward religion. Each of the human phenomena listed earlier is guilty-by-association. Religion is the social institution developed to nourish, manage, or control how persons explore spirit. To be blind to the phenomena of spirit because of their frequent cohabitation with religion is naïve and grossly unscientific. Indeed, a significant and relatively unexplored research question is how religious doctrine, ritual, and participation actually inform the process of spiritual development. But spiritual development happens with or without religion. It's as central to and universal in human life as any of the other streams of development (e.g., cognitive, social, moral).

We still face a key definitional challenge. And that is to define spiritual development in a way that breaks the link to religion without denigrating the latter. Here's one way of thinking about it. (It is offered and offered gently.) Spirit is an intrinsic, animating force that gives energy and momentum to human life. It also propels us to look inward to create and re-create a link between "my life" and "all life." Spiritual development, then, is a constant, active, and ongoing process to create and re-create harmony between the "discoveries" about the self and the "discoveries" about the nature of life-writ-large. The two journeys (inner and outer) constantly inform each other and are always brought back into balance. Within this understanding are three major dynamics in spiritual development:

Becoming aware of or awakened to the essence of the self and the essence of life-writ-large;

Seeking, accepting, and/or experiencing an interconnection between one's inward and outward journeys in a way that brings harmony or coherence;

Living life in accord with one's essence and one's understanding of or connection to life-writ-large (Roehlkepartain and Benson, in press).

Though spirit has been on the back burner of scholarship, a new story is being written. We may only be in Chapter 1, but that's progress. The idea of spirit is emerging in a variety of disciplines, particularly in those that are looking for new principles to understand movement, growth, harmony, unity, and interdependence. Physics, cosmology, and biology are beginning to play. And spirit is slowly working its way into the fields of youth development, positive psychology, and human development, and in particular, at those conceptual spaces where attention is being given to the dynamics of human thriving and flourishing.

In the developmental sciences, these are important signs of an awakening. Spiritual development sessions have recently been on the programs of the Society for Research in Child Development and the Society for Research in Adolescence. Sage has published, just in the last two years, both *The Handbook of Spiritual Development in Childhood and Adolescence* and *The Encyclopedia of Spiritual and Religious Development*. For the first time in its long and illustrious history, the 2006 version of the multivolume *Handbook of Child Psychology* included a review chapter on spiritual development. And, thankfully, the John Templeton Foundation is supporting new work at Tufts, Stanford, Bowling Green State University, and the Search Institute.

It is as though the soul of William James, the great pioneer of psychology, has been reborn and dwells among us. To paraphrase him, "the love of life, at any and every level of development," is the *spiritual impulse*. *Love of life* is a marvelous lens for exploring many of the vital questions of our time. It could be the wellspring for altruism, social justice, and stewardship of the earth. And when it is individually or collectively manipulated, it can be the trigger for our darkest side (e.g., genocide, terrorism, slavery).

The love of life. Spirit. It is our thinnest heritage of theory and research. And it is the line of inquiry—if we are bold enough to clarify it—that is needed to heal a broken planet.

What we need now are pioneers willing to break new ground and build the intellectual scaffolding for a science of spirit. *Positive Youth Development and Spirituality* fits the bill—and then some. This is a field building work of majestic proportions. In its multidisciplinary scope and quality, this volume advances our understanding of the ideas that should link many disciplines in a shared journey of discovery. And it powerfully advances our understanding of what it means to be fully human.

It seems like a good time to offer a benediction: may *Positive Youth Development and Spirituality* trigger a tsunami of new theory and research. It is time.

Peter L. Benson
Minneapolis, MN
January 2008

REFERENCES

Coles, Robert. 1990. *The spiritual life of children*. Boston: Houghton Mifflin.

Benson, P. L., E. C. Roehlkepartain, and S. P. Rude. 2003. Spiritual development in childhood and adolescence: Toward a field of inquiry. *Applied Developmental Science* 7, no. 3: 205–13.

Roehlkepartain, E. C., and P. L. Benson. in press. Children, religion and spiritual development: Reframing a research agenda. *Handbook of Child Research*.

Preface

This book initiates a new phase in the study of religious and spiritual development during adolescence. First, the chapters in this book are derived from an interdisciplinary network of biologists and social and behavioral scientists. Together, these colleagues provide a holistic, systems perspective about positive youth development and spirituality. This book reflects, then, an exciting and important time in the history of the scientific study of spirituality or religiosity. It is a marker of the fact that a growing number of scholars from a variety of scientific disciplines are seeking to explore the role of religion and spirituality in adolescent development.

Second, the focus of this book is on positive development during adolescence—on what goes right during the teenage years—and how spirituality and spiritual development can contribute to the positive development of diverse adolescents. Although the study of positive youth development had been a burgeoning area of scientific study for approximately fifteen years, only a relatively few scholars to date have explored the relation of youth spirituality and religiosity to positive development, despite its salience in the lives of millions of youth (e.g., Roehlkepartain et al. 2006).

Our intent is that the present book will frame—theoretically and methodologically—how the connection between spiritual development and positive youth development may be explored at the levels of biology, the individual, social settings, and culture. The chapters in this book provide heretofore unavailable discussions of conceptual/theoretical, definitional, and methodological issues that need to be addressed in understanding the connections between spirituality and positive youth development and the conditions of the individual and context enabling spirituality to promote or reduce positive youth development or to result in positive outcomes such as civic engagement and generosity (contribution).

The book advances the scientific study of spiritual development by detailing the key methodological challenges and choices that must be addressed to elucidate the spirituality–positive youth development relation. The contributors propose new ways of using qualitative data, physiological and brain imaging data, and a variety of quantitative techniques to address this important aspect of human development. Furthermore, across the chapters, recommendations are made about the ways in which research about the spirituality–positive youth development connection may be integral in building the larger field of spiritual development as a legitimate and active domain of developmental science.

The chapters in this book frame both this emerging field and serve as guideposts in a larger intellectual agenda for collaborative pilot research currently being conducted by the contributors to this book (a research project that is funded by the John Templeton Foundation). As such, the book serves as the intellectual forerunner of a series of new research findings, information that we hope will elicit the attention, interest, and additional research of scientists around the world. At the least, we know that our readers will be stimulated by the exciting and innovative theory and research they will find in the chapters of this book.

We would like to express our gratitude for the creativity and knowledge of the scientists who have contributed to this book. We thank all the colleagues whose have worked so hard to craft such useful and engaging chapters. It is their expertise that has made this book possible. We are grateful as well that Dr. Peter Benson, president of Search Institute, wrote such a thoughtful and supportive foreword to the book.

We appreciate also the important contributions to this book that have been made by Jennifer Davison, managing editor at the Institute for Applied Research in Youth Development, and Lauren White, assistant editor at the institute. Their expertise and impressive productivity in guiding the development of this work through all phases of the manuscript development and production process were invaluable to us.

We are grateful also for the support of and the commitment to quality scholarship of our publisher, Templeton Foundation Press. Its enthusiasm for, and expertise in, publishing high-quality work elucidating the links between human developmental science and spirituality have been vital in pursuing and completing our work.

We also are deeply appreciative of the support provided to us by the John Templeton Foundation. The collaborations among the scholars contributing to this volume, and the science that has been produced, could not have happened without the vision and support of the Foundation.

Finally, our work on *Positive Youth Development and Spirituality: From Theory to Research* has been framed by the intellectual leadership of Sir John M. Templeton. In writings such as *The Humble Approach: Scientists Discover God* (1995) and *Possibilities for Over One-Hundredfold More Spiritual Information* (2000), Sir John stressed the importance of generating new spiritual knowledge to help direct the lives of youth along positive life paths. Inspired by his vision, we undertook the scholarly work represented in this volume. As such, we dedicate this book to him.

R. M. Lerner, R. W. Roeser, & E. Phelps
Medford, MA
December 2007

REFERENCES

Roehlkepartain, E. C., P. E. King, L. M. Wagener, and P. L. Benson, eds. 2006. *The handbook of spiritual development in childhood and adolescence.* Thousand Oaks, CA: Sage Publications.

Templeton, J. M. 1995. *The humble approach: Scientists discover God.* Philadelphia: Templeton Foundation Press.

———. 2000. *Possibilities for over one-hundredfold more spiritual information.* West Conshohocken, PA: Templeton Foundation Press.

Positive
Youth
Development
&
Spirituality

1

Positive Development, Spirituality, and Generosity in Youth

An Introduction to the Issues

Richard M. Lerner, Robert W. Roeser, & Erin Phelps

The purpose of this book is to explore the study of spiritual development during the adolescent period (the second decade of life; Lerner and Steinberg 2004) and to ascertain the possible links among spirituality and the healthy, positive development of youth. Adolescence is a time of life when young people are prototypically engaged in finding a self-definition—an identity (e.g., Erikson 1959, 1968; Harter 2006)—that enables them to matter to self, family, and society, both in the teenage years and in their future adult life. The search for such an identity impels the young person to transcend a cognitive and emotional focus on the self (Elkind 1967) and to seek to contribute in important, valued, and even noble ways to his or her world. We believe that generosity that is derived from such transcendence and noble purpose is the essence of spirituality (see, too, Damon 2004; Damon, Menon, and Bronk 2003) and may provide a key foundation for positive youth development (Lerner 2008).

To frame our consideration of the links among positive development, spirituality, and generosity (or contribution) among youth, we use an approach to theory that, today, is at the cutting edge of developmental science, i.e., developmental systems theory (Damon and Lerner 2006; Lerner 2002, 2006). Developmental systems theory is an ideal model within which to explore the integration of self and context that we believe is involved in these links and, as well, to consider the potential impact on positive development and spirituality of the numerous biological, psychological, behavioral, and sociocultural changes converging within the adolescent period. In fact, in affording an integrative perspective to biological through sociocultural levels of organization, developmental systems theory allows science to consider within common research programs the emerging findings pertinent to neural, socioemotional, cultural, and historical influences on adaptive (healthy, positive) adolescent development that, today, are at the fore of the study of this pivotal period of life.

Accordingly, in this book we explore key conceptual and definitional issues useful in framing the understanding of the association between positive development in adolescents, spiritual development, and the attainment of a sense of self that moves the young person to make contributions to (or, in other words, be generous toward) self, family, community, and society. In addition, we discuss the biological covariates of these links among positive youth development (PYD), spirituality, and generosity and, as well, the individual-level, social-level, and cultural-level covariates of this linkage. All chapters in the book focus as well on the research that needs to be done to advance understanding of these linkages.

To prepare for these discussions, the present chapter presents a developmental systems model of the relations among the positive development of adolescents and the development of their spirituality and generosity. In the context of the model, we describe the extant, and admittedly limited, neuropsychological, behavioral, and social-relational data pertinent to the covariation among positive development, spirituality, and generosity. Because of the limits of existing data, we specify also some of the key features of to-be-conducted developmental research that is needed to elucidate these relations; for instance, we discuss the importance of longitudinal research with youth from diverse religious, racial, ethnic, socioeconomic, and cultural backgrounds. The chapter ends with a brief overview of the plan of, and the chapters in, this book.

POSITIVE YOUTH DEVELOPMENT AND SPIRITUALITY: A DEVELOPMENTAL SYSTEMS MODEL

How does human development, in general, and positive and healthy human development, in particular, happen? What are the processes that enable humans to adapt and thrive across their life spans? Are the bases of positive human development ones shared completely by other organisms, or are there features of development that are unique to humans? Are answers to these questions the same across the human life span, or are the explanations of positive development at least in part discontinuous across the course of ontogeny?

Answers to these questions address the foundations of developmental science. We believe that theoretically useful and empirically rich answers can best be derived from models derived from developmental systems theories of human development (Lerner 2002, 2006).

Within the frame of developmental systems theories, positive human development involves adaptive (i.e., health promoting) regulations (Brandtstädter 1998, 2006) involving individuals and the ecology of human development. Represented as individual ⟷ context relations, these associations involve mutually influential exchanges between person and context that are beneficial to both entities. Humans' evolutionary heritage established mutually supportive individual ⟷ context relations as integral for human survival (Gould 1977; Johanson and Edey 1981), and this phylogeny is repeated in ontogeny, where individuals must support a context that supports them if humans are to survive and prosper across life (Lerner 2004).

Adaptive developmental regulations involve, for instance, changing the self to support the context and altering the context to support the self. Such efforts require the individual to remain committed to contributing to the context and to possess, or to strive to develop, the skills for making such contributions. In turn, there is a requirement that the institutions of society support people in their individual attempts to find the means across life to thrive. A commitment to maintaining the social institutions that, in turn, provide the person with the opportunity to flourish as a healthy individual is not only the operationalization of adaptive developmental regulation; as well, the continuity of such structural relations across life is the process of positive human development, of "thriving" (Lerner 2004).

Derived from the neotenous phylogenetic history of humans (Gould 1977), that is, the slowing down or "retardation" of the rate of development in comparison to ancestral species, adaptive developmental regulations provide an ontogenetic basis for postulating that, in life periods of marked individual and social change—for instance, in human adolescence (Lerner and Steinberg 2004)—the positive development of people involves convergences in neural, emotional, cognitive, behavioral, and social developments. In adolescence, for instance, when there is new brain growth, as well as qualitative and quantitative changes in individual characteristics (involving affective, cognitive, and behavioral attributes) and social relations (with family members, peers, and the institutions of society), the positively developing young person must use his or her emergent affective and cognitive attributes to integrate all the inner and outer changes being experienced through formulating a sense of self (an identity) that involves a rationale for engaging in mutually beneficial individual ↔ context relations. In order for such relations to function in a manner supporting the adaptation of individuals and context, an individual must define the self as a person who works to enhance entities (people, institutions, the ecology) beyond the self.

For instance, the young person may formulate the belief that one should contribute *both* to self and civil society, to have a generosity of spirit that integrates self-regarding virtues and other-regarding virtues. In other words, in youth who manifest exemplary, positive development, who are thriving, there should be an integration of moral and civic identity that promotes adolescents' contributions to mutually beneficial relations between themselves and their social worlds.

Civic participation, civic engagement, and civic contribution (Lerner, Alberts, and Bobek 2007) are associated, then, with a young person developing along a life path marked by what are termed the "Five Cs" of positive youth development: character, competence, confidence, connection, and caring (or compassion). Such youth will pursue the noble purpose of becoming a productive adult member of their community (Damon, Menon, and Bronk 2003), a person showing generosity toward, or contributing positively to, self, others, and the institutions of civil society.

In other words, such youth will develop an integration of generosity to self and to others, or what researchers in the youth development field (e.g., Lerner 2004) would term the "Sixth C" of contribution—of making life better for the self, family, community, and, ultimately, civil society. For instance, a youth showing such generosity or contribution would manifest thriving by attending to both (a) his or her physical and psychological health (e.g., as in Scales et al. 2000), so that he or she would not be a burden on others and, as well, so that he or she could, in fact, have the energy to contribute; and (b) the "healthy" well-being of his or her community (e.g., as in Scales et al. 2000).

Indeed, a commitment to such contribution rests on defining behavior in support of mutually beneficial individual ↔ context exchanges as morally necessary. Individuals' moral duty to contribute exists because, as citizens receiving benefits from a social system supporting their individual functioning, it is necessary to be engaged actively in maintaining and, ideally, enhancing that social system (Youniss, McLellan, and Yates 1999). This type of developmental regulation—between thriving individuals and their civil society—is the essence of a system marked by liberty (Lerner 2004).

In short, adaptive developmental regulation results in the emergence among young people of an orientation to transcend exclusive self-interest and place value on, and commitments to, actions supportive of a social system promoting equity, democracy, social justice, and personal freedoms. The integration of individual and ecological assets (Benson 2003)—for example, support for positive development from family members and community organizations (e.g., youth-serving programs, faith institutions)—that occurs through adaptive developmental regulation provides the developmental "nutrients" (resources) requisite for thriving. We hypothesize that spirituality is the emotional "fuel" energizing the thriving process.

Thriving and Spiritual Development

Arguably, spirituality and religiosity are the only mental and behavioral characteristics that are distinctly associated with humans (Kalton 2000; Lerner et al. 2004). Characteristics of psychological and behavioral functioning, such as emotions, language, caregiving, cognition, temperament, personality, and goal setting, can be found or operationalized in other species. However, such generalizability does not appear to exist for spirituality and religiosity; these attributes may be the key characteristics that make humans human.

Following Reich (1998) and reflecting one "of the core themes of the monotheistic traditions, and many Asian traditions as well" (Goodenough 2001, 21), we conceptualize spirituality as fundamentally involving the concept of transcendence. Reich notes that transcendence represents a commitment to ideas or institutions that go beyond the self in time and place. He believes that transcendence involves viewing life in new and better ways, adopting some conception as transcendent or of great value, and defining one's self and one's relation to others, to nature, and to the universe in a manner that goes beyond provincialism or materialism and expresses authentic concerns about the world.

Kalton (2000) terms such transcendence of self "horizontal transcendence," and comments that it represents a "radically non-anthropomorphic spirituality" (193) and "a form of transcendence that is characteristic of degrees of abstraction rather than a movement towards some kind of Absolute metaphysical dimension. There is no cosmos posited apart from the historically ongoing one within which we find ourselves . . ." and "the movement of this kind of spiritual cultivation is horizontal, perfecting our relationship with the world of life about us" (195). Kalton contrasts horizontal transcendence with "vertical transcendence," by which he means "a metaphysical structure grounding the contingent in the Absolute" (2000, 190), in the "infinite, eternal, personal creator by whose will we may live" (192). Goodenough (2001) indicates that both forms of transcendence are essential to the full religious life. Here, concepts of grace and of the Divine are integral. Similarly, Haight (1999) presents a theology of evolution that integrates both horizontal and vertical concepts of transcendence.

We agree that the subordination of self to institutions that are believed to have relations to the Divine through vertical transcendence is the essence of religiosity. Reich (1998), in fact, operationalizes this hierarchical instantiation of transcendence (Goodenough 2001; Haight 1999) as religiosity and notes that it involves a relationship with a particular institutionalized doctrine about a supernatural power. He indicates that this relationship occurs through affiliation with an organized faith and participation in its prescribed rituals.

There are as yet unaddressed empirical questions about the development of these dimensions of human life that are of paramount importance in providing new spiritual knowledge (Templeton 1995). In addition, while we believe that spirituality and religiosity—or, in the terms of Kalton (2000) and Goodenough (2001), horizontal and vertical transcendence—may be singularly human characteristics, the neural, cognitive, emotional, and behavioral characteristics that operationalize spirituality are, nevertheless, not fully present in the newborn (Lerner et al. 2004). These characteristics *develop* across the life span. Their development may be reflected in the changing neural (brain), cognitive, emotional, personality, behavioral, and social relationship characteristics of the developing person (e.g., d'Aquili and Newberg 1999). Even more so than infancy, adolescence is the ontogenetic period within the life span within which there is the most profound convergence of quantitative and qualitative changes in these dimensions. As such, it is the ideal portion of the life span within which to seek this new spiritual knowledge.

Moreover, consistent with developmental systems theory, (d'Aquili and Newberg 1999) explain that human adaptability and survival are linked to human religiosity and spirituality, in that these domains of functioning involve both self-maintenance and self-transcendence. In our terms, this dual focus means engagement in the mutually beneficial individual ↔ context relations that define adaptive developmental regulations. d'Aquili and Newberg (1998) hypothesize that, across human evolution, the neural mechanisms that underlie spirituality and religiosity "appear to have become thoroughly ingrained in the human gene pool" (187). Thus, key questions that remain to be addressed are how these two facets of human functioning, which together may mark what is uniquely human about humans, emerge and evolve over the course of adolescence and what may be their biological (genetic, hormonal, and neuronal) covariates.

Theoretical and Empirical Ideas about the Links between Positive Youth Development and Spiritual Development. Despite the need for new knowledge regarding the contributions of the developmental system to the growth in adolescence of adaptive individual ←→ context functioning and spirituality, existing data indicate that both spirituality and religiosity are theoretically and empirically identifiable as important influences on human development across much of the life span (e.g., Koenig and Lawson 2004) and are transformed in personal, cognitive, and adaptive salience across the course of life. There is considerable theoretical reason, but as yet only some indirect empirical evidence, to believe that this transformation is a key feature of the adolescent period (Dowling et al. 2003, 2004; Keating 2004; Lerner, Dowling, and Anderson 2003; Spear 2000).

In regard to theory, Erikson (1959) discussed the emotional "virtues" that were coupled with successful resolution of each of the eight psychosocial crises he included in his theory of ego development. He specified that fidelity, defined as unflagging commitment to abstract ideas (e.g., ideologies) beyond the self, was the virtue associated with adaptive resolution of the identity crisis of adolescence and thus with the attainment of a socially prescribed, positive role (cf. Youniss, McLellan, and Yates 1999). Commitment to a role was regarded by Erikson (1959) as a means for the behaviors of youth to serve the maintenance and perpetuation of society; fidelity to an ideology coupled with a role meant that the young person would gain emotional satisfaction—which, to Erikson (1959), meant enhanced self-esteem—through contributing to society by the enactment of role behaviors (Lerner 2002).

One need not focus only on crisis resolution to suggest that behaviors attained during adolescence in the service of identity development may be coupled with an ideological "virtue," that is, with a sensibility about the meaningfulness of abstract ideas that transcend the self (Youniss, McLellan, and Yates 1999). From a perspective that focuses on adaptive developmental regulation within the developmental system, it is possible to suggest that spirituality is the transcendent virtue that is coupled with the behaviors (roles) reflecting an integrated moral and civic identity—or the character that is manifested by noble purpose and generosity (contribution) to self and context.

Contemporary researchers (e.g., Flanagan et al. 1998; Lerner, Alberts, and Bobek 2007; Sherrod, Flanagan, and Youniss 2002; Youniss, McLellan, and Yates 1999) show increasing interest in addressing the impact of community contributions and service activities on healthy identity development. Erikson (1959) proposed that, when young people identify with ideologies and histories of faith-based institutions, identities can be placed within a social-historical framework that connect youth to traditions and communities that transcend the immediate moment, thereby providing young people with a sense of continuity and coherence with the past, present, and future (e.g., see Kerestes and Youniss 2003). Consistent with Erikson's prescription, youth-service programs sponsored by faith-based institutions, such as the Catholic Church, are embedded in interpretive values and historical meaning (Kerestes and Youniss 2003).

For example, a parish that sponsors a highway cleanup activity for its youth will likely rely on a moral and value-laden framework to explain its involvement, describing that involvement in religious traditions and stories (Youniss, McLellan, and Yates 1999). Youth who take part in service activities are likely to "reflect on these justifications as potential meanings for their (own) actions. These established meanings, with their historical richness and picturing of an ideal future may readily be seen as nourishment for youths' identity development" (Youniss, McLellan, and Yates 1999, 244).

Developmental systems models of adolescent development suggest also that spirituality is important for the healthy, positive development of an adolescent's sense of self—his or her identity—and for enabling this identity to frame the individual's pursuit of a life path that involves mutually beneficial and socially positive relations between the individual and his or her world (Lerner 2004). For instance, as already suggested, we believe that exemplary positive youth development—thriving—is marked by having a noble purpose (i.e., a goal that transcends the self in time and place and that is valued by others as mattering for the positive and healthy maintenance of the social context; Damon, Menon, and Bronk 2003), one that is linked to generosity—contributions—to self, family, community, and civil society. The relationship between purpose and contribution is supported by character development among youth and, as well, by the development of other attributes that constitute a thriving young person, that is, competence, confidence, caring, and healthy social connections to others. Healthy and productive adolescents make transitions into adulthood that maintain and potentially enhance civil society. Thus, because of the intertwined personal and social characteristics associated with purpose, generosity, and healthy development, adolescence may be the time when humans are most questioning of and open to ideas about horizontal and vertical transcendence.

Accordingly, a key task of future research is to generate new information about spirituality, specifically the development of horizontal and vertical spirituality in youth. What would be invaluable would be generating empirical data through longitudinal research that is theoretically framed by a model of positive youth development and that is illuminated by a wide-ranging, multidisciplinary collaboration among scholars in the fields of psychology, theology, sociology, education, anthropology, medicine, and biological sciences. To our knowledge, an effort of this sort—a major longitudinal study of the development in young people of spirituality, purpose, generosity, and positive functioning that includes measures ranging from the structure and function of the brain to the role of community assets in promoting these developments—has not been conducted.

Nevertheless, some data sets suggest the importance of conducting such work. For instance, Smith and his colleagues at the University of North Carolina have been doing a large-scale study of the important role of organized religion in the lives of American youth and families (e.g., Smith 2003; Smith et al. 2002; Regnerus, Smith, and Smith 2004). In turn, King and Furrow (2004) report cross-sectional data about the contribution of youth religiosity to social capital and positive development. Such data sets suggest that youth whose exchanges with their contexts (whose developmental regulations) are marked by functionally valued behaviors should develop integrated moral and civic identities and a transcendent, or spiritual, sensibility (Benson 2003; Youniss, McLellan, and Yates 1999). In fact, there is evidence that adolescents' sense of spirituality is linked to thriving.

In two studies that used an archival data set from Search Institute (1984), entitled "Young Adolescents and Their Parents" (YAP), Dowling et al. (2003, 2004) studied the links among spirituality, religiosity, and thriving among a subsample of one thousand adolescents randomly selected from the larger cross-sectional group of 8,165 youth, ranging from fifth through ninth grades. From an item pool of more than three hundred items in the YAP survey, Dowling et al. (2003) found ninety-one items that pertained to religiosity, spirituality, and positive youth development or thriving (e.g., social competence, self-esteem, and respect for diversity). The results of orthogonal factor analyses for each of the three constructs of religiosity, spirituality, and thriving provided evidence for three spirituality factors, four religiosity factors, and nine thriving factors.

In turn, Dowling et al. (2004) explored the role of spirituality in thriving by using the one thousand adolescents from the YAP data set studied by Dowling et al (2003). Among the three models that were estimated, the one that included the paths from spirituality to religiosity, from religiosity to thriving, *and* from spirituality to thriving was a better fit to the data set than hierarchically related, reduced models that either lacked the direct effect of spirituality on thriving or the mediating effect of spirituality on thriving through religiosity. The fit of the comparison models was significantly worse than the fit of the complete mediation model. In short, more so than a path from spirituality to thriving that involved the mediation of religiosity, among the YAP youth, the strongest influence on their thriving was a path directly from spirituality.

These findings illustrate how contemporary scholars of adolescent development are pointing to the implications of spirituality on positive youth development (see, for example, Lerner 2004; Youniss, McLellan, and Yates 1999) and are conceptually differentiating the role of spirituality from the role of religiosity in such development. However, there has been no attempt to date to establish developmentally, i.e., through longitudinal research, the differential links in adolescence among spiritual development, religious development, and thriving, particularly in a data set having sufficient statistical power and a measurement model seeking to assess the neural, psychological, cognitive, affective, moral, and contextual bases of this developmental process.

Moreover, as illustrated in the above-cited research, although the psychosocial correlates of adolescent spirituality and positive youth development have been assessed cross-sectionally, the neural covariates of these constructs among youth remain a still largely neglected focus of even cross-sectional research. Nevertheless, there is reason to believe that such a focus is important.

Neuropsychological Bases of the Spirituality ←→ Thriving Relation. Humans are the most slowly developing (neotenous) species. In their extended postnatal developmental period, wherein their childlike (paedomorphic) features are retained well into the second decade of life, their slowly developing brain emerges in a context that provides them with rich opportunities to learn about, and become contributing members to, a social and cultural context (involving sustained and protective adult–child and peer-group relations). Development in such a social context was requisite for our relatively defenseless hominid ancestors (lacking sharp teeth and claws) to survive (Gould 1977; Johanson and Edey 1981). Accordingly, adaptive brain development meant in human evolution the development of a social brain. There had to emerge neural pathways and cognitive characteristics that enabled the developing young person to contribute to a social context whose support was critical to his or her survival (Gould 1977).

Spear (2000) notes that

among the brain areas prominently remodeled during adolescence is the prefrontal cortex, a brain region thought to be involved in various goal-directed behaviors (including rule learning, working memory, and spatial learning) and in emotional processing, particularly of aversive stimuli. Along with a decline in the relative size of the prefrontal cortex during adolescence, there is a substantial remodeling of connections between neurons—with some connections lost and others added. . . . To the extent that adolescence is associated with the developmental alterations in prefrontal cortex, limbic brain areas, and the dopamine input to these regions, concomitant developmental alterations in various motivated behaviors might also be expected. (112–13)

Building on such evidence of developmental changes in the adolescent brain, d'Aquili and Newberg (1999) hypothesize that such transformations provide a basis of the complex psychological and affective developments involved in the transcendence of self that is requisite for a sustained, self-defining investment of the individual in the "beyond-the-self" social world.

The possible neural substrates of this transcendence may be the emotional (i.e., spiritual or transcendent) and cognitive (i.e., self-definitions involving integrative ideas of moral and civic action) bases of the links among spirituality, purpose, generosity, and thriving. The importance of these neural bases for emerging links between the adolescent and his or her social world is underscored by recent findings in developmental neuroscience, involving dramatic advances in imaging technologies (e.g., Paus 2001). The data derived from these techniques point to the importance for adolescent development of both prefrontal and parietal cortex areas and maturation of cortico-cortical connectivity (Keating 2004; Paus 2001).

Supported by increasingly rapid transfer of information due to continued myelination, and perhaps, by localized synaptic pruning in the prefrontal cortex, the specific features of brain development in adolescence correspond precisely to the most marked differences between adult humans' brains and the brains of nonhuman primates (Keating 2004). In essence, then, brain development in adolescence may covary with the emergence of a spiritual sense that affords the self-transcendence that enables adaptive individual ↔ context relations. Thus, this brain development may be precisely what makes human beings human (Goodenough 1998, 2000, 2001; Kalton 2000; Lerner et al. 2004).

In other words, the continuous cortical maturation in adolescence, beginning in the 10- to 14-year-old age period and extending at least to age 18, provides youth not only the capacity for integrative cognitive functions but also with the neural networks enabling the integrative regulation of emotion, attention, and action (Keating 2004) and perhaps for positive development as well (Paus et al., in press). For instance, development of the anterior cingulated cortex may enhance the adolescent's ability to implement desired goals in the face of conflict (Paus 2001) and the development of the orbito-frontal cortex moderates the processing of emotions and making decisions in the absence of clear information (Keating 2004; Paus et al., in press).

Having a brain capable of understanding (cognizing) the importance of a mutually beneficial relationship between self and context and having a brain with neuronal networks where emotions "energize" the linkage of such cognitions with actions adaptive in support of the world beyond the self (e.g., Spear 2000) were critical for the survival of humans, both individually and collectively (d'Aquili and Newberg 1999; Fisher 1982; Gould 1977; Johanson and Edey 1981). In turn, and as we have already noted, the emergence of a sense of the importance of "transcending self"—and contributing to the world that extends beyond the self in time and place—is the essence of the conception of horizontal and vertical spirituality (Goodenough 1998, 2000, 2001; Kalton 2000; Reich 1998).

For instance, we have noted that Kalton (2000) conceived of horizontal transcendence of self as an emotional valence to act in ways that contribute to the social and cultural world, a world that extends beyond the self in time and place. When coupled with the cognitive understanding of the self (a self-concept or identity) that defines such contributions as one's moral requirement and civic duty, the developing person will be impelled to contribute actively to the social world in productive and positive ways (Lerner 2004; Sherrod, Flanagan, and Youniss 2002; Youniss, McLellan, and Yates 1999).

The emergence in adolescence of a press for identity development and the concomitant emergence of emotional energizers of such self-knowledge, whether they reflect the "virtue" of fidelity (Erikson 1950, 1959) or the "virtue" of spirituality (Lerner, Dowling, and Anderson 2003), may both reflect and contribute to the brain maturation that continues in the early portion of the second decade of life (Keating 2004). Indeed, it is reasonable to hypothesize that those maturing emotional and cognitive networks (e.g., in the prefrontal cortex, in the parietal lobe, and in their connection) that may involve the adolescent's sense of transcendence of self—his or her spirituality—should be the most slowly developing (neotenous) brain areas. Moreover, these networks also may be the most functionally active areas of the brain among those adolescents who are most likely to show noble purpose and generosity of spirit toward self, family, community, and civil society, e.g., through positive civic engagement.

Whatever the empirical fate of these specific ideas, theory and research combine to suggest that the development of new spiritual knowledge may advance significantly if research within the adolescent period explores the confluence of biology (brain maturation, as well as genetic and hormonal changes), identity development, spirituality, and youth engagement with and contributions to the context beyond the individual. This book is aimed at enhancing the theoretical and empirical rationale for such scholarship and, as such, for setting the stage for a new, major scientific enterprise elucidating the connections between positive development and spirituality in adolescence.

CONCLUSIONS

The theoretical model that we have forwarded to frame the scholarly agenda we envision reflects the cutting-edge of scholarship in human development (Damon and Lerner 2006). This theory specifies that spirituality may foster an integrated moral and civic identity within a young person and may lead the individual along a path to becoming an adult who contributes to self, family, community, and civil society (Lerner 2004; Lerner, Dowling, and Anderson 2003).

Given the theoretically engaging and empirically generative scholarly agenda that is linked to the study of adaptive individual ↔ context relations and spirituality in adolescence, it is reasonable to expect that a growing numbers of scientists may be drawn to this new domain of inquiry. The subsequent chapters of this book may be seen as evidence for the validity of this expectation.

As such, it is useful to describe the foci and key ideas presented in these chapters so that a frame for future research linking positive youth development, spirituality, and generosity may be envisioned. As will be clear as we note the substance of these chapters, the scholars contributing to this book stand with Sir John Templeton (1995) in believing that "no one can foresee exactly which research projects for spiritual progress should be undertaken or even the specific form that empirical inquiry may take in this realm. Nor can anyone foresee which experiments will prove fruitful" (70).

THE PLAN OF THIS BOOK

Formidable conceptual issues arise in operationalizing the constructs of interest in the study of linkages across ontogeny among thriving, spirituality, and generosity. These issues are addressed across the chapters in the first section of this book.

Johnson seeks to ground the study of spiritual development within the concept of spirit. Noting that the concept of spirit has not been a core part of the literature of contemporary developmental science, he discusses spirit as a dynamic reality, a force, that integrates other domains of development—for example, the physical, cognitive, and emotional. Scarlett conceptualizes spirituality in adolescence as emerging faith in something greater than one's self. His discussion focuses on the problem of how to evaluate emerging faith when it, in fact, occurs. For instance, should structural criteria, evidence of transcendence, or the strength of faith be the marker of emerging faith? King provides a framework for understanding how spirituality may be a unique and robust catalyst for positive youth development. She explores how the ideological, social, and transcendent dimensions of spirituality provide a productive means for promoting positive development among young people. Roeser, Issac, Abo-Zena, Brittian, and Peck discuss also the links between the development of spirituality and the development of self. They argue that spiritual development and religious development are intimately linked in human development and that there are important distinctions that exist between spiritual identity and religious identity; they point to the unique role that the former facet of identity may play in positive youth development.

Consistent with a developmental systems model linking the multiple levels of organizations within the human development system, the changes involved in processes of self and spiritual development affect and are affected by the changing biological features of the adolescent. Accordingly, chapters in the next section of the book are discussions of the potential associations between biological changes—and, in particular, the genetic, neuronal, and physiological—and the PYD ↔ spirituality ↔ generosity linkage. Grigorenko considers the role of genes in faith, spirituality, and religiosity. She discusses the magnitude of genetic effects on these constructs. Through drawing on literature pertinent to heritability, molecular genetics, and population genetics, she points to the potential mechanisms that might engage genes in the formation of spirituality, religiosity, and related characteristics. Paus, Gächer, Starmer, and Wilkinson discuss the possible relations between economic inequality and a key indicator of positive youth development, prosocial behavior. They outline several potential neural bases of cooperative behavior, conflict resolution, and resistance to peer influence and, as such discuss potential links about the brain-behavior-context relations that may be involved in positive youth development and in young people's transcendence of self and contributions to others. Urry and Poey address the question of whether emotion regulation might play a role in mediating associations between religious/spiritual beliefs and practices and psychological well-being. They frame their discussion within the context of a model that links brain structure and function with religious/spiritual practices, psychological functions such as emotion regulation, and well-being.

The chapters in the second section of the book underscore the intimate, mutually influential relations between human biology and the individual's cognitive, emotional, and social behavior and development. In the third section of the book, authors focus on various instances of these individual-level variables and how they may represent covariates of the PYD, spirituality, generosity linkage. Feldman explores how an understanding of developmental change processes may contribute to the understanding and promotion of spiritual development among adolescents. He discusses the meaning of human-developmental processes and the links between developmental and spiritual-change processes and considers the nature of fixed beliefs versus spiritual growth in the context of an analysis of the nature of developmental change. Templeton and Eccles examine spirituality in relation to the development of an ethic of care and of related actions aimed at making positive contributions to other people. They argue that spirituality includes not only individuals' personal reflections about what to believe and how to behave but, as well, their most cherished values, beliefs, and life purposes. Mariano and Damon note that the literature on youth development has rarely focused on the intentions of young people to contribute to the world beyond the self, and, as such, they discuss the nature of "youth purpose" and suggest that one way of understanding the relationship between spiritual development and positive youth development is to examine the role of religion and spirituality in influencing young people's purposes in life. Arnett uses the lens of a study of the afterlife beliefs of emerging adults to discuss the links between spirituality and positive development in adolescence. He notes that the emerging adulthood period, the time between adolescence and adulthood, may provide especially fertile ground for an assessment of how beliefs in an afterlife provide insight into how the young person establishes his or her own spiritual beliefs (i.e., beliefs independent of parental influences) and, as such, the association between these beliefs and the individual's positive development.

The social and cultural covariates and settings of the PYD, spirituality, generosity linkage are the focus of the chapters in the final section of the book. Several chapters discuss the central role of culture in framing the associations among PYD, spirituality, and generosity, often through the lens of racial and ethnic differences and in regard to religious institutions. For example, Jensen focuses on the development of immigrant youth and considers the bases of their civic engagement. She discusses the individual motives and institutional contexts that are linked to the positive engagement of youth with their new national context and focuses on the specific role of religious motives and organizations in moderating the civic engagement of these young people. Juang and Syed explore the connection between ethnic identity and religious/spiritual identity. They illustrate how youth think about these identities by drawing upon data derived from a study of ethnic identity and discuss future directions for research about how these intersecting identities are negotiated and how they contribute to positive youth development. Nasir discusses the interrelations of context, culture, and development in shaping the relation between spirituality and positive youth development. She argues that the intersection between positive youth development and spirituality is both an important scientific endeavor and an arena of scholarship that affords a re-envisioning of our human potential for happiness, wellness, wholeness, and greatness. Nicolas and DeSilva provide an overview of the ways that researchers may conduct culturally sensitive research on spirituality and positive youth development. They argue that such research must be informed by a comprehensive understanding of the sociocultural factors in adolescents' lives. They indicate that without such a frame, it is not possible to understand adequately the impact of spirituality on the development of youth. Sherrod and Spiewak examine the potential relationships among spirituality/religiosity, civic engagement, and positive youth development. They note that all of these concepts have multiple meanings in the literature, and, as such, in clarifying this variation and discussing the possible relations among these constructs, they demonstrate that each construct links values, identity, and behavior. Bers provides an overview on the positive role that technology can play in the spiritual development and the religious education of young people. She discusses the history of identity construction technological environments and presents results of several psychoeducational interventions aimed at enhancing the links between technology and positive youth development and spirituality.

Across the chapters in this book, the reader will encounter a rich and varied array of theoretical ideas, methodological approaches to the study of youth and spiritual development, and recommendations for subsequent research. Nevertheless, across this variation, several common perspectives exist, ones that—together—organize an agenda for new, programmatic scholarship aimed at producing new spiritual information about adolescent spirituality and adaptive individual ↔ context relations.

First, in the midst of theoretical variation there is also theoretical commonality. All contributors to this book embed their ideas within developmental systems notions of human development. All share the view that multiple levels of organization are fused across the life span and that variables for each level are reciprocally integrated. All understand the basic process of development as a mutually influential one within and across levels, and, as such, all believe that research must include multivariate, multilevel designs in order to capture the complexity of the change processes within the developmental system.

Second, all contributors emphasize that research must be attentive to diversity—to the variation in gender, race, ethnicity, religion, family structure, and culture that makes each person an individual. All people possess generic characteristics, and, as well, all people possess group-specific attributes. However, each person also possesses specific characteristics of individuality, and the contributors share the view that, unless research is sensitive to the distinctive characteristics of people, essential features of human functioning will be missed.

Third, the contributors note that no one method can appraise adequately the generic, group, and individual characteristics of people. As such, multiple methods of data collection must be used, and, ideally, each method must be triangulated with other methods in order to identify what is unique and what is common about the associations in adolescence among positive development, spirituality, and generosity.

Finally, the fundamental point of scientific agreement among the contributors to this book is that all of the variables involved across the multiple, fused levels of organization *develop*. The links among adaptive individual ↔ context relations, spirituality, and generosity are not static. They change dynamically over the course of the second decade of life. Accordingly, all the contributors to this book believe that the full advancement of this quest for new spiritual information rests on the design and implementation of longitudinal research.

Indeed, as will be evident across the chapters of this book, there are substantial theoretical and empirical reasons to believe that such longitudinal work can create a new era in the study of spiritual development. The chapters in this book make clear that science and scientists are poised to engage the difficult conceptual and methodological issues involved in the search for new spiritual realities about positive development in adolescence. As such, this book may mark a watershed event in the crystallization of a new domain of scientific activity.

ACKNOWLEDGMENTS

The preparation of the chapter was supported in part by a grant from the John Templeton Foundation to Richard M. Lerner.

REFERENCES

Benson, P. L. 2003. Developmental assets and asset-building community: Conceptual and empirical foundations. In R. M. Lerner and P. L. Benson, eds., *Developmental assets and asset building communities: Implications for research, policy, and practice,* 19–46. New York: Kluwer Academic Press.

Brandtstädter, J. 1998. Action perspectives on human development. In W. Damon and R. M. Lerner, eds. *Theoretical models of human development,* 807–63. Vol. 1: *Handbook of child psychology.* 5th ed. New York: Wiley.

———. 2006. Action perspectives on human development. In R. M. Lerner, ed., *Theoretical models of human development,* 516–68. Vol. 1: *Handbook of child psychology.* 6th ed. Hoboken, NJ: Wiley.

Damon, W. 2004. What is positive youth development? *The Annals of the American Academy of Political and Social Science* 591: 13–24.

Damon, W., and R. M. Lerner. 2006. *Handbook of child psychology.* 6th ed. Hoboken, NJ: Wiley.

Damon, W., J. Menon, and K. C. Bronk. 2003. The development of purpose during adolescence. *Applied Developmental Science* 7: 119–28.

D'Aquili, E., and A. B. Newberg. 1999. *The mystical mind: Probing the biology of religious experience.* Minneapolis, MN: Fortress Press.

Dowling, E. M., S. Gestsdottir, P. M. Anderson, A. von Eye, J. Almerigi, and R. M. Lerner. 2004. Structural relations among spirituality, religiosity, and thriving in adolescence. *Applied Developmental Science* 8: 7–16.

Dowling, E. M., S. Gestsdottir, P. M. Anderson, A. von Eye, and R. M. Lerner. 2003. Spirituality, religiosity, and thriving among adolescents: Identification and confirmation of factor structures. *Applied Developmental Science* 7: 253–60.

Elkind, D. 1967. Egocentrism in adolescents. *Child Development* 38: 1025–34.

Erikson, E. H. 1950. *Childhood and society*. New York: W. W. Norton and Company.

———. 1959. Identity and the life-cycle. *Psychological Issues* 1: 18–164.

———. 1968. *Identity, youth, and crisis*. New York: Norton.

Fisher, H. E. 1982. Of human bonding. *The Sciences* 22: 18–23, 31.

Flanagan, C. A., J. M. Bowes, B. Jonsson, B. Csapo, and E. Sheblanova. 1998. Ties that bind: Correlates of adolescents' civic commitments in seven countries. *Journal of Social Issues* 54: 457–75.

Goodenough, U. 1998. *The sacred depths of nature*. New York: Oxford University Press.

———. 2000. Reflections on scientific and religious metaphor. *Zygon: Journal of Religion and Science* 35: 233–40.

———. 2001. Vertical and horizontal transcendence. *Zygon: Journal of Religion and Science* 36: 21–31.

Gould, S. 1977. *Ontogeny and phylogeny*. Cambridge, MA: Harvard University Press.

Haight, J. E. 1999. *God after Darwin: A theology of evolution*. Boulder, CO: Westview Press.

Harter, S. 2006. The self. In N. Eisenberg, ed., *Social, emotional, and personality development*, 505–70. Vol. 3: *Handbook of child psychology*. 6th ed. Hoboken, NJ: Wiley.

Johanson, D. C., and M. A. Edey. 1981. *Lucy: The beginnings of humankind*. New York: Simon and Schuster.

Kalton, M. 2000. Green spirituality: Horizontal transcendence. In M. E. Miller and P. Young-Eisendrath, eds., *Paths of integrity, wisdom and transcendence: Spiritual development in the mature self*, 87–200. London and Philadelphia: Routledge.

Keating, D. P. 2004. Cognitive and brain development. In R. M. Lerner and L. Steinberg, eds., *Handbook of adolescent psychology*, 2nd ed., 45–84. New York: Wiley.

Kerestes, M., and J. E. Youniss. 2003. Rediscovering the importance of religion in adolescent development. In R. M. Lerner, F. Jacobs, and D. Wertlieb, eds., *Applying developmental science for youth and families: Historical and theoretical foundations*, 165–84. Vol. 1: *Handbook of applied developmental science: Promoting positive child, adolescent, and family development through research, policies, and programs*. Thousand Oaks, CA: Sage Publications.

King, P. E., and J. Furrow. 2004. Religion as a resource for positive youth development: Religion, social capital, and moral outcomes. *Developmental Psychology* 40: 703–13.

Koenig, H. G., and D. M. Lawson. 2004. *Faith in the future: Health care, aging, and the role of religion*. Radnor, PA: Templeton Foundation Press.

Lerner, R. M. 2002. *Concepts and theories of human development*. 3rd ed. Mahwah, NJ: Erlbaum.

———. 2004. *Liberty: Thriving and civic engagement among America's youth*. Thousand Oaks, CA: Sage Publications.

———. 2006. Developmental science, developmental systems, and contemporary theories of human development. In R. M. Lerner, ed., *Theoretical models of human development*, 1–17. Vol. 1: *Handbook of child psychology*. 6th ed. Hoboken, NJ: Wiley.

Lerner, R. M. 2008. Spirituality, positive purpose, wisdom, and positive development in adolescence: Comments on Oman, Flinders, and Thoresen's ideas about "integrating spiritual modeling into education." *The International Journal for the Psychology of Religion*, 18, no. 2:108–18.

Lerner, R. M., A. E. Alberts, and D. Bobek. 2007. Thriving youth, flourishing civil society—How positive youth development strengthens democracy and social justice. In *Civic engagement as an educational goal*, 21–35. Guterslöh, Germany: Verlag Bertelsmann Stiftung.

Lerner, R. M., P. M. Anderson, A. E. Alberts, and E. M. Dowling. 2004. *On making humans human: Spirituality and the promotion of positive youth development*. Invited presentation to the Mid-Winter Meetings of Division 36 (The Psychology of Religion) of the American Psychological Association. Loyola College, Columbia, MD. March.

Lerner, R. M., E. M. Dowling, and P. M. Anderson. 2003. Positive youth development: Thriving as a basis of personhood and civil society. *Applied Developmental Science* 7: 172–80.

Lerner, R. M., and L. Steinberg, eds. 2004. *Handbook of adolescent psychology*. 2nd ed. New York: Wiley.

Paus, T. 2001. Primate anterior cingulate cortex: Where motor control, drive and cognition interface. *Nature Reviews Neuroscience* 2: 417–24.

Paus, T., R. Toro, G. Leonard, J. V. Lerner, R. M. Lerner, M. Perron, G. B. Pike, L. Richer, L. Steinberg, S. Veillete, and Z. Pausova. 2007. Morphological properties of the action-observation cortical network in adolescents with low and high resistance to peer influence. *Social Neuroscience* DOI: 10.1080/17470910701563558; First published on September 20, 2007.

Regnerus, M., C. Smith, and B. Smith. 2004. Social context in the development of adolescent religiosity. *Applied Developmental Science* 8: 27–38.

Reich, K. H. 1998. Psychology of religion: What one needs to know. *Zygon: Journal of Religion and Science* 33: 113–20.

Scales, P.C., P. L. Benson, N. Leffert, and D. A. Blyth. 2000. Contribution of developmental assets to the prediction of thriving among adolescents. *Applied Development Science* 4: 27–46.

Search Institute. 1984. *Young adolescents and their parents*. [Unpublished raw data archive]. Minneapolis, MN.

Sherrod, L., C. Flanagan, and J. Youniss. 2002. Dimensions of citizenship and opportunities for youth development: The what, why, when, where, and who of citizenship development. *Applied Developmental Science* 6: 264–72.

Smith, C. 2003. Theorizing religious effects among American adolescents. *Journal for the Scientific Study of Religion* 42: 17–30.

Smith, C., M. L. Denton, R. Faris, and M. Regnerus. 2002. Mapping American adolescent religious participation. *Journal for the Scientific Study of Religion* 41: 397–612.

Spear, L. P. 2000. Neurobehavioral changes in adolescence. *Current Directions in Psychological Science* 9: 111–14.

Templeton, J. M. 1995. *The humble approach: Scientists discover God*. Philadelphia: Templeton Foundation Press.

Youniss, J., J. A. McLellan, and M. Yates. 1999. Religion, community service, and identity in American youth. *Journal of Adolescence* 22: 243–53.

Conceptual Issues in Operationalizing Spirituality in Positive Youth Development

2

The Spirit of Spiritual Development

Carl N. Johnson

The object of this chapter is to ground spiritual development in "spirit." Although this root connection seems obvious, it has been little considered in the recent developmental literature.[1] The term *spirit* does not even appear in the index of the *Handbook of Spiritual Development in Childhood and Adolescence* (Roehlkepartain, King, Wagener, and Benson 2005).

To be recognized as a core domain, spiritual development must be about something real. Physical development is about the development of bodies, cognitive development is about development of minds, and social development is about the development of social behavior. Development obviously divides into these parts. But what exactly is spiritual development about? The answer is that spiritual development is not about any part of development. It is about a dynamic reality that is greater than these parts—the vital force that organizes the whole.

Spirit is the vital force of life—the creative source of every kind of thing. It is what generates physical, biological, psychological, and social kinds, organizing the world as we know it. As the universal ordering force, this idea is the foundation of process metaphysics, emerging in ancient cosmology and elaborated in contemporary process philosophy (Bergson 1911; Rescher 1996).[2] It is also essential to the ordinary way we think about life.

We commonly think about life as a whole. In this sense, we recognize that life, or spirit, can be full or empty, deep or shallow, heavy or light. Life can be good or bad, true or false, beautiful or ugly. It can be ordered or fragmented, purposeful or aimless, free or strained. Reality is thus framed in structural, modal, and directional dimensions of intrinsic value (Nozick 1989).

In this sense, we commonly think about the spirit of everything. We readily refer to the spirit of kinds, individuals as well as collective life—the spirit of childhood or youth, the spirit of a nation or a school, the spirit of a historical time or place, the spirit of religion, art, or science. Spirit is not just about human minds. Every matter of human concern, natural and cultural, is framed in terms of dynamic qualities of life (cf. Nozick 1989; Alexander 2002).

Reality in this sense is full of life. More than mere objective existence, reality is a vital matter. We know that young children distinguish real identifiable objects from figments of the imagination (Wellman and Estes 1986; Wellman 1990; Wellman and Johnson in press). But being "real" is not just a matter of objective existence; it is a matter of vital existence. Thus, we recognize that a person may be actually alive and yet have "no life." Such a person needs to "get a life," one that is "more real."

From this standpoint, spiritual development rests upon a distinct human capacity to become aware of what is truly vital in life. From the outset, human beings are adaptively oriented to attend to qualities of dynamic organization (cf. Hunt 1995). What develops is an increasingly conscious appreciation of what it is that genuinely fills reality with life. The idea of "spirit" distinctly orients attention to this vital matter, framing what it means to thrive (Lerner 2004, 2008), be vitally engaged (Nakamura and Csikszentmihaly 2003), or uplifted (Haidt 2003).

Spiritual development is about orienting life toward what most vitally matters. It is about fostering richer, deeper, fuller life by carefully attending to its spirit. William James (1990) describes the spirit of religion in these terms: "Not God, but life, more life, a larger, richer, more satisfying life, is, in the last analysis, the end of religion. The love of life, at any and every level of development, is the religious impulse" (453). The theologian Jurgen Moltmann (1997) comes to the same conclusion: "True spirituality is the rebirth of the full and undivided love of life. The total Yes to life and the unhindered love of everything living are the first experiences of the Holy Spirit" (85).

This chapter offers a sketch of the development of this love of life from the ground up. The ontology of the idea of spirit is initially sketched, framing the domain of spirit. A developmental framework is then described, followed by illustrations of development of spirit at each level.

THE ONTOLOGY OF SPIRIT

How does the idea of spirit develop and provide the ground for a distinctly spiritual way of life? The answer, proposed here, builds upon recent thinking about how cognition, evolution, and culture are intertwined in human development. The guiding idea is that evolution and culture work together in the design of a human organism that prepared to attend to what is adaptively relevant in the human life-world.

We now know that children are adaptively prepared to distinguish mental from biological and physical kinds of things (Bloom 2004; Wellman and Johnson in press; Bering and Bjorklund 2004). These ontological distinctions notably coincide with elements in the "great chain of being," common in traditional human cosmologies (see Wilbur 2006). Physical reality is on the bottom of this chain, next is biology, then mind and soul. The most inclusive reality, however, is spirit. While bodies, minds, and souls are understood to be parts of the natural order, spirit accounts for how this natural order itself comes into being.

Spirit in this sense is a higher-order reflective idea (see Johnson 2000). It is the ordering force that brings all order into being. The idea of spirit naturally arises with children's ability to reflect upon the natural-order world, as it is intuitively appears to be. The metaphysical question is how the apparent existence of physical, biological, psychological, and social order comes into being; the natural inference is that there must be some universal ordering force.

Ontologically, it is important to emphasize that spirit is not any substantial kind of thing—not physical or mental, biological or social. While young children intuitively distinguish physical and mental entities, as well as natural and intentional causes, spirit stands out as an adualistic force. Spirit is boundless, as it is both immanent in all things and transcendent in possibility.

Spirit is a metatheoretical idea that underlies reality as it intuitively appears to be (Johnson 2000). As an essentially unbounded, adualistic force, spirit cannot be defined as any kind of natural or supernatural thing. In this regard, the spiritual domain can be distinguished from the domain of supernatural religious belief (cf. Boyer 1994; Bloom 2004; Bering 2006). Spiritual development is not about acquiring beliefs about otherworldly minds; it is about being oriented to what is truly vital in life.

Are young children biologically prepared to discern the reality of spirit? Admittedly, spirit is a comparatively difficult reality to grasp. But, before knowing anything about the world, infants appear to be oriented to centers of organizing energy. The human spirit literally rises as an infant's whole being is energized, focused, and organized in dynamic relationship to the positive organizing energy of an Other (cf. Trevarthen 2001). Subsequently, the whole of reality opens up as a toddler gains the capacity to share, show, and talk about reality with others.

The human spirit grows as with the developing capacity of children to communicate and participate in the collective reality of human life. With reflection, children become able to frame life in a world that consists of institutional entities (school, science, religion, sports, occupations). In turn, recursive reflection on this world opens consideration of possible ways of life, such as a life of spirit, a life of materialism, or a life of mind. From this standpoint, spiritual life exists as a possibility of recursively framing a finite life in relation to the infinite spirit of life as a whole. Among possible ways to frame a life, a spiritual way is distinctively attentive to the spirit of life itself. In its fullest expression, a spiritual life organizes a person's whole being (mind, heart, body, and soul) in relation to life in its fullest sense.

FRAMING DEVICES

Spirit is an abstract device that directs attention to organizational processes of life. Development is about the abstraction of this distinct framing device. The general idea, grounded in recent cognitive-development theory, is that the human cognitive system gets off the ground with simple devices that function selectively to pick up units of information that adaptively orient the organism to world. There are a number of such devices, selected through evolution, that attune the human organism to what is adaptively relevant, especially to the shared intentionality that constitutes the human cultural life-world (Tomasello and Rakoczy 2003).

Three rudimentary framing devices have been well described. From early on, infants begin to attend selectively to mechanical, intentional, and vital forces in the world. Thus, infants begin to discern that some kinds of things in the world are moved by external force (mechanical), while other kinds of things appear to be moved on their own (agency). In addition to these internal and external forces, infants are also attuned to the peculiar way in which certain external things (food, water, air) are vitally necessary for life, as they must be literally assimilated into living bodies. (See Wellman and Gelman 1998; Inagaki and Hatano 2004.)

At the outset, framing devices simply function to orient the infant to relevant information in the world. Importantly, however, this information is only significant in the context of a higher-order collective enterprise, as children come to understand reality in the context of sharing, showing, naming, and communicating about everyday things in the world.

From this ground, development proceeds with increasing capacities to reflect upon reality in higher-order ways. These successive levels are summarized in Table 1, below. At each level, framing devices function to transform incoming information into elements (schemes/concepts) that then self-organize into more coherent systems (frameworks/theories). At the primary level, intuitive devices provide the input for the construction of intuitive theories (Wellman and Gelman 1998; Wellman and Johnson in press). This constitutes the everyday reality of individual actions and objects in the world, as it appears to be given directly in experience. At the secondary level, framing devices are abstracted from the intuitive level in the form of reflective ideas that are the input for the construction of first-order reflective "theories." Thus, children acquire collective ideas about natural and artificial "kinds" of things that constitute a human life-world (Habermas 1987).

At this reflective level, children enter into the collective reality of human life that consists of all sorts of collective ideas that extend beyond boundaries of intuition. These ideas distinctly depend on trust in the testimony of others (Harris and Koenig 2006; Harris 2007), constituting "reflective beliefs" (Sperber 1996, 1997). Epistemologically, as Kuhn and Dean (2004) describe it, these life forms are absolutist. While reality is initially experienced as intuitively given (realist), it is now experienced as collectively given (trust). While initially encoded in isolation, these reflective beliefs gradually become organized into frameworks that constitute the life forms of culture.

Together, the primary and secondary frames constitute the ordinary way people make sense out of the world, first, in relation to things that are directly experienced and, then, in relation to the kinds of things that frame collective real-

TABLE 1. **Levels of Framing**

	Devices	*Frameworks*	*Epistemology*[1]	*Experience*
Primary	Intuitive Devices	Intuitive Frameworks	Realist	Intuitively Given
Secondary	Reflective Ideas	Reflective Frameworks	Absolutist	Collective Trust
Tertiary	Meta Ideas	Meta-Reflective Frameworks	Multiplist	Personally Chosen
Quadri	Critical Ideas	Critical Frameworks	Evaluativist	Critically Chosen

1. From Kuhn & Dean, 2004

ity. In either case, reality appears to be given directly in experience, first, in the given reality of intuition, then, in given reality of collective ideas. (See Table 1.)

From the reflective level upward, human life is framed by "transcendental" kinds of things that extend beyond the boundaries of intuitive reality. School, science, and religion; language, mathematics, and art—these are real, transcendental kinds, as they exist beyond the boundaries of subjective/mental and objective/physical kinds. There is nothing otherworldly about this transcendence, even though there is a natural tendency to interpret it as such. The simple truth is that human life is framed in terms of collective ideas. Mostly, these ideas simply serve to orient human beings to what is collectively relevant in actual life. However, such ideas can also function to direct attention to otherworldly possibilities, including the existence of spirits, deities, angels, and the afterlife.

The tertiary level emerges with the capacity to reflect consciously upon alternative ways to frame the life-world. Essentially, this marks the emergence of true philosophical thinking, or meta-thinking, which distinctly requires the capacity to think about ideas as ideas (possibility) separate from the actual reality that they frame. Instead of accepting the life-world as a given, this level brings the realization that reality can be framed in radically different ways. Most strikingly, it is now possible to imagine that reality is just lifeless matter, completely devoid of spirit. As Michael Chandler describes it, this opens the possibility of existential loneliness and the devastating loss of faith in the life-world (Chandler 1987; Chandler, Lelonde, Sokol, and Hallett 2003). Youth is a transitional period in identity development, posing risk and opportunity, entertaining radically different forms of life, against the terrible prospect that existence by itself, apart from cultural framing, is entirely devoid of vital significance (cf. Hart, Shaver, and Goldenberg 2005). At this juncture in development, the risk is that adolescents will tend either to cling to absolutism or to reduce everything in life to a matter of personal choice.

The fourth level marks an additional capacity to evaluate life-worlds critically. For the first time in this developing ontological edifice, cognition is able to evaluate human life-worlds as life-worlds. Beyond recognizing that people have different subjective views, this late-developing capacity involves evaluating the adaptive function of life-worlds, insofar as they effectively serve to energize real, valuable human lives.

Within this scheme, we can trace the developing idea of spirit. At the outset, preconceptually, infants are prepared to be attuned to the collective reality of human life. The spirit of the human infant is literally energized and organized as it is attuned to the energy of a more highly organized human being. This budding spirit grows, as toddlers naturally seek, share, show, and talk about what they are prepared to experience as being vitally significant. Thus, children acquire an intuitive understanding of the world, vitalized in the context of interpersonal sharing. The idea of spirit, however, arises with reflection, as children are engaged in thinking about the reality of collective kinds of things. Spirit arises as an implicit idea that explains order in the world, as it is conceptually given. This spontaneous idea is culturally framed in terms of conventional ideas about spirit. In leaving childhood, youths, in turn, gain the capacity to think about spirit in a subjective sense. In this regard, adolescents can reflect upon the state of their spirit, experiencing existential emptiness and thinking about what it means to "get a life." In this sense, adolescents can think about possible lives, thinking about them as personal matters of choice. Finally, with maturity, young adults may come to appreciate more fully and deeply how human life is actually a dynamic reality, a concept that is wholly dependent on life-worlds, a transpersonal kind.

For the most part, spirit is a completely ordinary idea, framing our common conception of life. However, with reflection, extraordinary ideas about spirits are easily framed. And, absent high-order awareness, there is a natural transcendental tendency to reify spirit as an otherworldly kind of thing. At first, beliefs about such things are primarily a matter of collective trust (in what you're being told or what is commonly assumed to be the case). With meta-reflection, however, such beliefs can gain the force of personal conviction, as spirit may reflectively appear to be grasped in its pure, abstract form. With critical awareness, human beings gain the capacity to think about the actual ground of spirit, leading to an integral spirituality (Wilber 2006).

THE SPIRIT OF INFANCY

The framework described above rests on the idea that human infants are prepared to be oriented to the human spirit at birth. Before knowing anything at all about the actual kinds of things that exist in the world, human infants are attracted to distinctive centers of the organizing energy (Hunt 1995; Mandler 1992, 2004). Indeed, an infant's whole perceptual-motor-emotional-cognitive system is oriented, energized, and organized in dynamic transactions with a more highly organized human being (see Trevarthen 2001). In playful interaction, the infant literally comes alive as an organized human being in relation to a Being of higher organizational value. Engaged in playful interaction, the infant's spirit rises in ecstasy. Faced with an unresponsive other (a "still face"), the infant's spirit falls into despair; life disorganizes. Responding to this distress, the Other quells the chaos, bringing peace, order, and tranquility.

The whole spiritual drama of life is evident in the dynamic interplay of infant-parent interaction. From the outset, it seems, the human spirit is wholly dependent upon connection to something of higher value. The human spirit emerges in the dynamic flow of affective attunement. From this standpoint, it becomes evident how the dynamic systems that underlie attachment frame the later development of religious ideas (Kirkpatrick 2005).

Effectively, the infant comes into the world with powerful, dynamic systems-detection devices. From the outset, the infant is vitally oriented to dynamic centers of value in the world. Before knowing anything definitely about self or the world, the infant's whole perceptual, motor, affective, and cognitive being is prepared to respond to organization of a higher kind.

Drawing on the work of Hess, Trevarthen (2001) describes how the vital existence of the infant is framed by a two-dimensional affective field. One dimension is termed *energy* (ranging from strong, attentive, forceful to weak and inattentive); the other is *trophic* or *nurturing* (from positive and attractive to negative and avoidant). Here is spirit in its basic form: energy directed positively toward what nurtures life.

Everything in the infant's life arises from this spiritual field. Before knowing anything about the world, the infant organizes itself in spirit. When something attractive enters its perceptual field, the infant's whole perceptual-emotional-cognitive-motor system is oriented toward engaging with it. Hence, the infant's spirit develops in dynamic intersubjective relation to others, driven by an intrinsic drive to connect to higher-source energy.

It is important to recognize that spirit is an adualistic dynamic reality from the very outset. The infant's spirit, as organizing energy, is not an autonomous private mental state but a wholly interdependent dynamic relationship. The infant's spirit exists only insofar as it is dynamically engaged.

THE SPIRIT OF CHILDHOOD

As infants develop into toddlers, they develop an intuitive conceptual understanding of objects and agents in the world. Presumably, they also develop an intuitive understanding of spirit, especially with regard to what is vitally engaging. For example, it seems likely that children have intuitions about how cultural practices like ritual, dance, and meditation genuinely serve to elevate, organize, or focus life. While this seems obvious enough, we know almost nothing about what children understand about the ordinary ups and downs of the human spirit.

Absent a systematic look at this domain of understanding, the case of Helen Keller will be used to illustrate dramatically insights that are common to all children: how participating in the collective naming of things brings an intuitive understanding of spirit into the world, which subsequently provides the ground for a reflective idea of spirit.

Keller (1996) describes the sudden, conscious enchantment with life that occurred when she acquired the capacity to name objects and verbally communicate at the age of 7. Having developed normally up to the age of 20 months, Helen suddenly lost both her sight and hearing due to an illness. Helen describes the awaking of her soul, at age 7, when her teacher Anne Sullivan suddenly opened her world to a fusion of language, love, and reality. Before Anne's intervention, Helen describes herself as being like a ship lost in a dense fog, with no tools—no compass or sounding line—to give a sense of direction. Language was the key to her enlightenment. Her discovery that everything had a name opened an infinite kinship with everybody and everything in the world. As Helen explains, "I did nothing but explore with my hands and learn the name of every object that I touched; and the more I handled things and learned their names and uses, the more joyous and confident grew my sense of kinship with the rest of the world" (Keller 1996, 12).

At first, Helen's sense of spiritual kinship was simply intuitive, but, soon after, she describes the emergence of the distinctively reflective idea of love. She got this idea from her teacher. At first, she was baffled by the lack of any concrete referent of this term. But with some guidance by her teacher, Helen came to this realization: "The beautiful truth burst upon my mind—I felt that there were invisible lines stretched between my spirit and the spirits of others" (16).

Thus, cued by her teacher's reference to the abstract idea of love, Helen Keller spontaneously constructed a reflective idea of spirit. This idea effectively served to account for the actual way in which Helen's life was interpersonally connected. Spirit, thus, appears as a real, transpersonal force in the world.

Beyond this example, there is clear evidence that young children are prepared to think about life in vital terms, at least in the domain of biology (Inagaki and Hatano 2004). Evidence in this case shows that children infer that there is some vital power of life force in substances like food and water that makes humans able to act, grow, and fend off disease. These naïve, vitalistic biological intuitions are striking because they clearly demonstrate that young children are prepared to think about energy transfer across body boundaries. Just as children infer that there is something in food (or other substances) that vitalizes the body, presumably they also understand that there is something in language, dance, and music that vitalizes the human spirit. It is not clear whether intuitions about biological vital force are to be related to intuitions about spirit. Spirit distinctly has to do with energy that comes from participating in collective life. We have overlooked this domain of spirit.

A small body of recent work on children's concept of soul is notable insofar as it essentially draws on the idea of spirit (see Bering 2006; Richert and Harris 2006). Just as children develop reflective ideas about natural kinds (human kind, dog kind, and the like), so too they develop ideas about individual human kinds. The reflective idea of soul apparently arises from the simple inference that, because person is identified as an individual kind, there is some organizing force that creates and maintains this kind. Thus, just as the human spirit accounts for why human beings are the way they are, a person's individual spirit accounts for his or her soul. Research in this field shows that school- age children are beginning to frame an idea of soul that is distinct from the idea of mind. Instead of existing as a mental substance, the soul refers to the organization of an individual human spirit, as a whole, living, cognitive-emotional-moral being.

THE SPIRIT OF YOUTH

As children enter adolescence, they become capable of reflecting on alternative metaphysical possibilities in an explicit way. In this sense, they can imagine alternative ways of life—a life of spirit, a life of pleasure, a life of mind—and think explicitly about what it means to "get a life" or suffer from having "no life." Just as we know little about children's implicit understanding of life in this vital sense, so too we know little about adolescents' explicit ideas of this sort.

One might think that ideas of spirit would be evident in adolescents' religious views, but the opposite may often be the case. While early religious ideas commonly emphasized the spirit in all things, modern religion is characteristically dispirited (Moltmann 1997; Weber 2003). A recent survey of American adolescents indicates the persistence of this disenchanted religious worldview. Smith (2005) reports that American teenagers commonly share a collective creed, described as "Moralistic Therapeutic Deism," which consists of the following tenets:

A God exists who created and orders the world and watches over human life on earth.

God wants people to be good, nice, and fair to each other, as taught in the Bible and most world religions.

The central goal of life is to be happy and to feel good about oneself.

God does not need to be particularly involved in one's life, except when God is needed to resolve a problem.

Good people go to heaven when they die.

What is strikingly absent in this creed is any appreciation that spirit is immanently present in the world. God exists exclusively as an autonomous, supernatural agent watching over the world from the outside. The goal of life is personal satisfaction, not a fuller life.

Of course, there are exceptions to this dispirited worldview: postmodern spirituality is largely an effort to reclaim spirit in an immanent sense (Goodenough 1998; Moltmann 1997, 2001; Wilber 2006). The example of Piaget's own adolescent spiritual development illustrates how one modern youth achieved this end.

As Piaget (1971) saw it, the turning point of his life was the discovery of spirit at the age of 15. His godfather, concerned about the narrowness of Piaget's scientific interests, invited the teenager to his summer home where he guided him in reading Henri Bergson's *Creative Evolution*. Piaget describes this experience as opening his eyes to larger philosophical questions, of which he was previously unaware. These philosophical questions led him to rethink the conventional frames of religion and science that he acquired as a child.

Growing up, Piaget's life was exposed to vitally conflicting frames. His mother was a Christian believer. His father, a nonbeliever, successfully encouraged Piaget to narrowly pursue scientific truth. His godfather, in contrast, encouraged Piaget to reflect on reality as a whole. In reading Bergson, Piaget was struck by the idea of unifying these framing influences together. As he explains it, "in a moment of enthusiasm close to ecstatic joy, I was struck by the certainty that God is life under the form *élan vital*, and my biological interests provided me at the same time with a small sector of study. Internal unity was thus achieved in the direction of an immanentism, which has long satisfied me" (Piaget 1971, 5).

The idea of life-spirit redirected Piaget's thinking about value, away from a supernatural God outside the world, toward an immanent ordering force in nature itself. Thus, he was able to view scientific interests as directed toward something of ultimate reality and value. While the idea of life-spirit changed the direction of Piaget's thinking, this was only the beginning of a prolonged developmental struggle, extending through his college years, culminating in his autobiographical novel *Recherche*. Importantly, Piaget's development during this period was scaffolded by the Swiss Christian Students' Association (ACSE), an evangelical Protestant organization expressly designed to help youth find Christ as a "living personality," not merely as "theological argument." The organization saw an inherent link between religion and youth, especially as it appealed to their noblest aspirations: "heroism of the will, intellectual emancipation and generosity of the soul" (Flournoy 1905, 46, cited in Vidal 1994, 107).

The ACSE fostered spiritual development through what was taken to be a normal period of crisis in the college years, toward a period of resolution in early adulthood. As a member of this group, Piaget worked deliberately on his development, a period that culminated in writing a "formation novel," which was considered to be an ideal format for reflecting upon one's spiritual quest. In his novel, through the character of Sebastian, Piaget describes his spiritual journey through a period of "preparation" to "crisis" ending finally in "reconstruction." The crisis, as Piaget described it, was propelled by a profound "disequilibrium," a conflict between science and religion, truth and value, which had to be resolved, given Sebastian's "thirst for absolute truth."

The "disequilibrium" that propelled Piaget's development was not merely a cognitive disequilibrium; it was a vital disequilibrium that had a profound bearing on everything in his life. In this struggle, Piaget's own life was totally bound up with an effort to both know and love reality in its fullest sense. What drove the crisis and its resolution was the very real spirit of life.

What can we learn from Piaget's example, as it compares to the conventional religious ideas of the typical American adolescent? First, certain conventional religious ideas are remarkably persistent. In this regard, it is especially important to consider whether contemporary youth are developing ideas about spirit outside the conventions of religion. Second, the intense, vital conflict that drove Piaget's development depended on the existence of salient life forms in his personal life. Piaget's spiritual struggle grew out of his father's passionate love of scientific truth, his mother's conflicting Christian values, and his godfather's encouragement to think about ultimate matters. Moreover, once this crisis was triggered, development was extensively scaffolded by a

Christian community that framed youth as a profound period of spiritual crisis and development. Given these conditions, however, the key organizing idea in Piaget's metaphysical development was spirit, as this simple idea oriented Piaget to thinking about how organization arises from immanent processes.

VITAL REALITY

Piaget's vital spiritual crisis essentially framed the metatheory of his science. In resolving his spiritual crisis in *Recherche*, Piaget proposes a "research program" based on a "science of genera" with a theory based on the dynamic equilibrium of wholes and parts (Vidal 1994, 183). Importantly, this "science of genera" is a reconstruction of ancient philosophical ideas about vital order. In this, Piaget draws on Bergson's insight that ancient philosophers were distinctly concerned not with causal mechanisms but with explaining why the natural order is as it is. Aristotle, for example, regarded every kind of thing as having a natural way of being. Each individual tends to realize its place in the natural order. A stone tends to be stonelike, just as a horse tends to be horselike or a human tends to be humanlike. In every case, spirit is the vital force that makes things as they are. In these spiritual tendencies, reality and value are unified in a conception of the natural order of being. Reality, in this sense, is vitally alive.

As Piaget's thinking matured, he became able to evaluate critically the value and limitations of "spirit" as a framing device. While he gave up the idea of *spirit* as a scientific term, he continued to appreciate the essential value of a vital framing of life. Distinguishing between cognitive and vital meaning, he regarded science as being properly limited to determining actual, objective matters, whereas vital meaning has to do with the essential matter of life, which, he explains,

remains central from the point of view of human existence and the thinking subject, for we have a choice between a life without values, a life with relative and unstable values, and a life involving values experienced as absolute, engaging one's whole being. To deny such a problem because it is a vital one and without positive cognitive solutions is plainly absurd, since it constantly occurs and forces itself on us in the form of "engagement," even if we do not know how to formulate it intellectually. (Piaget 1971, 43)

Piaget's stance is bold: the meaning of a life depends upon the degree to which a person's "whole being" is vitally engaged in values experienced as "absolute." In other words, the meaning of life depends upon being fully committed to ideals. In his maturity, Piaget certainly was aware of the limits of the vitalistic frame, but he nevertheless saw that it was essential to the most fundamental question in life.

The vital disequilibrium that drove Piaget's own development fell into the background, as his science focused almost exclusively on cognitive meaning alone. Removed from the vital ground that propelled his own development, Piaget framed cognitive development as a matter of increasing abstraction, free of its affective, embodied ground. In this regard, he neglected the spirit of human knowing, as it aims to embrace reality in a fuller sense.

While Piaget (1971) appreciated the vital question of the meaning in life, he did not explicate how deeply reality is framed in a vital sense. The philosopher Robert Nozick (1989) points in this direction. Contrary to the idea that human beings primarily seek pleasure, Nozick argues that human beings inherently seek lives that are "more" real—lives that are actually more deeply, more fully connected to greater reality. This reality, however, is not just an abstract ideal; it is full of qualities like weight, depth, intensity, and texture. Whether something actually exists in the world (cognitive/scientific reality) is a minimal condition of reality. Reality in its full sense is framed in dimensions and degrees that are grounded in bodily experience and end-anchored in absolute ideals. In this sense, things are more real if they are more integrated, focused, energetic, full (rich), and substantial, as these mark fundamental dimensions of reality (see Figure 1).

Framed by these dimensions, every given matter becomes more real insofar as its embodied reality is framed by the ideal. Nozick was not a developmental-

FIGURE 1. Nozick's Dimensions of Reality. Reprinted with permission from Nozick 1989.

	Inherent	*Relational*	*Fulfillment or Telos*	*Ideal Limit*
Integration	Value	Meaning	Completeness	Perfection
Substance	Weight	Importance	Greatness	Omnipotence
Light	Truth	Goodness	Beauty	Holiness
Scope	Depth	Amplitude	Height	Infinitude
Energy	Intensity	Vitality	Creativity	Infinite Energy
Focus	Sharpness	Vividness	Individuality	Sui Generis
Fullness	Texture	Richness	Wholeness	All-Encompassing

ist, but it is not hard to appreciate that the dimensions he frames arise in the embodied experience of infancy. Reality comes into consciousness not merely as an abstraction but as richly vibrant matter. As Trevarthen (2001) explains, reality is originally framed by dynamic, affective qualities of force, effort, pressure, intensity, focus, freedom, and the like. Thus, reality is embodied, relational, and dynamic (Overton 2003).

From this standpoint, the driving force, metric, and ultimate end of reality becomes clear. From the outset, human beings are intrinsically driven toward a positive interpersonal engagement in the world. Reality grows with the human capacity to expand this positive engagement to the most encompassing ideals.

Spirit is real. It originates in the dynamic interchange with a significant other and develops with increasing participation in the collective realities of human life. Spiritual development is about attending to this reality, as it genuinely fosters the flourishing of life.

ACKNOWLEDGMENTS

Thanks to my wife, Deborah Landen, whose spirit was essential for writing this chapter. Thanks also go to Paul Harris, Melanie Nyhof, and Rob Roeser for kindly indulging and encouraging nascent ideas in early states of inchoate formulation.

NOTES

1. This is not to say that the subject of spirit has been generally neglected. Pneumatology, the study of spirits or spiritual beings, has waxed and waned especially with regard to the modern emergence of the "spiritual" domain (see Moltmann 2001).

2. As A. N. Whitehead saw it, this generic creative force, which brings all novelty into being, is a "category of the ultimate," the "universal of universals," the ultimate essence of reality (cited in Rescher 1996, 80).

REFERENCES

Alexander, C. 2002. *The nature of order. Book 1: The phenomenon of life.* Berkeley, CA: The Center for Environmental Study.

Bergson, H. 1911. *Creative evolution.* Trans. Arthur Mitchell. London: Macmillan.

Bering, J. M. 2006. The folk psychology of souls. *Behavioral and Brain Sciences* 29: 453–98.

Bering, J. M., and D. F. Bjorklund. 2004. The natural emergence of reasoning about the afterlife as a developmental regularity. *Developmental Psychology* 40: 217–33.

Bloom, P. 2004. *Descartes' baby: How the science of child development explains what makes us human.* New York: Basic Books.

Boyer, P. 1994. *The naturalness of religious ideas.* Berkeley: University of California Press.

Chandler, M. J. 1987. The Othello effect: An essay on the emergence and eclipse of sceptical doubt. *Human Development* 3, no. 3: 137–59.

Chandler, M. J., C. E. Lalonde, B. W. Sokol, and D. Hallett. 2003. *Personal persistence, identity, and suicide: A study of native and non-native North American adolescents.* Monographs for the Society for Research in Child Development, serial no. 273, 68 (2).

Goodenough, U. 1998. *The sacred depths of nature.* New York: Oxford University Press.

Habermas, J. 1987. *The theory of communicative action.* Boston: Beacon Press.

Haidt, J. 2003. Elevation and the positive psychology of morality. In C. L. M. Keyes and J. Haidt, eds., *Flourishing: Positive psychology and the life well-lived*, 83–104. Washington, D.C.: American Psychological Association.

Harris, P. L. 2007. Trust. *Developmental Science* 10, no. 1: 135–38.

Harris, P. L., and M. A. Koenig. 2006. Trust in testimony: How children learn about science and religion. *Child Development* 77: 505–24.

Hart, J., P. R. Shaver, and J. L. Goldenberg. 2005. Attachment, self-esteem, worldviews, and terror management: Evidence for a tripartite security system. *Journal of Personality and Social Psychology* 88: 999–1013.

Hunt, H. T. 1995. *On the nature of consciousness.* New Haven, CT: Yale University Press.

Inagaki, K., and G. Hantano. 2004. Vitalistic causality in young children's naïve biology. *Trends in Cognitive Sciences*, no. 8: 356–62.

James, W. 1990. *The varieties of religious experience.* New York: Random House. Originally published in 1902.

Johnson, C. N. 2000. Putting different things together: The development of metaphysical thinking. In K. Rosengren, C. N. Johnson, and P. Harris, eds., *Imagining the impossible: Magical, scientific, and religious thinking in children.* New York: Cambridge University Press.

Keller, H. 1996. *The story of my life.* New York: Dover Publications.

Kirkpatrick, L. A. 2005. *Attachment, evolution and the psychology of religion.* New York: Guilford Press.

Kuhn, D., and D. Dean. 2004. Metacognition: A bridge between cognitive psychology and educational practice. *Theory into Practice* 43, no. 4: 268–73.

Lerner, R. 2004. *Liberty: Thriving and civic engagement among America's youth.* Thousand Oaks, CA: Sage Publications.

Mandler, J. M. 1992. How to build a baby II: Conceptual primitives. *Psychological Review* 99: 587–604.

———. M. 2004. *The Foundations of mind: Origins of conceptual thought.* New York: Oxford University Press.

Moltmann, J. 1997. *The source of life.* Minneapolis, MN: Fortress Press.

———. 2001. *The spirit of life.* Minneapolis, MN: Fortress Press.

Nakamura, J., and M. Csikszentmihalyi. 2003. The construction of meaning through vital engagement. In C. L. M. Keyes and J. Haidt, eds., *Flourishing: Positive psychology and the life well-lived*, 83–104. Washington D.C.: American Psychological Association.

Nozick, R. 1989. *The examined life.* New York: Simon and Schuster.

Overton, W. F. 2003. Development across the life span: Philosophy, concepts, theory. In R. M. Lerner, M A. Easterbrooks, and J. Mistry, eds., *Comprehensive handbook of psychology: Developmental psychology*, Vol. 6. 13–42. Editor-in-chief: Irving B. Weiner. New York: Wiley.

Piaget, J. 1971. *Insights and illusions of philosophy.* New York: World Publishing Company.

Rescher, N. 1996. *Process metaphysics.* Albany: State University of New York Press.

Richert, R. A., and P. L. Harris. 2006. The ghost in my body: Children's developing concept of the soul. *Journal of Cognition and Culture* 6: 409–27.

Roehlkepartain, E. C., P. E. King, L. M. Wagener, and P. L. Benson, eds. *The handbook of spiritual development in childhood and adolescence.* Thousand Oaks, CA: Sage.

Smith, C. 2005. *Soul searching.* New York: Oxford University Press.

Sperber, D. 1996. *Explaining culture: A naturalistic approach.* Oxford: Blackwell.

———. 1997. Intuitive and reflective beliefs. *Mind and Language* 12, no. 1: 67–83.

Tomasello, M., and H. Rakoczy. 2003. What makes human cognition unique? From individual to shared to collective intentionality. *Mind and Language* 18, no. 2: 12147.

Trevarthen, C. 2001. *Infant Mental Health Journal* 22, nos. 1–2: 95–131.

Vidal, F. 1994. *Piaget before Piaget.* Cambridge, MA: Harvard University Press.

Weber, M. 2003 *The Protestant ethic and the spirit of capitalism.* Mineola, NY: Dover Publications.

Wellman, H. M. 1990. *The child's theory of mind.* Cambridge, MA: MIT Press.

Wellman, H. M., and D. Estes. 1986. Early understanding of mental entities: A reexamination of childhood realism. *Child Development* 57: 910–23.

Wellman, H. M., and S. A. Gelman. 1998. Knowledge acquisition in foundational domains. In D. Kuhn and R. S. Siegler, eds. *Handbook of child psychology.* Vol. 2: Cognition, Perception and Language, 5th ed. New York: John Wiley, 523–74.

Wellman, H. M., and C. N. Johnson. In press. Developing dualism: From intuitive understanding to transcendental ideas. In A. Antoinette, A. Corradina, and E. J. Lowe, eds., *Psychophysical dualism today: An interdisciplinary approach.* Lanham, MD: Lexington Books.

Wilber, K. 2006. *Integral spirituality.* Boston and London: Integral Books.

3

Spirituality and Positive Youth Development

The Problem of Transcendence

W. George Scarlett

Spirituality in late adolescence and early adulthood may be defined as emerging faith in something greater than oneself. For some, this emerging faith means faith in God. For others, it means a nonreligious faith in a way of life. Whatever form emerging faith may take, there is the sense that something transcends, even if what transcends is something within the self.

The problem focused on in this paper is the problem of how to evaluate emerging faith when it happens. Do we evaluate using structural criteria only? Is transcendence all that matters? Is measuring how strong faith becomes the only or main measure of development? Do we need to consider the content of faith? To answer these and other questions, this paper discusses three well-known individuals: Jean Piaget, Abraham Lincoln, and Adolf Hitler—because each meets the criterion of undergoing a spiritual transformation to develop a strong faith; each represents a kind of thinking common to modern, post-Enlightenment society; and each adopts a very different understanding of transcendence. The comparison shows, therefore, that, in evaluating spiritual development, what matters is not so much whether transcendence figures in an individual's life but which transcendence figures. A discussion of Piaget's late adolescence and early adulthood provides our first example.

JEAN PIAGET

In his late adolescence and early adulthood, Piaget focused on the problem of how to reconcile religion with science. For him, the problem was intensely personal, so much so that it seemed he needed to resolve this problem before he could move on in life.

Piaget's resolution was to challenge orthodox, Calvinist faith in a personal, transcendent God, a God who is outside of natural law and a God who is a cause beyond the causes studied by science. During the decade that he focused on this problem, Piaget developed a new understanding of God. God, for Piaget, went *inward* and became a purely spiritual force, at once real and mystical. For Piaget, this mystical but real force was experienced when listening to conscience, when being active in following the norms of reason, and when being successful in aligning conscience and reason with the model of Jesus. Conscience, reason, and embodied spirituality, then, connected Piaget to an immanent God, one who (that) freed rather than limited and one who (that) subordinated the ego to that which is universally good (Reich 2005; Vidal 1987).

However, Piaget went further than simply working through for himself a personal solution to the science versus religion problem. He also took his own development as indicating religious-spiritual development in general. He makes this clear in a lecture he gave to members of the Swiss Christian Students Association:

If, beyond men, one examines the currents of thought that propagate from generation to generation, immanentism appears as the continuation of the impulse of spiritualization that characterizes the history of the notion of divinity. The same progress is accomplished from the transcendental God endowed with supernatural causality to the purely spiritual God of immanent experience, as from the semi-material God of primitive religions to the metaphysical God. Now—and this is the essential point—to this progress in the realm of intelligence corresponds a moral and social progress, that is, ultimately an emancipation of inner life. (Piaget 1928, 22)

This story about Piaget's own spiritual development illustrates two important points about the possibilities for spiritual development in late adolescence and its relation to thriving. First, the transformation in Piaget's thinking about God and spirituality came about through his own active participation with not only his specific and local communities and groups but also with his culture and historical period. As already indicated, he was an active member of the Swiss Christian Students Association, a group of young people who, like many other Christian groups in Europe at that time, were trying to make sense of how Christianity had failed to prevent World War I. The Swiss Christian Students' Association, with Piaget being the prime example, responded by, at least implicitly, blaming orthodox theology for the churches' failures to prevent the war.

Second, the development of his spirituality was crucial to Piaget's later achievements and partially explains the energy, persistence, commitment, and independence in his work. Even though he rarely mentions theology and his ideas about immanentism, Piaget kept essentials in his theology as undergirdings of his genetic epistemology (Vidal 1987). Anyone reading his groundbreaking work on the development of children's moral judgment can easily detect the same general themes as are outlined in his developmental scheme for religious and spiritual development. For Piaget, morality, like theology, develops insofar as the individual takes rules, which are initially experienced as sacred and outside the self (the analogue to a transcendent God), and brings them inside by treating them as creations of the self and by subordinating rules to conscience, the norms of reason, and consideration for multiple points of view (the analogue to an immanent God). The parallels, then, between the transformations in the young Piaget's theology and the transformations he describes much later in children's thinking about reality in general, but morality in particular, are striking—which is why it is reasonable to say that the development of the youthful Piaget's spirituality partially explains his later success and thriving.

So, is Piaget's solution the only solution to spiritual development leading to thriving? Must we conclude that good spiritual development rests on individuals' replacing a transcendent God with a God who is immanent? Not at all. Immanentism is not the only answer. Indeed, positive development is more than possible when an individual holds on to even a Calvinist image of God. Lincoln serves as a prime example.

ABRAHAM LINCOLN

Hein (1983) claims that the religious faith expressed in Lincoln's second inaugural address was there, in Lincoln, from very early on:

The central elements of Lincoln's mature religious faith were already present in that of the youthful Lincoln. In 1842 he wrote a letter to his closest friend, Joshua F. Speed, in which he discussed Speed's engagement to be married and stated his belief that "God made me one of the instruments of bringing your Fanny and you together, which union, I have no doubt He had fore-ordained." Here we have clearly expressed a strong belief in God's overruling providence and a conviction that Abraham Lincoln might be employed by God as an "instrument" to bring about the specific good of reconciliation. Both of these would be among the wartime themes of Lincoln's faith. (Hein 1983, 113)

But does the religious faith expressed in this letter truly have the same central elements as Lincoln's faith later on? I don't think so.

The faith expressed in the letter to Speed has all the characteristics of what Lincoln referred to as "the doctrine of necessity," whereby "the human mind is impelled to action, or held in rest by some power, over which the mind itself has not control" (Donald 1995, 15). This faith has none of the trust and devotion to a just and caring, albeit mysterious, God that so characterized Lincoln's faith later on.

More importantly, the doctrine of necessity has humans acting as God's puppets, their autonomy minimized, at least during critical moments. In contrast, Lincoln's faith later on always emphasized both God's omnipotence and humans' autonomy. These are not the days of miracles, Lincoln reminded others; they are the days for struggling to interpret cold, hard facts with the limited aid of reason.

The younger Lincoln's religious faith also lacked the visions of justice and caring that characterized the faith of the presidential Lincoln. In the 1850s, the possibility that slavery would spread throughout the country roused Lincoln to make explicit his previously intuitive grasp that the Declaration of Independence's principle of equality was the defining principle of American democracy. In having to make explicit his beliefs concerning the nation's moral mission, Lincoln also had to make explicit his mature faith in a just and caring God. This was never more evident than in his reaction to the news that most of Springfield's clergy and prominent church members were going to vote against him in the 1860 presidential election. To this news, Lincoln became deeply sad and replied:

I am not a Christian. God knows I would be one; but I have carefully read the Bible.... These men well know that I am for freedom in the territories, freedom everywhere, as free as the Constitution and the laws will permit, and that my opponents are for slavery. They know this, and yet with this book [the Bible] in their hands, in the light of which human bondage cannot live a moment, they are going to vote against me. I do not understand it at all. I know there is a God, and that He hates injustice and slavery. I see the storm coming and I know that His hand is in it. If He has a place and work for me, and I think He has, I believe I am ready. I am nothing, but truth is everything. I know I am right because I know that liberty is right, for Christ teaches it, and Christ is God. I have told them that a house divided against itself cannot stand, and Christ and Reason say the same, and they will find it so. (Macartney 1951, 43)

In this statement and many that followed, we see Lincoln connecting religion to a vision of a just and caring community of humankind, a vision outlined in both the American Declaration of Independence and in the Bible. In this connecting, Lincoln begins his tenure as a spiritual president working tirelessly to do what he can to make that vision reality.

The sequence in Lincoln's religious development was, then, as follows: As a young man, Lincoln acquired a good deal of religious knowledge, largely through his reading the Bible but also through participating in the religious conversations common in Protestant America during the first half of the nineteenth century. That knowledge spawned a faith in an impersonal, all-powerful God but not a strong faith in the sense of its being a continuous presence that affected Lincoln's daily discourse and actions. During the 1850s, when the issue of slavery brought out Lincoln's passionate faith in the principle of equality, Lincoln began to make explicit the link between his religious faith and his moral vision for the country. That transition ended with Lincoln's being elected president and with the realization that the weight of two issues, union and liberty, rested most heavily on his shoulders.

Lincoln's mature religious faith in a just and caring God and his identity as someone constantly striving to be an instrument of God's will, even as God's will remains forever veiled, are evident in his farewell address to his friends in Springfield. Just before he took office, in his farewell speech at Springfield, he said,

I now leave, not knowing when or whether ever I may return, with a task before me greater than that which rested upon Washington. Without the assistance of that Divine Being who ever attended him, I cannot succeed. With that assistance, I cannot fail. Trusting in Him who can go with me, and remain with you, and be everywhere for good, let us confidently hope that all will yet be well. To His care commending you, as I hope in your prayers you will commend me, I bid you an affectionate farewell. (Basler 1969, 568)

What we see in Lincoln's spiritual development, then, is not a rejection of a transcendent God but a deepening connection to a transcendent God through the principles and guidelines that Lincoln took to be sacred—in particular, Jesus' "summary of the Law" (to love God and neighbor as self) and the Declaration of Independence's principle of justice expressed as equal opportunity for all. The connection to a transcendent God was, then, indirect or filtered. God's will was defined neither in terms of elaborate dogmas nor in terms of prescriptions written in a sacred text, but in terms of guidelines and principles. For Lincoln, then, there was a normative ambiguity that called for the same kind of approach as advocated later by Piaget, an approach directed at integrating conscience and reason and the model of Jesus.

As indicated in the example of Lincoln, it is not transcendence that is the problem, but rather *which* transcendence and whether transcendence is used for good or for evil. Lincoln retained an essentially Calvinist view of transcendence. However, he avoided the problems associated with a Calvinist view, the problems alluded to in Piaget's rejection of a transcendent God. By emphasizing the essential mystery of God and God's will, Lincoln could provide an equal emphasis on personal autonomy, reasoning, and principles.

Other views of transcendence have yielded very different faiths that have led in very different directions than those led by Piaget and Lincoln's faith. Nowhere do we see this more clearly than in the pathological spiritual development of Adolf Hitler.

ADOLF HITLER

As a young man, Hitler underwent a spiritual conversion. Lukacs (1997) describes Hitler before his conversion:

There is a drastic change in both the content and the tone of [Hitler's] personal documents in 1919, but not before. The impression one receives of the younger Hitler is that of a loner, a dreamer, and a German idealist—an impression that accords with the reminiscences of those who knew him during the war. In Hitler's extant notes, letters, and postcards to his friends, whether sent from the front or on leave, the expressions are often childlike, showing a doglike loyalty and deference to his officers and his country. With one exception . . . they are also apolitical. (Lukacs 1997, 59)

After 1919, everything about Hitler—his thinking and manner of speaking, his passion and persistence amid tremendous adversity, his supreme confidence in himself and his cause—all pointed to one thing, namely, to his having a faith that was both powerful and terrible. From 1919 on, Hitler as a man of faith lived precisely as Paul Tillich (1957) and W. C. Smith (1998) describe—at a different level and as if guided always by what he took to be ultimate, transcendent, and sacred. After 1919, he was a man with a growing underlying spiritual purpose to save not only Germany but also the world.

What happened around 1919 to transform Hitler from a young man with no faith into a man whose faith defined his identity? First, there was Germany's defeat at the end of World War I. That defeat proved devastating for most Germans, but for Hitler it was more so. Before the war, Hitler's frustrated career as an artist and his near-vagabond existence in Vienna came close to supporting what Redlich referred to as a "malignant identity diffusion" (Redlich 1999, 306). The war years saved Hitler by giving him a temporary and functional identity, but, once the war ended, he was no further along in forming a lasting identity than he had been before the war. As Fest explained, when Hitler heard the news of Germany's defeat and cried, "I knew that everything was lost," he was expressing more a personal than a national sense of loss (Fest 1973, 79). So, the events following Germany's defeat proved crucial to Hitler's developing a faith. He was "lost" and needed to be "found," "re-born" and "saved." In short, he was ready for a conversion experience, for a new and (for him) saving faith.

Then, too, there was the near state of anarchy following Germany's defeat— the constant and bloody battles between right-wing and leftist groups. Hitler became a member of a right-wing paramilitary group and received special training in "civic thinking," the object of which was to indoctrinate soldiers with historical, economic, and political theories that would foster nationalism and counter the Communist revolution. For Hitler, then, this time encouraged him to organize his thinking about broad issues having to do with economics, history, and nationalism.

However, by far the most important development at this time was Hitler's discovering his special talent for public speaking. The first recorded account of this talent came from Alexander von Muller, who came across a gathering of students from one of the courses in "civic thinking." Fest records Muller's experience as follows:

He [Muller] found his way blocked by a group that stood fascinated around a man in their midst who was addressing them without pause and with growing passion in a guttural voice. "I had the strange feeling that the man was feeding on the excitement, which he himself had whipped up. I saw a pale, thin face beneath a drooping, unsoldierly strand of hair, with close-cropped mustache and strikingly large, light blue eyes coldly glistening with fanaticism." (Fest 1973, 113)

This was the beginning of Hitler's amazing career as a public speaker and demagogue.

Hitler's talent for demagoguery provided an important mechanism for the development of his faith. Hitler systematized ideas from his days in Vienna, from his war days, and from his courses in "civic thinking," not mainly for the purpose of organizing his thoughts but for the purpose of winning over his followers. More than anything else, it was Hitler's success as a public speaker before crowds that grew ever larger and ever more responsive that transformed his system of ideas into a faith. The crowds confirmed his identity as a spiritual person and spoke back to him with the overriding message that he was the man to lead them out of the wilderness, that he was their prophet.

All this was not immediately evident to Hitler. Nevertheless, even from the beginning, it must have occurred to him at some level that this incredible turnabout in his life, from being a miserable adolescent nobody to being a charismatic leader, all this must have meant that he had a much greater identity than could fit within a bourgeois role, that in some sense he was a prophet. Years later, before huge party rallies, Hitler would speak of the spiritual relationship between himself and his followers. At the 1937 party rally, he said, "That you found me long ago and that you believed in me has given your lives a new meaning, posed a new task. That I have found you is what has made my life and my struggle possible" (Fest 1973, 515). Therefore, in the case of Hitler, his prophetic faith developed as much through the reactions of his followers as through any revelations that arose in solitude.

Hitler's faith emerged, then, as a resolution to a personal identity crisis. The role of prophet not only filled a vacuum created by his having no role at the end of the war, but it also satisfied a long-standing desire to be great. In 1905, at the age of sixteen, Hitler convinced his mother that he should quit school to prepare himself for a career as an artist. For two years, he lived mostly in his dreams of becoming a great artist. Those dreams were destroyed when the Viennese Academy of Art twice rejected him. Over a decade later, they were revived in a new form—when the militant crowds of Munich made him their fuehrer. The lesson here is an old one. Faith is not just about ideas and ideals. Faith is also about the practical matter of finding a workable identity.

From 1919 on, Hitler's emerging faith was a faith focused on nature and biology. Later, in 1942 and during one of his "table talk" conversations, he summarized his faith as follows: "If I can accept a divine Commandment, it's this one: Thou shalt preserve the species" (Hitler 1953, 116). Hitler's faith, then, was expressed by a particular image of nature. Everything about Hitler's faith follows from this image. Trevor-Roper (1953) described Hitler's image of nature best when he said:

[Hitler] had views indeed of Nature and often spoke of his "communion" with it, but it was a hideous Nature, the devouring Nature whose cruelty justified his own; not a sociable pagan Nature of nymph-haunted woods and populated streams, but a romantic Wagnerian Nature of horrid Alps in whose intoxicating solitude he could best hatch his own equally violent and implacable interventions. (Trevor-Roper 1953, xxviii)

Throughout his speeches, his table talk, and *Mein Kampf*, Hitler made clear that nature is cruel, not nurturing. In *Mein Kampf*, he wrote (or rather dictated):

In the struggle for daily bread all those who are weak and sickly or less determined succumb, while the struggle of the males for the female grants the right or opportunity to propagate only to the healthiest. And struggle is always a means for improving a species' health and power of resistance and, therefore, a cause of its higher development. (285)

There is, then, nothing aesthetic or beautiful in Hitler's image of nature, nor, certainly, is there a deep and accurate understanding of nature's workings. Furthermore, there is nothing in Hitler's image of nature to foster and sustain spiritual values. For Hitler, throughout the grim and gruesome drama of constant struggle, "Nature looks on calmly, with satisfaction, in fact" (Hitler 1971, 285).

Since Hitler conflated creature and creation, he did not view humans as having dominion over nature. In *Mein Kampf*, he said, "man does not dominate Nature" (287); and much later, in his table-talk, he said, "It is senseless to encourage man in the idea that he's king of creation" (Hitler 1953, 71). In Hitler's image of nature, then, there is nothing to suggest that humans are responsible for nature. In fact, in his image, humans dominate nature by being smarter than other creatures. Hitler's image of nature and the human-nature relationship provides, then, no support for being responsible in general—except if one adopts his specious argument that following the law of "claw and fang" is being responsible.

Furthermore, from Hitler's "hideous" image of nature came his equally hideous images of man, woman, and community. Out of Nazi education, he envisioned a "new man" as well as a "new woman" emerging:

My pedagogy is hard. The weak must be hammered away. In my castle of the Teutonic Order a youth will grow up before which the world will tremble. I want a violent, domineering, undismayed, cruel youth. Youth must be all that. It must bear pain. There must be nothing weak and gentle about it. The free, splendid beast of prey must once more flash from its eyes. (Fest 1973, 233)

A girl's object is, or should be, to get married. Rather than to die an old maid, it's better for her to have a child without more ado! Nature doesn't care the least bit whether, as a preliminary, the people concerned have paid a visit to the registrar. Nature wants a woman to be fertile. (Hitler 1953, 75)

Following his image of nature, Hitler's image of community was not that of the just and caring community. Rather, it was a "blood" community. In Fest's words:

[Hitler's] sincerest and most solemn thought, . . . was this: to gather again the Aryan blood that had wasted itself on all the seductive Klingsor gardens of this world and to guard the precious grail for all time in the future, thus becoming invulnerable and master of the world. All the calculations of power tactics and all cynicism stopped short of this vision. (Fest 1973, 533)

This image of a community based on race or "blood" makes clear that Hitler was not the supernationalist he is often characterized to be. Rather, he was a superracist. It also makes clear that his hatred for democracy stems not so much from a political philosophy as it does from his faith and image of nature.

As for his image of himself, Hitler clearly took himself to be a prophet and instrument of providence. This self-image derived from Hitler's seeing himself as having discovered the key to history. What was that key? Once again, it had to do with nature and biology. Rome had built a great empire, and, at its height, Rome was, for Hitler, a model community. But why, he asked, had Rome declined? Why had the glory of Rome been replaced by the Dark Ages? The answer was race—the mixing of blood that led to the adoption of poisonous democratic ideologies, which, in turn, supported the weak and weakened the strong. In "discovering" race to be the key to history, Hitler came to see the world of his day as threatened on all sides by poisonous ideologies propagated by inferior races. He warned that "this planet once moved through the ether for millions of years without human beings and it can do so again some day if men forget that they owe their higher existence . . . to the knowledge and ruthless application of Nature's stern and rigid laws" (Hitler 1971, 288).

His mission, then, was no less than that of turning this worldwide decline around and saving the world from regressing beyond saving.

CONCLUSION

What are we to conclude from this discussion of faith development in the lives of Piaget, Lincoln, and Hitler? I think several things in particular. First, late adolescence and early adulthood appear to be especially important times for the development of strong faith—either for their establishing the framework for an implicit faith that flowers later on (Lincoln's case) or for their establishing a faith that remains a central aspect of identity for years to come (Piaget's and Hitler's cases). The reasons for this have much more to do with transactions occurring between youth and their immediate circumstances and culture than they have to do with maturation and cognitive structure. Piaget took on his contemporaries' questions of how science and religion can be reconciled and why the churches failed to prevent World War I. Lincoln took up a legal and political career that eventually confronted him with the limitations of law and politics for resolving the moral problem of slavery. Hitler found his calling as a demagogue because of the need for Germans to find their way out of anarchy. Strong faith, then, develops from constructing a fit between person and context.

Second, the fit manufactured by faith has both short- and long-term implications, making it important to consider spiritual development's relation to thriving and positive youth development in terms of both and not in terms of the short term only. As Fest pointed out and as virtually every major Hitler historian has agreed, if Hitler had died in 1938, he would have gone down in German history as a great statesman:

> If Hitler had succumbed to an assassination or an accident at the end of 1938, few would hesitate to call him one of the greatest German statesmen, the consummator of Germany's history. The aggressive speeches and *Mein Kampf*, the anti-Semitism and the design for world dominion, would presumably have fallen into oblivion, dismissed as the man's youthful fantasies. . . . (Fest 1973, 9)

If he had died in 1938, Hitler would, then, have been an exemplar of positive development—at least to the majority of Germans—and so his faith had short-term implications for his thriving.

The opposite can be said about Lincoln. Lincoln's spirituality led initially not to his thriving so much as to his being misunderstood, especially misunderstood by both sides of the slavery issue. As his most beloved biographer, Lord Charnwood, put it:

> . . . a patient being, who, long ago in his youth, had boiled with anger against slavery, but whose whole soul now expressed itself in a policy of deadly moderation towards it . . . In almost every department of policy we shall see him watching and waiting while blood flows, suspending judgment, temporizing, making trial of this expedient and of that, adopting in the end, quite unthanked, the measure of which most men will say, when it succeeds, "That is what we always said should be done." Above all, . . . we shall witness the long postponement of the blow that killed negro slavery, the steady subordination of this particular issue to what will not at once appeal to us as a larger and a higher issue. All this provoked at the time in many excellent and clever men dissatisfaction and deep suspicion; they longed for a leader whose heart visibly glowed with a sacred passion; they attributed his patience, the one quality of greatness which after a while everybody might have discerned in him, not to a self-mastery which almost passed belief, but to a tepid disposition and a mediocre if not a low level of desire. . . . [But] perhaps the sense will grow upon us that this balanced and calculating person, with his finger on the pulse of the electorate as he cracked his uncensored jests with all comers, did of set purpose drink and refill and drink again as full and fiery a cup of sacrifice as ever was pressed to the lips of hero or of saint. (Charnwood 1916/1996, 115)

Presumably, then, if the South had won the Civil War, Lincoln would have gone down in history not the way he has, as a political saint, but as a failure.

Third, these three examples indicate that spiritual development cannot be measured by strength of faith or structural development alone. By Fowler's

(1995) and Oser and Gmunder's (1991) structural criteria, Hitler's strong faith was a developed faith. We need different criteria if development is to retain its useful meaning as "movement towards perfection" (Kaplan 1983).

These three cases leave us, then, with a greater appreciation of the complexities in assessing spiritual development and its relationship to youth development. To make accurate and complete assessments of spiritual development, we need details, details that can only come from case material and details that are not only about context and circumstances but also about an individual's thinking and imagining. These details are essential for determining whether an emerging faith supports a moral life or its opposite—and whether a faith is for positive development in the long term or in the short term only.

The question of transcendence and its relation to youth development is, then, ultimately a question of which transcendence and how individuals construe their connection to transcendence. For Piaget, transcendence was a product, not a given—a product of integrating conscience, reason, and the example of Jesus. For Piaget, there was no personal relationship with transcendence, for transcendence was, for him, purely spiritual and in the person. For Hitler, there was providence and the laws of nature as he construed those laws. For Hitler, transcendence was outside and impersonal, and his connection was similar to that of a prophet, someone to whom truth and mission had been revealed. For Lincoln, transcendence also was outside and impersonal, but his connection to transcendence was indirect and mediated by the sacred guideline laid down by Jesus' summary of the Law (to love God and neighbor) and by the sacred principle of justice outlined in the Declaration of the Independence. He saw himself as an instrument of God, but not in the same way that Hitler saw himself as an instrument of providence.

The comparisons of Piaget, Lincoln, and Hitler bring us to a final point. Evaluating spiritual development and its relation to thriving should include an evaluation of how youth develops convictions while dealing with ambiguity. There is little in Hitler's communications with others that indicates he felt or dealt with the ambiguities inherent in life's many moral dilemmas. As a result, Hitler lacked humility. The opposite was the case with Lincoln, as is evident in his frequently pointing out that good men disagree. Humility, then, may turn out to be a more critical virtue in the development of faith and spirituality than research has previously indicated. Perhaps, then, we should take Reinhold Niebuhr's assessment of Lincoln as a way to define what develops in positive-adaptive spiritual development in general:

This combination of moral resoluteness about the immediate issues with a religious awareness of another dimension of meaning and judgment must be regarded as almost a perfect model of the difficult but not impossible task of remaining loyal and responsible toward the moral treasures of a free civilization on the one hand while yet having some religious vantage point over the struggle. . . . Abraham Lincoln . . . was . . . that rare and unique human being who could be responsible in executing historic tasks without equating his interpretation of the task with the divine wisdom. (Niebuhr 1952, 87)

Strong, positive faith, then, is neither the opposite of doubt nor a synonym for belief. Rather, strong, positive faith refers to the quality of persons who have become exemplars, exemplars that help define what it is that develops when we speak of spiritual development.

REFERENCES

Basler, R., ed. 1969. *Abraham Lincoln: His speeches and writings.* New York: The World Publishing Co.

Charnwood, G. 1996. *Abraham Lincoln.* New York: Madison Books. Originally published in 1916.

Donald, D. H. 1995. *Lincoln.* New York: Simon & Schuster.

Fest, J. 1973. *Hitler.* New York: Harcourt Brace.

Fowler, J. 1995. *Stages of faith: The psychology of human development and the quest for meaning.* 2nd ed. San Francisco: Harper Collins.

Hein, D. 1983. Abraham Lincoln's theological outlook. In K. Thompson, ed., *Faith and politics,* 105–79. New York: University Press of America.

Hitler, A. 1953. *Hitler's secret conversations.* New York: Farrar, Straus, & Young.

———. 1971. *Mein kampf.* New York: Houghton Mifflin Co.

Kaplan, B. 1983. Genetic-dramatism: Old wine in new bottles In S. Wapner and B. Kaplan, eds., *Toward a holistic developmental psychology,* 53–74. Hillsdale, NJ: Lawrence Erlbaum.

Lukacs, J. 1997. *The Hitler of history.* New York: Random House.

Macartney, C. 1951. *Lincoln and the Bible.* New York: Abingdon-Cokesbury Press.

Niebuhr, R. 1952. *The irony of American history.* In K. Thompson, ed., *Faith and politics,* 1–104. New York: University Press of America.

Oser, F., and P. Gmunder. 1991. *Religious judgment: A developmental approach.* Birmingham, AL: Religious Education Press.

Piaget, J. 1928. Immanence et transcendence: Deux types d'attitude religieuse. *de L'Association Chretienne D'Studiants de Suisse Romande., Depot: Geneva: Labor.* Vol. 29.

Redlich, F. 1999. *Hitler: Diagnosis of a destructive prophet.* New York: Oxford University Press.

Reich, H. 2005. Jean Piaget's views on religion. University of Fribourg (Switzerland).

Smith, W. C. 1998. *Faith and belief: The difference between them.* Oxford, England: Oneworld Publications.

Tillich, P. 1957. *Dynamics of faith.* New York: Harper Brothers.

Trevor-Roper, H. R. 1953. The mind of Adolf Hitler. In A. Hitler, *Hitler's secret conversations,* New York: Farrar, Straus, & Young.

Vidal, F. 1987. Jean Piaget and the liberal Protestant tradition. In M. Ash and W. Woodward, eds., *Psychology in twentieth-century thought and society,* 271–94. Cambridge: Cambridge University Press.

4

Spirituality as Fertile Ground for Positive Youth Development

Pamela Ebstyne King

This volume is evidence of the growing interest in adolescent spirituality within the positive youth development (PYD) movement. In particular, the current emphasis within PYD on the role of civic engagement as an indicator or outcome associated with positive development or thriving has led scholars and practitioners to look at spirituality and religion as potential resources for promoting such social contribution (Benson, Scales, Sesma, and Roehlkepartain 2005; Damon 2004; King et al. 2005; Lerner, Dowling, and Anderson 2003; Lerner et al. 2005). Although there is burgeoning interest and a mounting number of relevant publications, many theoretical and methodological issues remain to be addressed in order to elucidate more fully the relationship between positive youth development and spirituality. Based on two foundational theoretical underpinnings of PYD, *plasticity* and *context*, this chapter aims to provide a framework for understanding how spirituality may be a unique and robust catalyst for positive development in young people. Specifically, this chapter will explore how the ideological, social, and transcendent dimensions within spirituality provide fertile ground for promoting positive youth development. In addition, the chapter will discuss methodological issues pertinent to building the study of spirituality and positive youth development.

Central to the optimistic view of youth within the positive youth development approach are the roles of plasticity and context (Lerner 2002; Lerner, Almerigi, Theokas, and Lerner 2005). Within PYD, *plasticity* refers to the view that a young person has the capacity to change over time. Human development is not stagnant, but a person will continue to change in both positive and negative ways throughout his or her life. All people have the capacity to develop along a vast array of possible trajectories. The PYD movement emphasizes the potential for positive growth, building on strengths and promoting character and competence through life. Plasticity legitimates the optimistic search for characteristics of people and their contexts that promote positive development.

Also foundational to a PYD approach is the significance of *context*. A young person does not develop in a test tube. Rather PYD emphasizes the reciprocal relations between a young person and the environments in which he or she lives (Bronfenbrenner 1979; Lerner 2002). A PYD approach optimistically looks to the systems in which a young person is embedded for potential resources or assets that may promote optimal development within the young person (Benson, Leffert, Scales, and Blyth 1998). Positive development occurs when the mutual influences between person and environment maintain or advance the well-being of the individual and context.

Based on the concepts of plasticity and context, this chapter explores how spiritual contexts may promote optimal development of young people. Specifically, this chapter takes spirituality under consideration as a contextual asset that may serve as a resource for positive development. Although spirituality may be considered a human capacity (see Roehlkepartain, King, Wagener, and Benson 2006), this chapter will focus on how it might serve as a resource for positive change within a young person's developmental system. From this standpoint, a spiritual context refers to opportunities that provide experiences of transcendence, moving the self beyond its own and creating a sense of meaning through connectedness with a divine, human, or natural "other." This may occur through beliefs, practices, relationships, experiences, worship, and service.

Given the current lack of clarity of this abstract, transcendent domain, this chapter offers specific hand-holds to provide researchers a place to begin to get a grip on the study of spirituality and positive youth development. Specifically, this chapter focuses on three salient aspects of spirituality that may promote a particularly "fertile ground" for nurturing positive development. In particular, the chapter focuses on the ideological, social, and transcendent contexts within spirituality that enable spirituality to be particularly effective in promoting positive development in such a way that young people are motivated and committed to contributing back to the greater good (King 2003). It is noted that, from a PYD and thriving perspective, individual development emerges with a commitment to making social contributions (Damon 2004; Lerner, Dowling, and Anderson 2003). Consequently, the following framework will consider how these three contexts or dimensions of spirituality promote both individual development (i.e., identity, purpose, character, and competence) and a contribution to the greater good (i.e., caring, civic engagement, and contribution).

CONCEPTUAL FRAMEWORK

In order to add insight into the potentially rich resources available to youth through spirituality, this chapter sets forth a conceptual framework proposing that spirituality offers young people ideological, social, and transcendent resources. Although spirituality has been more recently added to the academic study of the ecology of youth development, this chapter identifies two dimensions of spirituality that are more familiar to developmentalists, namely, the potentially rich ideological and social resources available to youth and one dimension that has been less explored in the social sciences, transcendence.

Spirituality as Ideological Context

Different forms of spirituality, whether individual experiences of spirituality or organized spiritual traditions, have the potential to serve as rich ideological resources. For example, more organized forms of spirituality, such as religious traditions, have specific beliefs, values, and morals that are foundational for a young person to generate a sense of meaning, order, and prosocial identity. Erikson (1964, 1965) recognized religion's potential in development. He pointed to religion as an important aspect of the sociohistorical matrix in which identity takes shape. He argued that religion is the oldest and most lasting institution that promotes fidelity, the commitment and loyalty to an ideology that emerges upon the successful resolution of the psychosocial crisis of identity formation (Erikson 1968). Religious beliefs, values, and morals enable youth to make sense of the world and understand their place in it.

Although Erikson (1964, 1965) specifically wrote about ideology and religion, his thinking may be applied to less-institutionalized forms of spirituality as well. In less-organized forms of spirituality, ideology is present and formative. Benson (2006) states that the function of spiritual development is to make sense of one's life by weaving the self into a larger tapestry of connection and meaning. Spirituality entails the intentional identification and integration of beliefs, narrative, and values in the process of making meaning. Whether this process is one of personal construction or socialization, the intentional act of relying on personal, religious, or cultural ideology is central to spirituality and crucial to the development of identity, meaning, and purpose—all foundational to positive youth development.

Ideology gives meaning to events and experiences. Particularly in a culture where youth face an ever-changing and cluttered present, Erikson (1964) writes that transcendent meaning is imperative to adolescent identity formation and well-being. Without such a worldview to give meaning and guide behavior, the choices and options before youth are likely to lead to confusion and despair. Spirituality offers ultimate answers and perspective about the larger issues in life. In addition, the morals and worldview present in most forms of spirituality are prosocial and encourage youth to make actual contributions to their society. For example, Youniss and colleagues (Kerestes, Youniss, and Metz 2004; Youniss, McLellan, and Yates 1999) found that religious youth are more apt to participate in various forms of civic engagement than their less-religious peers. Furthermore, they found that students who participated in community service within a religious context adopted a religious rationale for their action, demonstrating how ideology provided interpretive value for their life experiences.

Spirituality as Social Context

In addition to providing an ideologically rich context, spirituality may offer a social context helpful for positive development. More often than not, spirituality is practiced in community or with others—whether that be a part of a religious tradition and include a youth group, the practice of various forms of yoga, or less-organized gatherings of like-minded people who spur one another on in their spiritual journeys. These fellow travelers play an important role in enabling young people to internalize beliefs, values, and morals. For example, Erikson (1968) pointed to the salience of religion because it not only provides a transcendent worldview but because religious tradition also exemplifies these principles and behavioral norms in actual historical events and in the lives of fellow believers. Erikson wrote that it is the embodiment of these ideologically based principles and behavioral norms that enables religion to be so effective in the development of a prosocial identity. During adolescence, personal integration is facilitated not only by abstract ideology but by having that ideology lived out in the flesh. Spirituality often provides opportunities for adolescents to interact with peers and build intergenerational relationships as well.

One way of understanding the potential social impact of spirituality is through the concept of *spiritual modeling* (Lerner, in press; Oman, Flinders, and Thoresen, in press; Oman and Thoresen 2003). Spiritual modeling refers to emulating another in order to grow spiritually. This occurs through observing and imitating the life or conduct of one or more spiritual exemplars. Spiritual exemplars are living or historic examples of spiritual ideology and values that serve as role models. Spiritual modeling is founded on social modeling and observational learning in the acquisition and maintaining of human behaviors (Bandura 2003), illustrating the potential social influences of spirituality on positive youth development.

Organized forms of spirituality offer many opportunities for such examples. Models can be spiritual leaders such as swamis, priests, rabbis, or spiritual directors; fellow worshippers; or even a late saint or founder of the faith documented in sacred texts. For example, faith communities provide rich opportunities for intergenerational relationships to occur. King and Furrow (2004) found that religiously active adolescents report higher levels of social interaction and trust with their parents, immediate friends, and the significant adults in their lives. Further analysis revealed that these intergenerational relationships mediated the influence of religion on moral outcomes. These findings suggest that religious youth are embedded in a social context that is characterized by interactive, trustworthy relationships in which they share common goals, beliefs, and values and lead to prosocial behaviors and actions. In addition, studies of moral exemplars have suggested that trusting and sharing relationships have been shown to promote self-reflection and internalization of values, beliefs, and commitments that constitute identity vital to sustained moral, caring, and civic behaviors (Colby and Damon 1995).

In addition to encouraging relationships and role models, the spiritual context also provides social experiences that nurture positive youth development. Many spiritually based organizations offer service opportunities for young people, whether in local, national, or international settings. For example, interviews with a woman recognized for her extraordinary commitment to serving the poor revealed that her strong sense of moral identity was shaped within her religious context (Colby and Damon 1995). Through her experiences of serving the children in her congregation and working with her pastor, a vision emerged to expand her ministry to homeless people living around a garbage dump in Mexico. Such experiences may promote skills, competencies, and motivation in young people as well as provide opportunities to experience the value and fulfillment of service.

Spirituality as Transcendent Context

Perhaps the most unique aspect of spirituality is the potential for transcendence. Many youth programs and organizations offer ideology and rich social environments, but not many intentionally promote experiences of transcendence, where young persons acutely experience a reality beyond themselves. The Fetzer Institute (1999) describes spirituality as that which "is concerned with the transcendent, addressing ultimate questions about life's meaning, with the assumption that there is more to life than what we can see or fully understand" (2). Engaging in the spiritual may provide connectedness with a divine, human, or natural other, giving a young person an opportunity to experience himself or herself in relationship to God, a community of believers, the essence of all Being, or nature, for example. In this sense, spirituality and transcendence are not limited to a connectedness or awareness to other than the Divine. As Lerner (1996) emphasized, spirituality includes the connection between the self and human other. He describes spirituality as "the awareness of the fundamental unity of all being and of our connectedness to one another and the universe" (Lerner 1996, 56).

Such experiences of transcendence have a rich potential for promoting positive development in young people. The following section highlights various experiences of transcendence available through spirituality and a discussion of how these might contribute to positive youth development.

Experiences of transcendence can affirm one's own sense of identity and self-worth through a profound sense of connection to a divine or human other. For example, within Judeo-Christian traditions, believers experience themselves as being in a special relationship with God. Such traditions both teach and provide opportunities through spiritual practices or rituals, where believers not only learn about their sense of belonging to God but experience themselves in relationship to God. For example, being the covenant people of God has always been at the core of Jewish identity, and staking their claim as the chosen people of God has been central throughout history from the ancient Israelites to contemporary Jews (Furman 1987). Within the Christian tradition, believers are to understand themselves as "sons or daughters of God" (Galatians 3:26). The New Testament also refers to believers being chosen by God (Ephesians 1:4). Both traditions acknowledge the uniqueness and inherent value of each person. Understanding oneself as a beloved or chosen one by the Creator may have profound implications for identity. Within some Eastern traditions, intrinsic value is found in understanding people and the Divine to be of the same essence of all being. For example, in the Hindu tradition, the *Atman* (the individual soul) and *Brahman* (the cosmic spirit) are one and the same. Salvation is found through a deepening one's understanding of unity of Self, Universe, and Spirit (Roeser 2005). Spiritual traditions that promote an awareness of the connection between individuals and something greater than themselves, like a group, all of humanity, the universe, or a divine being, potentially affirm the worth of that young person.

In addition, youth may experience transcendence through connection to specific others. For example, as young people participate in religious congregations, they can locate themselves as members of a historic tradition. Religion provides both a past community of believers who have gone before them as well as a present body of believers that live alongside them, giving youths a sense of being a part of something greater than themselves. Participating in such a historic and current community of believers or followers can offer young people a profound sense of being a part of something greater than themselves. Not only may this nurture strong self-worth based on a sense of belonging but also a sense of commitment and responsibility to another.

Not limited to religious traditions are experiences of transcendence that occur in nature. Young people may find a strengthened sense of identity as being part of creation. Being confronted with the majesty of creation or being sensually aroused through the aesthetics of nature—whether the enormousness of mountains, the intricacies of a flower, or the power of a storm—may inspire perspective on one's sense of self or one's life. For example, awareness of creation may inspire a sense of purpose or meaning. For some, outdoor experiences may inspire young people to take seriously the call to stewardship of the environment. Spiritual experiences in nature may promote positive development by nurturing identity formation, a sense of purpose, and/or well-being.

Spirituality also promotes transcendence through ritual, whether that be through worship, spiritual practices, or rites of passage. Ongoing worship rituals may promote one's awareness of the divine or human other as well as confirm one's place in a community. For instance, one of the foremost religious practices within Islam is *salat* or prayer. According to this tradition, believers pray five specific times a day. In this repetitive act, believers experience themselves in solidarity with other Muslims worldwide prostrating themselves toward Mecca. The Lord's Supper or Communion provides Christians with a consistent experience of transcendence with the "communion of saints," which refers to all believers past and present. Within the practice of the Lord's Supper, it is understood that, by the taking of the bread and wine, one not only experiences union with Christ but one also participates in this act with the communion of saints.

Rituals of rites of passage also promote rich opportunities to confirm identity and engender a sense of obligation. For example, within Judaism, *bar* or *bat mitzvahs* serve to recognize a boy or a girl in their transition to manhood or womanhood as members of a synagogue, confirming their unique place in the body of believers. Within the Christian tradition, a young person's confirmation takes place within a parish or congregation. Confirmation serves as a time to confirm a young believer's commitment to beliefs and his or her religious tradition, as well as for the faith community to covenant, to nurture and care for the spiritual formation of the young person. These religious rites of passage are unique events that intentionally celebrate and affirm a young person's sense of identity as a believer as well as recognize their place within their faith community, potentially contributing to both personal development and a commitment to something beyond themselves.

Spiritual practices or disciplines also promote experiences of transcendence. For example, the practice of meditation promotes the realization of higher states of awareness and eventual enlightenment, yielding complete understanding of one's own unity with the essence of being (Roeser 2005). In addition, fasting from food allows an individual to heighten his or her awareness and concentration on the spiritual. Spiritual practices including service to the poor provide the opportunity to learn the intrinsic value of the other, gratitude, and self-sacrifice. In addition to promoting character, they allow youth to explore different competencies and skills. Experiences of answered prayer may generate a sense of commitment or devotion to one's practice of spirituality (Smith 2003). These practices may provide young people with an experience of something beyond themselves, which not only affirms their uniqueness but might engender a commitment to others.

These experiences of transcendence may promote key facets of positive development. This moving beyond the self provides the opportunity for the search for meaning and belonging that is central to positive youth development (Benson 1997; Damon, Menon, and Bronk 2003). Awareness that stems from this search provides ultimate answers and perspective in the larger issues of life that are crucial to the resolution of the adolescent identity crisis (Erikson 1964, 1965). Devotion, responsibility, and commitment inspired

FIGURE 1. Spirituality's Potential Influence on Positive Youth Development

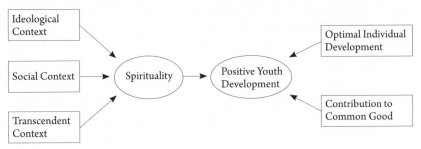

by transcendence may play an important role in both motivating and sustaining an altruistic or generative lifestyle. Spirituality provides opportunities to experience a profound sense of connectedness with either the supernatural or human other that invokes a sense of awareness of self in relation to the other. This heightened consciousness of others often triggers an understanding of self that is intertwined and responsible to the other (King 2003). This attentiveness usually promotes a manner of living that is carried out with the highest regard to the life of self, others, and/or the Divine (Miller and Thoresen 2004). As the Fetzer Institute (1999) wrote, "Spirituality can call us beyond self to concern and compassion for others" (2). Consequently, the transcendent quality of spirituality is especially pertinent to shaping a commitment to contributing to the common good within young people.

Figure 1 illustrates this conceptual model, suggesting that spirituality promotes positive development in young people. The model depicts spirituality as a latent variable consisting of ideological, social, and transcendent dimensions. In addition, positive youth development is also depicted as a latent construct comprising optimal individual development and commitment to the common good.

SPIRITUAL ANCHORS

This framework suggests that spirituality potentially offers ideological, social, and transcendent experiences that may promote positive youth development. Given that spirituality and religiosity have neither been a major focus of theory nor conceptualized as a basic human capacity until recently (Kerestes and Youniss 2003; King and Boyatzis 2004; Roehlkepartain, King, Wagener, and Benson 2006), the field of psychology lacks the terminology and conceptual understanding to explain thoroughly the mechanisms within spirituality that might influence positive youth development. This article has attempted to provide psychological insight into a young person's potential experience of spirituality, suggesting that spirituality may play a valuable role in society—providing youth with an environment of intergenerational support that can foster beliefs, values, meaning, identity, and a sense of belonging and connectedness beyond themselves. As such, spiritual institutions serve as what Garbarino (1995) refers to as *spiritual anchors,* "institutions of the soul that connect children and teenagers to the deeper meanings of life and provide solid answers to the existential questions: Who am I? What is the meaning of life?" (150). Youth need contexts in which to grapple with the spiritual issues of purpose in life, what they believe, and their place in the world. Spirituality may provide a distinct context in which a young person can explore these issues, gain support, and have experiences that are critical to the formation of self and commitment to the common good.

Is spirituality different from other organizations that have shared values and purpose? American society provides a rich tapestry of opportunities for youth development. Youth today have a myriad of options for finding a sense of belonging, for affirming a unique sense of self, and for providing opportunities to grow. Young people participate in families, sports, entrepreneurial enterprises, employment opportunities, philanthropy, and different forms of civic engagement. Personal endeavors such as pursuing art, academics, and sports also offer pathways for development. Does spirituality play a unique role in this social fabric? If so, how does it differ from other opportunities and experiences available to American young people?

The framework set forth in this chapter suggests that spirituality may offer opportunities for development through providing ideological, social, and transcendent contexts. As such, spirituality may provide an especially rich environment for development. Spirituality provides an intentional and coherent worldview that offers prosocial values and behavioral norms that are grounded in an ideology. Spirituality intentionally offers these values, and members may embody these ideals and values and serve as role models. In addition, when practiced in a community, spirituality provides an intergenerational network of enduring, caring relationships through which youth may wrestle with issues pertinent to development and also offers experiences in which they can grow in competency and character. Finally, spirituality promotes transcendent experiences that enable youth to move beyond their daily concerns and potentially encounter the supernatural or human other in a meaningful way that simultaneously promotes optimal individual development and inspires commitment to the common good.

Although other institutions and activities offer a wide range of opportunities for youth, they rarely offer the breadth and depth of developmental resources that foster positive development as spirituality does at its best. Rarely do organizations intentionally offer ideological cohesiveness; supply an intergenerational social network that nurtures and sustains beliefs, meaning, and values; and provide opportunities for sacred and communal transcendence. The confluence of these three dimensions enables spirituality to serve as a potentially potent resource for positive youth development.

SPIRITUALITY GONE AWRY

The intention of this chapter is not to promote spirituality as a social panacea; rather, it is to recognize the potency of spirituality and its potential to help or hinder youth development. Just as spirituality has the potential to promote positive development through ideology, social support, and experiences of transcendence, it also has the capacity to do harm. As stated earlier, positive youth development involves the optimal development of the individual and a contribution to the common good. At its best, spirituality promotes both self and commitment to the other. What occurs when spirituality is not at its best? What if the self is sacrificed for the sake of the other? What if the other is neglected for the sake of the self?

Let us consider. The optimal spiritual context affirms both individual development and engenders a commitment to the greater good. This balance is important, for, if one violates the other, healthy development does not occur. For example, if a religious tradition emphasizes the faith community, without valuing the uniqueness of its members, youth may not have the necessary opportunities to explore different aspects of identity. When youth are not given the freedom to explore and are either forced or pressured into adopting a specific ideology, social group, or expression of spirituality, identity foreclosure is at risk.

Taken to the extreme, cults can be understood from this perspective as spiritual expressions that devalue the individuality of their members in order to elevate the ideology and group. This is graphically illustrated by spiritual groups that demand that their members dress alike. In addition, recent current events such as suicide bombings illustrate the devastation caused by religious groups that value the goals and ends of the religion more than an individual life itself.

On the other hand, some traditions might not leverage their potential as being conducive to promoting positive development because they emphasize the individual, over and above promoting a sense of community and belonging. For example, some conservative traditions within Christianity emphasize the individual believer's relationship with God to the extent that they do not expend time or resources on promoting a sense of community or contribution to larger society. When this occurs, although youth are reinforced of their personal worth, they lose out both on the support and accountability of a faith community and the value of learning what it means to belong and to contribute to a greater good. In addition, individual forms of spirituality that are not connected with a group of followers also have the potential to leave youth without the web of support present in spiritual traditions associated with an intentional group of followers. These manifestations of spirituality are not necessarily deleterious for youth development or for society; rather, they lack the rich social context that is so effective for optimal development. Forms of spirituality that do not connect youth with a social group or a transcendent experience of other may not promote a self-concept that fully integrates a moral, civic, and spiritual identity. However, taken to the extreme, forms of religion and spirituality that exalt the individual over a greater good can promote a sense of narcissism and entitlement and a lack of connectedness and contribution to society.

Figure 2 presents a circumplex model that depicts the significance of spirituality nurturing both individual development and a commitment to something beyond the self. The vertical axis, labeled "Self," indicates higher or lower levels of individual development, where the horizontal axis, labeled "Other," represents high and low levels of commitment to the greater good. The labels in the

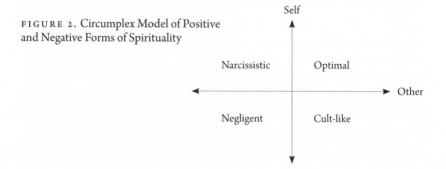

FIGURE 2. Circumplex Model of Positive and Negative Forms of Spirituality

four quadrants represent extremes for illustrative purposes. For example, the upper-left quadrant is labeled "Narcissistic" because this quadrant is indicative of forms of spirituality that promote individual development over and above nurturing a commitment to contributing to the greater good. Conversely, the lower-right quadrant is labeled "Cult-like," to represent at the extreme oppressive forms of spirituality that mandate serving the greater good or at least the spiritual community to the detriment of individual personal development and fulfillment. The upper-right quadrant, labeled "Optimal," represents the outcome of spirituality, where individual development and commitment to the common good are both promoted. In contrast, the lower-left quadrant, labeled "Negligent," represents contexts that do not nurture the self or a commitment to the common good.

THE EMPIRICAL STUDY OF SPIRITUALITY AND POSITIVE YOUTH DEVELOPMENT

Understanding the complex dynamics of spirituality holds considerable promise, not only for offering insight into the internal workings of spiritual and religious traditions but also for maximizing the practical relevance of spirituality and religion for strengthening positive youth development. At the core of the increasingly perceived relevance of spirituality for individual development and for society is a growing scientific recognition that religion and spirituality are worthy domains of study. Using the current understanding of development to examine the potential resources embedded within spirituality reveals unique contexts and opportunities that may promote optimal development. The conceptual framework presented in this chapter provides direction for further empirical work. In order to understand more fully the dynamics behind the association, a more thorough inquiry is necessary. The current model suggests that ideology, social context, and transcendent experiences available through spirituality need to be explored.

If the assertions of this framework are true, then integrating spirituality with one's personal ideology may be important in development. In order to explore more fully spirituality's role in adolescent development, the extent to which spiritual beliefs are integrated into personal ideology needs to be examined. Corroborating data from the adolescent, other members of a spiritual community, clergy, family, and peers would inform the extent to which a young person internalized a set of spiritual beliefs, values, and morals embedded in the spiritual context. Also worthy of asking is whether the clarity and explicitness of the teaching of ideology affect the extent to which beliefs and worldviews are internalized. Does the degree of organization of various forms of spirituality influence the degree to which ideology is internalized? What are the associations between the content of ideology and aspects of identity? For example, is prosocial ideology correlated with a prosocial identity? What are the effects of exploring multiple spiritual ideologies on one's identity during adolescence? Do the cognitive abilities required to evaluate and assimilate beliefs, meaning, and values have developmental implications for exposing youth to multiple spiritual systems?

Empirical study of the spiritual social context will need to bring understanding to the role of interpersonal and group relationships on adolescent development and explore the effects of social experiences available through spirituality. Spiritual modeling raises many questions, such as:

Whom do youth consider as spiritual role models?

How, if at all, do young people intentionally learn from role models?

Does the proximity of relationship influence the impact on development?

How do parents, peers, leaders, or historic heroes of a tradition impact youth differently?

How do trust and frequency of social interaction mediate or moderate the influence of the relationship on a young person?

How do negative or hypocritical examples influence youth behavior, identity, and beliefs?

How does spirituality bring meaning to different social activities carried out in the spiritual context?

How do individual social skills influence the way one accesses resources for development present through their spiritual context?

In order to gain more comprehensive insight into the relationship between spirituality and adolescent development, the role of transcendence must be considered. For example, norming existing measures that assess transcendence, such as the Spiritual Transcendence Index (Seidlitz et al. 2002) on adolescent populations, would enable social scientists to begin to understand the role of transcendence in adolescent development. Exploring how adolescents conceptualize and experience transcendence is also needed to answer questions such as,

Do adolescents experience transcendence as an awareness or connectedness to the Divine and/or to humanity and/or nature?

How, if at all, do young people intentionally seek out such experiences?

How do such experiences provide a sense of meaning, perspective, and values that inform identity?

How do such experiences engender a commitment to the common good?

How do spiritual and religious rituals influence identity and commitments?

Do initiation rites predict character, identity, contribution? If so, under what circumstances?

What cognitive skills and affective awareness are necessary for registering transcendent spiritual experiences?

Furthermore, it is important to explore how aspects of ideology, social context, and transcendence interact. Among the questions raised in this area are

Are spiritual role models more effective when ideology is shared?

Do spiritual groups that have more shared transcendent experiences exert more influence on adolescents?

How do ideological, sociological, and transcendent dimensions interact to promote positive development?

Does spirituality practiced outside the context of a community promote social contribution to the same extent that spiritual communities might?

How can spiritual groups or faith-based organizations leverage potential resources inherent in their beliefs and traditions?

Only after these questions and others are explored will a more thorough understanding of the role of spirituality in adolescent development emerge.

Finally, testing the proposed model in its entirety (see Figure 1) would determine if, in fact, the ideology, social context, and transcendent experiences proposed to be embedded within spirituality make up a latent variable "spirituality." In turn, it would answer how such a latent variable might be associated with another latent variable—positive youth development—that might comprise two endogenous variables of optimal individual development and contribution to the common good. Sufficient data collection and analysis are important next steps to further an understanding of the relationship between the complexities of spirituality and youth development.

In addition, further exploration would reveal if, in fact, discrete categories exist, such as are proposed in Figure 2. Are there significant differences between experiences of spirituality that are rich in ideological, social, and transcendent resources? Do experiences of spirituality that encourage both self-awareness and connecting to others promote higher levels of positive development for young people than do experiences of spirituality that emphasize just personal growth or awareness, without encouraging connection or service to others? Are experiences of spirituality that promote experiences of community and connection without emphasizing the individual associated with different developmental outcomes? It would be helpful to explore if there are certain thresholds of emphasis on the individual and emphasis on connection that become less helpful to a young person's developmental journey.

In addition to these research questions, several methodological issues need to be addressed in order to expand an understanding of the relationship between spirituality and positive youth development. Studies need to be designed that allow for an understanding of how spirituality develops over time. This demands high-quality longitudinal data with diverse samples that employ both qualitative and quantitative methods. This approach with diverse samples studied longitudinally will allow scholars to begin to understand what are the universal and the particular forms of spiritual development. Diverse samples need to include different spiritual or religious groups, cultures, and socioeconomic statuses in order to avoid a limited understanding of how people develop spiritually and how that impacts their general development. Large qualitative datasets would allow researchers to generate descriptives that help identify norms of spiritual concepts and practices at different ages and stages and for different traditions. Such datasets would also allow for the exploration of relationships between factors of spiritual development and positive youth development. On the other hand, in-depth qualitative data would allow scientists to come to a deeper understanding of the dimensions of spirituality present in different contexts.

In addition, crucial to building a field of study is the need to create interdisciplinary methods that build on the wisdom and experience of social scientists, anthropologists, theologians, religious scholars, and religious and spiritual leaders. In addition, youth need to be engaged as active participants in research design in order to ensure the validity of this research. Involving such a diverse array of participants is no small task. Establishing a common language among such an interdisciplinary group is the first challenge. Case studies or the use of focus groups might provide helpful methodologies in order to bridge cross-disciplinary boundaries.

CONCLUSION

Fortunately, there is renewed interest in the study of spiritual development. Hopefully, the not-too-distant future will see the rise of creative and rigorous methodologies that will begin to answer some of the questions raised in this chapter regarding the relationship between spirituality and positive youth development. As the field continues to explore this dynamic relationship, considering the ideological, social, and transcendent dimensions of spirituality will be helpful to elucidate how spirituality promotes a fertile ground for positive youth development. At its best, spirituality provides an environment where youth can experience the self embedded within a larger context that simultaneously validates the inherent value of the self and promotes a sense of belonging and connectedness beyond the self. The young person can gain a sense of self as a unique individual, as well as a self that is a contributing member to a larger whole. When all three dimensions are activated, spirituality can serve as a potentially potent aspect of the developmental system, where young people can experience the "self" embedded within a larger context that enables the "caring, compassion, and variants of 'we' [to] temper the rampant pursuit of 'me'" (Benson 1997, 8).

REFERENCES

Bandura, A. 2003. On the psychosocial impact and mechanisms of spiritual modeling. *International Journal for the Psychology of Religion* 13, no. 3: 167–73.

Benson, P. L. 1997. Spirituality and the adolescent journey. *Reclaiming Children and Youth* 5, no. 4: 206–9.

———. 2006. The science of child and adolescent spiritual development: Definitional, theoretical, and field-building issues. In E. C. Roehlkepartain, P. E. King, L. M. Wagener and P. L. Benson, eds., *The Handbook of Spiritual Development in Childhood and Adolescence*, 484–98. Newbury Park, CA: Sage Publications.

Benson, P. L., N. Leffert, P. C. Scales, and D. A. Blyth. 1998. Beyond the "village" rhetoric: Creating healthy communities for children and adolescents. *Applied Developmental Science* 2, no. 3: 138–59.

Benson, P. L., P. C. Scales, J. Arturo Sesma, and E. C. Roehlkepartain. 2005. Adolescent spirituality. In K. A. Moore and L. H. Lippman, eds., *What Do Children Need to Flourish?* 29–40. New York: Springer.

Bronfenbrenner, U. 1979. *The ecology of human development: Experiments by nature and design.* Cambridge, MA: Harvard University Press.

Colby, A., and W. Damon. 1995. The development of extraordinary moral commitment. In M. Killen, D. Hart et al., eds., *Morality in everyday life: Developmental perspectives,* 342–70. New York: Cambridge University Press.

Damon, W. 2004. What is positive youth development? *Annals of the American Academy of Political and Social Science* 591: 13–24.

Damon, W., J. Menon, and K. Bronk. 2003. The development of purpose during adolescence. *Applied Developmental Sciences* 7, no. 3: 119–27.

Erikson, E. H. 1968. *Identity: Youth and crisis.* New York: Norton.

———. 1964. *Insight and responsibility.* New York: W. W. Norton.

———. 1965. Youth: Fidelity and diversity. In E. H. Erikson, ed., *The challenges of youth*, 1–28. Garden City, NY: Anchor Books.

Fetzer Institute. 1999. *Multidimensional measurement of religiousness/spirituality for use in health research.* Kalamazoo, MI: A John E. Fetzer Institute Publication.

Furman, F. K. 1987. *Beyond Yiddishkeit: The struggle for Jewish identity in a reform synagogue.* Albany: State University of New York Press.

Garbarino, J. 1995. *Raising children in a socially toxic environment.* San Francisco: Jossey Bass.

Kerestes, M., & Youniss, J. E. 2003. Rediscovering the importance of religion in adolescent development. In *Handbook of applied developmental science: Applying developmental science for youth and families: Historical and theoretical foundations*, ed. R. M. Lerner, F. Jacobs, and D. Wertlieb, eds., vol. 1, 165–84. Thousand Oaks, CA: Sage Publications.

Kerestes, M., J. Youniss, and E. Metz. 2004. Longitudinal patterns of religious perspective and civic integration. *Applied Developmental Sciences* 8, no. 1: 39–46.

King, P. E. 2003. Religion and identity: The role of ideological, social, and spiritual contexts. *Applied Developmental Sciences* 7, no. 3: 196–203.

King, P. E., and C. J. Boyatzis. 2004. Editors' introduction: Exploring adolescent spiritual and religious development; Current and future theoretical and empirical perspectives. *Applied Developmental Sciences* 8, no. 1: 2–6.

King, P. E., and J. L. Furrow. 2004. Religion as a resource for positive youth development: Religion, social capital, and moral outcomes. *Developmental Psychology* 40, no. 5: 703–13.

King, P. E., E. M. Dowling, R. A. Mueller, K. White, W. Schultz, P. Osborn et al. 2005. Thriving in adolescence: The voices of youth-serving practitioners, parents, and early and late adolescents. *Journal of Early Adolescence* 25, no. 1: 94–112.

Lerner, M. 1996. *The politics of meaning.* Reading, MA: Addison-Wesley.

Lerner, R. M. 2002. *Concepts and theories of human development.* 3rd ed. Mahwah, NJ: Lawrence Erlbaum Associates Publishers.

Lerner, R. M. In press. Wisdom, positive purpose, spirituality, and positive development in adolescence: Theoretical, empirical, and pedagogical bases; Comments on Oman, Flinders, and Thoresen's ideas about "Integrating Spiritual Modeling into Education." *The International Journal for the Psychology of Religion.* Lerner, R. M., J. B. Almerigi, C. Theokas, and J. V. Lerner. 2005. Positive youth development: A view of the issues. *Journal of Early Adolescence* 25, no. 1: 10–16.

Lerner, R. M., E. M. Dowling, and P. M. Anderson. 2003. Positive youth development: Thriving as the basis of personhood and civil society. *Applied Developmental Sciences* 7, no. 3: 172–80.

Lerner, R. M., J. V. Lerner, J. B. Almerigi, C. Theokas, E. Phelps, S. Gestsdottir et al. 2005. Positive youth development, participation in community youth development programs, and community contributions of fifth-grade adolescents: Findings from the first wave of the 4-H study of positive youth development. *Journal of Early Adolescence* 25, no. 1: 17–71.

Miller, W. R., and C. E. Thoresen. 2004. Spirituality, religion, and health: An emerging research field. *American Psychologist* 58, no. 1: 24–35.

Oman, D., T. Flinders, and C. Thoresen. In press. Integrating spiritual modeling into education. A college course for stress management and spiritual growth. *The International Journal for the Psychology of Religion.*

Oman, D., and C. E. Thoresen. 2003. Spiritual modeling: A key to spiritual and religious growth? *International Journal for the Psychology of Religion* 13, no. 3: 149–65.

Roehlkepartain, E. C., P. E. King, L. M. Wagener, and P. L. Benson. 2006. *The handbook for spiritual development in childhood and adolescence.* Newbury Park, CA: Sage Publications.

Roeser, R. 2005. An introduction to Hindu India's contemplative psychological perspectives on human motivation, self, and development. In M. L. Maehr and S. Karabenick, eds., *Advances in motivation and achievement,* 14: 297–345. New York: Elsevier.

Seidlitz, L., A. Abernethy, P. Duberstein, J. S. Evinger, T. H. Chang, and B. L. Lewis. 2002. Development of the spiritual transcendence index. *Journal for the Scientific Study of Religion* 41, no. 3: 439–53.

Smith, C. 2003. Theorizing religious effects among American adolescents. *Journal for the Scientific Study of Religion* 42, no. 1: 17–30.

Youniss, J., J. A. McLellan, and M. Yates. 1999. Religion, community service, and identity in American youth. *Journal of Adolescence* 22, no. 2: 243–53.

5

Self and Identity Processes in Spirituality and Positive Youth Development

Robert W. Roeser, Sonia S. Issac, Mona Abo-Zena,
Aerika Brittian, & Stephen C. Peck

The nature and number of biopsychosocial changes during adolescence have been hypothesized to precipitate a quest for a "sense of continuity and sameness" among young people who experience their bodies, minds, and social relationships as simultaneously in flux (Erikson 1950, 261). Erikson (1950, 1968) proposed that attaining this sense of self-sameness was the focal life task of the adolescent period, a quest resolved, ideally, by a conscious exploration of, and eventual commitment to, vitalizing identities, ideologies, ideals, and institutions. According to Erikson, identity development during adolescence involved a more-or-less conscious recycling through and reworking of prior developmental task resolutions from infancy, toddlerhood, and childhood. Erikson proposed that earlier task resolutions associated with security and attachment (e.g., trust versus mistrust), agency and will (e.g., autonomy and initiative versus shame, doubt, and guilt), and personal and social competence (e.g., industry versus inferiority) were reworked during adolescence in the process of identity exploration and commitment and in the context of (emerging) adult roles, relationships, institutions, and ideological systems. Specifically, earlier task resolutions around trust were said to be renegotiated during adolescence in terms of the kinds of people, role models, cultural ideals, and social institutions in which the growing young person could (or could not) have *faith*; earlier task resolutions around issues of autonomy and initiative were renegotiated in terms of the self-images, purposes, and corresponding activities and ideologies to which youth could (or could not) *freely choose to commit*; and previous task resolutions around issues of industry were renegotiated in terms of desired social and occupational roles in which youth could (or could not) expect to *excel*.

In this chapter, we propose that one fruitful way to study religion and spirituality in relation to development during adolescence and early adulthood is from the vantage point of identity and identity development (see also Benson, Donahue, and Erickson 1989; Elkind 1999; Furrow, King, and White 2004; Markstrom-Adams, Hofstra, and Dougher 1994; Templeton and Eccles 2006; Youniss, McLellan, and Yates 1999). For many adolescents, it appears that religion and spirituality represent important sources of self-images, role models, ideals, and ideologies that shape the course of identity development and related behavioral choices during adolescence. We know that religion and spirituality are important to many American adolescents ages 13–17 today (Smith and Denton 2005); that high levels of interest in spirituality among contemporary American college students exist (Higher Education Research Institute 2007); and that religion and spirituality address the very existential questions regarding identity, purpose, and meaning that are at the heart of development during adolescence and early adulthood. We believe a closer scientific understanding of the role of self and identity processes in religion, spirituality, and positive development during these stages is an ideal, if not heretofore relatively neglected, topic for investigation in the developmental sciences (e.g., Donelson 1999). Thus, our goals in this chapter are (a) to conceptualize *spirituality* from the perspective of human identity and human-identity development and, in so doing, to distinguish *religious identity* from *spiritual identity*; (b) to explore the potential role that spirituality, defined in terms of a unique domain of identity development, may play in positive development during adolescence and early adulthood; and (c) to suggest directions for future research on spirituality and positive youth development. We begin by providing a heuristic framework we have developed for conceptualizing human identity.

DEFINING HUMAN IDENTITY

Based upon the works of William James, Erik Erikson, and the many social scientists who subsequently elaborated upon their ideas (cf. Case 1991; Coté and Levine 2002; Harter 2006; Sheldon 2004), we have developed several conceptual models that describe and define the focal levels of human identity in context (Roeser, Peck, and Nasir 2006). The foundation of these models is James' (1890) classic distinction between the I and the ME.

According to James, "*the constituents of the Self* may be divided into two [different] classes [or levels], those which make up respectively—(a) The material Self; (b) The social Self; (c) The spiritual Self;[1] [on the one hand,] and [those that constitute] (d) The pure Ego" [on the other] (James 1890, 292). The material self was described in terms of the physical body and material possessions; the social self was described in terms of the "recognition" we get from other people; and the spiritual self was described as our "inner or subjective being, [our] psychic faculties or dispositions, taken concretely; not the bare principle of personal Unity, or 'pure' Ego" (296). Whereas the material, social, and spiritual selves (as the Empirical Me) refer to the *representational objects or contents* of identity-related experience, James viewed the I-self or "pure Ego" as the *subjective process of experiencing*.

To arrive at this conclusion, James (1890) undertook a lengthy review of the various philosophical positions on the "soul" and the "knower in consciousness" in his day. His aim was to arrive at a scientifically tractable, substantive definition of the subjective pole of experience he called the I-self. He concluded that the "I" is synonymous with "Thought" and that "it is enough to know that it exists; and that in everyone, at an early age, the distinction between Thought as such, and what it [Thought] is of or about, has become familiar to the mind" (James 1890, 296–97). James relegated any further conclusions to either metaphysical speculation or future scientific discoveries, though he did go on to describe how the I-self or Thought is experienced within the stream of consciousness.

James (1890) described the I-self as "the present mental state [that] binds the individual past facts with each other and with itself" (338). He maintained that "Thought is a perishing and not an immortal or incorruptible thing" (345) and that, despite its moment to moment, pulsating quality, the I-self fulfills several essential psychological functions for the individual. These include the "subjective synthesis" and the "appropriating and disowning" of experience. For James, the experience of particular ME-selves was dependent on distinguishing what was me from what was not me; therefore, "there must be an agent of the appropriating and disowning. . . . It is the Thought to whom the various 'constituents' are known. That Thought is a vehicle of choice as well as of cognition" (340). In sum, James viewed the I-self as synonymous with *phenomenal awareness*, its degree of clarity and concentration (Wallace and Shapiro 2006), and its inherent potentials for willful action and the conscious regulation of emotion, cognition, and behavior (Ryan and Deci 2006).

A key aspect of James' (1890) distinction between the ME and the I was that it afforded a way to understand the age-old question of human self-sameness and continuity. James posited that it was the intrapsychological relationship between the ME and I at any given moment that formed the basis of our *sense of personal identity*. Specifically, personal identity reflected that which is *sensed* as the "Empirical Me" (in this case, past and present material, social, and spiritual ME-selves) in relationship to that which *senses* (or "I", "Thought," and "pure Ego") in any given moment. In order to codify clearly the Jamesian distinction between these two focal levels of self and to update this conceptualization with evidence gleaned over the last one hundred years on the nature of self/identity, we developed the **B**asic **L**evels **of S**elf (BLoS) Model. This model describes the I and the ME in terms of their substance and functions (Roeser, Peck, and Nasir 2006). A version of the BLoS model is depicted in Figure 1.

FIGURE 1. Basic Levels of Self *(BLoS)* Model

(I) Body (ME)	
(II) Mental Representations *(ME-selves)*	
Temperamental representations	
Temperamental characteristics	Basic arousal
Discrete emotions	
Characteristic moods	
	Complex emotions
Iconic representations	
Affective-motor schemas	
Sensory-affective schemas	
Sensory-affective-motor scripts	Implicit motives
Symbolic representations	
Plans and strategies	
Goals and goal hierarchies	Explicit motives
Belief- and value-systems	
Self and world narratives	
Phenomenological representations	Objects of awareness
Feelings	
Ideas	
Thoughts	
	Sense of Self
	Self-reflection
	Stream of Consciousness
(III) Phenomenal Awareness *(I-self)*	
Alert, passive awareness	
Alert, active awareness	

ME-selves

The BLoS model characterizes the content of the ME-selves in terms of various kinds and domains of mental representations (Roeser, Peck, and Nasir 2006). These include temperamental representations (e.g., stimulus reactivity, basic and complex emotions, moods); affectively charged iconic representations (e.g., implicit motives, attachment schemas); valenced symbolic representations of self and world (e.g., beliefs, values, goals); and the phenomenological representations that constitute the stream of thoughts, feelings, and perceptions that characterize our conscious experience of daily living. The representational ME, as a higher-order self-system, develops through a characteristic set of stage structures across the lifespan (Case 1991; Cook-Greuter 2000; Damon and Hart 1988; Fisher and Bidell, 2006; Harter 2006; Loevinger 1978; McAdams 1997; Wilbur 2006). The BLoS model posits that the representational ME functions to afford relatively automatic, nonconscious, rapid, and efficient forms of meaning making, information-processing, motivation and (ME-) self-regulation. Socially, the representational ME, the content of which is largely culturally constituted and conditioned (Markus and Kitayama 1991), affords predictability in social relations and a means of social integration.

I-self

The BLoS model describes the I in terms of "awareness." Erikson once referred to the I-self as "an observing center of awareness and of volition" (1968, 135). Scientific research has now identified several distinct states and functions of I-self awareness. Damasio (1999) used the term *core consciousness* to describe the *alert passive state of awareness* that is present at or very near birth and that is a necessary condition for conscious experience. In these terms, awareness *is* and forms the ground of mental life (Tart 1975). The passive state of I-self awareness is associated with a fusion of I and ME and relatively nonaware, nonvolitional modes of motivation and self-regulation (Tart 1975).

A second mode, an *alert active state of awareness*, is associated with volitional action or agency. This mode involves the *willful directing and sustaining of attention through awareness* (cf. Lieberman et al. 2002; Posner and Rothbart 2000; Shiffrin and Schneider 1977) and affords myriad possibilities for the executive control of emotion, cognition, and behavior (Roeser and Peck 2007). The active state of I-self-awareness makes possible a different kind of relationship between I and ME, one in which awareness is relatively free from the contents of ME and can reflect upon and volitionally relate to and regulate mental states and behavioral acts. We propose that spiritual identities, by definition, involve active forms of I-self-awareness in that the domain is necessarily intentional and contemplative (i.e., reflective) in nature (e.g., Feldman 2008; Ho and Ho 2007; Templeton and Eccles 2006). The BLoS model is meant to advance a central proposition related to the various collection of identities that characterize individuals: *any* domain of identity, including religious identity and spiritual identity, can be conceptualized in terms of these basic aspects of self and their constituent elements: (1) I-states of awareness; (2) ME-self-representations; and (3) the embodiment of I/ME as a whole person engaging in particular kinds of behaviors and social relationships in particular physical and social worlds. BLoS is an organizational framework rather than a substantive theory and, in fact, can be applied to many different prominent theories of self (e.g., Epstein 1990; Markus & Kitayama 1991; Ryan and Deci 2006). In the next section, we use the BLoS model to develop a conceptualization of spirituality as a unique domain of identity development.

DEFINING SPIRITUALITY, SPIRITUAL IDENTITY, AND SPIRITUAL IDENTITY DEVELOPMENT

The study of spirituality in human health and development hinges upon whether it is possible to formulate a clear and scientifically tractable definition of what spirituality *is* substantively, what it *does* functionally, and how it *develops* systematically over ontogenetic time (Emmons and Paloutzian 2003; Feldman 2008; Ho and Ho 2007; Johnson 2008; Oser, Scarlett, and Bucher 2006; Roehlkepartain, Benson, King, and Wagener 2006; Weaver, Pargament, Flannelly, and Oppenheimer 2006). Furthermore, addressing the distinction, and mutual relation, between spirituality and religion in human health and development is also important for moving this field of inquiry forward (Zinnabauer and Pargament 1999).

Below, we review three different meanings of spirituality that are prevalent in the literature today: spirituality as states of consciousness; spirituality as a domain of development with its own stages and structures; and spirituality as the postconventional levels within any domain of development (cf. Wilbur 2006). We then relate these three meanings to the basic levels of human identity outlined in Figure 1. Spiritual development, we propose, reflects changes in meaning perspective associated with transformations in the degree and depth of awareness (I-self), changes in meaning perspective associated with transformations in the content and organization of identity representations, such as purposes and worldviews (me-self), and changes in meaning perspective associated with transformations in the relation between awareness (I-self) and the representational ME from fused to free forms.

Spirituality as States of Consciousness

In one usage of the term, spirituality refers to certain kinds of phenomenological states of awareness—those that transcend the limits of normal, everyday, waking, ego-consciousness. Ego-consciousness is, by definition, centered in the ego (ME). According to the contemplative traditions of the world, the state of ego-consciousness is said to be permeated with a sense of uniqueness, separation, lack, fear, desire, and division. States of awareness that transcend ego-consciousness are often labeled "religious," "spiritual," "mystical," or "non-ordinary" (e.g., Hay 2007; James 1890; Wilbur 2006) because they transcend this limited state and bring one temporarily into communion with "something more." Such states are often accompanied by particularly powerful emotions such as awe, wonder, elevation, and love, which signal experiences beyond the limits of ego (Haidt 2003; Shiota, Campos, and Keltner 2003). Conceptions of spirituality as "transcendence" often rest upon this notion of spirituality as "states of consciousness."

The capacity to turn transient states of ego-transcendence into enduring traits of awareness in which a stable and clear witness state is continually realized is a core goal of spiritual development from a contemplative perspective (Wilbur 2006). This is associated with the attainment of wisdom (Roeser 2005). The developmental line here runs from dull and passive forms of awareness in which much of life is lived in a reactive state, to clear and active forms of awareness that are more reflective and volitional in nature, to "witness forms of awareness" in which a lucid meta-awareness or "awareness of awareness" is attained (Lutz, Dunne, and Davidson 2007).

Spirituality as a Domain of Development

A second meaning of spirituality is that of a distinct representational domain of human development with its own focal psychological content, functions, and stage structures (see Oser, Scarlett, and Bucher 2006 for a review of such approaches). A defining feature of the spiritual domain is that it involves ideals and ideologies that individuals sanctify or construe as sacred, divine, or of ultimate value. By focusing on first-person perspectives on what, if anything, individuals construe as sacred, divine, spiritual, or of ultimate value in their lives, we bracket out questions about the reality of the phenomena and objects to which individuals attribute such significance (e.g., God, paradise, soul, grace, salvation, etc.). Rather, the focus is on the mental (representational) content and its functional implications for the motivation and regulation of positive and problematic behavior across development. This approach is consistent with a naturalistic and scientifically tractable study of religion and spirituality (Dennett 2006).

William James (1902) defined the substantive content of what he called *personal religion* (or what today is called *spirituality*) as "the feelings, acts, and experiences of individual men [*sic*] in their solitude, as far as they apprehend themselves to stand in relation to whatever they may consider the divine" (James 1902, 32). In our reading of James (1902), this suggests spirituality is substantively about *basic aspects of self* (e.g., feelings, thoughts, acts, and experiences); *spiritual ideals* (e.g., subjective conceptions of that which is divine, sacred, or of ultimate purpose and value in life), and *the centrality of such ideals to one's core identity* (e.g., a personal relationship between self and that which is considered divine, sacred, or of ultimate purpose and value).

William James (1902) went on to posit that the function of spirituality was to motivate individuals to realize a more satisfying existence: "Not God, but life, more life, a larger richer, more satisfying life, is, in the last analysis, the end of religion. The love of life, at any and every level of development, is the religious impulse" (453). This suggests that spirituality is fundamentally about being whole, being wholly human, and being part of the whole that is existence. This view accords well with the classic notion of *spirit* and its relation to notions of health, healing, wholeness, and salvation (Fields 2001; Johnson 2008; Smith 1992).

Similar to James, Pargament (2007) defines spirituality as "a search for the sacred." In this work, the sacred is defined as "concepts of God, the divine, and transcendent reality, as well as other aspects of life that take on divine character and significance by virtue of their association with, or representation of, divinity" (Pargament 2007, 32). This definition highlights how, in studying spirituality, researchers need not address themselves to the validity of individuals' claims about spirituality and sacred, divine, or transcendental visions of self, nature, and existence. Rather, the study of spirituality addresses itself to the meanings and practices that characterize individuals' spiritual lives, the social contexts that inform and are informed by such meanings and practices, and the consequences of individuals' spiritual lives for their personal identity development, social relationships, well-being, and physical health. So, for instance, if we know that individuals search for the sacred because they believe that the sacred is synonymous with the realization of a fuller happiness and a more meaningful and satisfying life, we can examine the functional significance of these beliefs, attitudes, and behaviors with regard to happiness, satisfaction, or any number of other possible indicators of healthy development.

James Fowler (1981) described "faith" in terms that are strikingly similar to those used in reference to spirituality today. His work is a good example of spirituality that develops through the forms of a religious tradition. For him, faith is defined as one's personal way of responding to "transcendent value and power as perceived and grasped through the forms of the cumulative tradition" (Fowler 1981, 9). Central to this notion of "faith" is a sacred ideal, value, or end to which one commits one's life and "sets one's heart upon." Fowler describes faith as an "orientation of total person" involving an "alignment of the will" and "a resting of the heart" in accordance with "a vision of transcendent value and power, one's ultimate concern" (14). Faith, as such, is "the human quest for relation to transcendence" (14) and to "that which is universal" (15). The relation between self and a cherished ideal or ultimate concern, Fowler hypothesized, functions to energize and direct, to "give purpose and goal to one's hopes and strivings, thoughts and actions" (14).

Consistent with these works, we consider spirituality as a unique domain of identity development in which the content of the domain is defined by issues of *ultimate* meaning, value, and purpose that transcend but include the wisdom and insights found in any particular cultural-religious tradition. This domain can evolve from, coevolve with, or evolve independent of the religious domain of identity development (Koenig, McCullough, and Larson 2001; Templeton and Eccles 2006). That is, individuals can self-identify as spiritual, religious, both, or neither. Furthermore, to be spiritual, in our view, means to develop from an exclusive identification with a particular cultural-religious tradition toward a transcultural outlook in which an identification with the whole of humanity is more and more salient and central. In addition, a respect for and openness to wisdom traditions other than one's own is central. The central representational (ME-self) contents of the spiritual domain of identity development, we propose, are (a) self-identification as a spiritual person; (b) a spiritual ideal and cardinal life purpose; (c) a spiritual worldview; (d) a "spiritual life story" that features spiritual experiences and associated emotions like awe, wonder, and gratitude that are encountered in pursuit of one's ideal and purpose; and (e) patterns of engagement in spiritual practices such as selfless service to others, private prayer and meditation, and scriptural study.

Spiritual Self-identification. A key characteristic of spiritual identities, we assume, is that these identities are intentional and reflective in nature (Ho and Ho 2007; Feldman 2008; Templeton and Eccles 2006). Through reflection and experience, individuals' identification as "spiritual" does not serve to tie them to a specific group only (though it does), but to that whole which is taken to constitute "Spirit"—humanity, life on earth, the universe itself. The internalization and personal transformation of beliefs about ultimate concerns, including those from sources beyond one's own tradition, along with the experiences and behavior corresponding to these beliefs, constitute the substance and development of individuals' spiritual identities and spiritual life stories. In addition, spiritual development often seems to involve the movement of individuals from a mode of a "received religious tradition" or no religion at all to engaging in reflection upon and dialogue with ultimate concerns that apply to all of humanity, including oneself, in a personal manner. This personalization and questioning of ultimate concerns that are transpersonal and transcultural in nature (e.g., the meaning of death), either through a religious tradition or outside their bounds, is in part what has historically been meant by "personal religion" or "spirituality" (James 1890).

The study of spiritual self-identification has led scholars to an interest in the intersectionality between "being religious" and "being spiritual" in the lives of young people. Smith and Denton (2005), for instance, asked U.S. adolescents if they considered themselves "spiritual but not religious." Results showed that 8 percent of all U.S. teenagers said this was "very true," 46 percent said this was "somewhat true," 43 percent said this was "not true at all," and 2 percent were uncertain. Another study of young people (18–25-year olds) found that, although a majority identified themselves as "religious" (44 percent), substantial percentages also identified themselves as "spiritual but not religious" (35 percent) or "neither spiritual nor religious" (18 percent; Greenberg 2007). In addition, young people in this cohort reported being comfortable with espousing more of a "personal faith" rather than one that was closely tied to a particular religious institution compared to those emerging adults of previous generations (Greenberg 2007). These results support the notion that the meanings of spirituality and faith are changing among newer generations of Americans. For instance, some sociologists, noting the declining importance of religious institutions in American cultural life, have argued that Americans have become increasingly private about their religion. This "privatization of faith" is viewed as evidence of a broader advance of individualism in American life (Bellah 1985).

Spiritual Ideals and Cardinal Life Purposes. Another important characteristic of the spiritual domain of identity development, we propose, is that its subjective content is not the real but the ideal. Key elements of this domain consist, in part, of individuals' beliefs about that which is of the highest value in life, the ultimate, the ideal. Another term for such ideals are *ultimate concerns* (Emmons 1999) or that which individuals consider *sacred* (Pargament 2007), *divine* (James 1902), or of any *ultimate value equivalent* (Fowler 1981). We refer to these as *spiritual ideals* and liken them to high-level organizing *purposes* (e.g., Damon, Menon, and Bronk 2003), *core life values* (Kasser and Ryan 1993), or *cardinal identities* (Peck, Eccles, and Malanchuk 2006) insofar as they can be expected to produce a consistency of intentional focus, energy investment, and behavior across developmental time and space (Ho and Ho 2007).

Two archetypal spiritual ideals or *visions of the ultimate* are useful to consider in research on self and identity processes in spirituality and positive youth development: that of an ultimate being, and that of an ultimate state of being (d'Aquili and Newberg 1999). Although the range of what is marked as "sacred" varies from person to person (Pargament 2007), these two ideals are among the most common among those who identify themselves as having a spiritual life. Often, these ideals are enshrined in images of divine figures, in the lives of great prophets and saints, and in other kinds of spiritual role models (Oman and Thoresen 2003). These common ideals, and their associated forms of theistic and contemplative spirituality, are described in Table 1.

The first spiritual ideal is that of an *ultimate being* such as God, Allah, or Yahweh. This ideal anchors theistic forms of spirituality. The aim of the one committed to this ideal is to attain closeness to and union with the ultimate being or other soteriological aims (e.g., going to heaven or paradise after death). The cardinal life purposes of those who are more or less committed to this ideal involve pleasing one's Deity through ethical conduct, devotion, love, and service to others. Spiritual practices associated with this ideal include devotional prayer, singing the divine name, scriptural recitation and study, and service to others. Many of the virtues associated with this form of spirituality are the same as those described in conceptualization of positive youth development—caring, compassion, character, connection, and contribution (e.g., Lerner et al. 2005).

TABLE 1. **Two Common Spiritual Ideals and Related Forms of Spirituality**

Features	*Type 1* *Theistic Spirituality*	*Type 2* *Contemplative Spirituality*
Orientation	God-centered	Consciousness-centered
Spiritual Ideal	Ultimate Being (e.g., God, Allah)	Ultimate State of Being (e.g., Enlightenment, Nirvana)
Aim	Closeness to/union with Ultimate Being Salvation Going to Heaven / Paradise	Realization of Ultimate State of Being Liberation Freedom from rebirth
Cardinal Life Purposes	Ethical conduct Devotion Love Service	Ethical conduct Wisdom Compassion Service
Signature Behaviors	Devotional practices Scriptural recitation and study Service to others Attendance at public worship services	Contemplative practices Scriptural recitation and study Service to others Training with an adept
View of Human Nature	Fallen	Contracted
View of Nature	Spirit—matter division	Spirit—matter continuum
View of Time	Arrow of time, historical Future-expectant	Circle of time, epochal Present-attendant
View of Spiritual Development	Sin to salvation Fall to redemption Bondage to freedom	Ignorance to wisdom Delusion to enlightenment Bondage to freedom

The second spiritual ideal is that of an *ultimate state of being* such as enlightenment or liberation. This ideal is characteristic of contemplative or consciousness-
centered forms of spirituality. The aim of one committed to this ideal is to realize
the ultimate state of being in this body and this lifetime—something referred
to as enlightenment or liberation (e.g., freedom from rebirth). The cardinal life
purposes of those who are more or less committed to this ideal include ethical
conduct, the realization of wisdom, and the manifestation of compassion and
service to all beings. Spiritual practices associated with this ideal include con-
templation and meditation, scriptural recitation and study, service to others, and
studying under an adept. A closely related spiritual ideal is that of an ultimate
reality such as Tao, Brahman, or "the kingdom of heaven." The main aim of the
one who is committed to this ideal is personal alignment or harmonization with

this ultimate reality. This is thought to herald peace, wisdom, and well-being. These common spiritual ideals and their attendant forms of spirituality, we propose, are central elements of individuals' spiritual identities. Again, many of the virtues associated with these forms of spirituality are the same as those described in conceptualizations of positive youth development and thriving (e.g., Benson, Roehlkepartain, and Rude 2003; Lerner et al. 2005; Sherrod and Spiewak 2008). Like all aspects of the representational ME, representational aspects of individuals' spiritual identities such as their conceptions of the sacred and identifications with larger and larger groups of "others" develop through various stage structures across the lifespan (Fowler 1981). One hallmark of spiritual identity development is the process by which individuals consciously and self-reflectively organize or reorganize their identity and motivation for living around a commitment to a spiritual ideal and orchestrate their goals and behaviors accordingly. A second hallmark of spiritual identity development with respect to such ideals involves changes in the nature of the motivation behind pursuit of a spiritual ideal. Is the ideal sought for intrinsic reasons or extrinsic reasons (Allport and Ross 1967; Batson, Schoenrade, and Ventis 1993)? How well integrated is the ideal in the individuals' overall sense of identity and behavior?

A third hallmark of spiritual identity development with respect to such ideals is whether individuals' commitment to them are associated with a growing and abiding concern for the welfare of all, particularly those beyond one's own "kith and kin" (Templeton and Eccles 2008). Unlike those who propose that spiritual identities reflect personal identity beliefs that distinguish oneself from other people (Templeton and Eccles 2006), we propose that spiritual identities are actually transpersonal in nature. *Transpersonal* means that the content of spiritual identities are not about one's personal uniqueness at all but rather about one's gradually developing understanding of what are really shared aspects of human experience across lines of creed, class, caste, race, and religion—the desire for happiness, questions about the meaning of life and human purpose, the existence of suffering, and the problem of death as the shared fate of all human beings. The intrinsic desire to gain wisdom into these ultimate concerns, regardless of one's particular cultural, religious, ethnic, or racial background, make such ultimate concerns universal and, therefore, spiritual.

In this sense, spiritual identity development is not so much about the personal but really about that which is transpersonal (Wilbur 2006). It involves a renunciation of the human impulse towards cruelty and violence that comes from a belief that one's in-groups are inherently superior to others—what Erikson (1985) called *pseudospeciation*. "Pseudospeciation, which is fueled by historical and cultural experience, creates a false sense of unique identity in groups and ignores the genetic integrity of the human species" (Erikson 1985, 212). The "exclusion clauses" at the heart of many traditions of Christianity, Islam, and Judaism in which some are said to be "saved," "chosen," or "superior" to others are certainly a problematic aspect of religion in the world today (Harris 2004). Whereas religious self-identifications, as collective identities, mark "my religious in-group" from others and promote cooperation with in-group members, such identifications can also inhibit cooperation with out-group members and, in some instances, foster conflict and mistrust between members of different groups (Bodenhausen 1992). Authentic spiritual development reflects a movement beyond basic forms of egocentricity, ethnocentricity, and pseudospeciation; and later beyond even reciprocal altruism (Fehr and Fischbacher 2003) toward acts of generosity, compassion, and contribution to those beyond our selves and our social groups (Dalai Lama 2001; Hanh 1995; Tutu 1999). Spiritual identity, in short, involves transformations in identifications with particular human collectives to the exclusion of others, toward an identification with the human collective as whole—with humanity and our shared values, aspirations, qualities, and experience. As a consequence, some suggest that another hallmark of spiritual-identity development is that the circumference of one's circle of empathic care widens beyond the egocentric and the ethnocentric to the worldcentric (Templeton and Eccles 2008; Wilbur 2006). This is another meaning of the word *transcendence* in reference to the spiritual: transcendence of concern only for oneself or one's in-groups.

Worldviews. A third key dimension of a spiritual identity is a worldview. Koltko-Rivera (2004) defines a worldview as:

a way of describing the universe and life within it, both in terms of what is and what ought to be. A given worldview is a set of beliefs that includes limiting statements and assumptions regarding what exists and what does not (either in actuality, or in principle), what objects or experiences are good or bad, and what objectives, behaviors, and relationships are desirable or undesirable. . . . A worldview defines what can be known or done in the world, and how it can be known or done. . . . In addition to defining what goals can be sought in life, a worldview defines what goals should be pursued. . . . Worldviews include assumptions that may be unproven, and even unprovable, but these assumptions are superordinate, in that they provide the epistemic and ontological foundations for other beliefs within a belief system. (Koltko-Rivera 2004, 4)

Bem (1982) referred to such interlocking sets of zero-order propositions as "nonconscious ideologies" and suggested these ideologies are tacitly socialized by our primary reference groups across development—family, friends, religious leaders, and society as a whole. Erikson (1968) referred to ideology as a cultural-level phenomena available for appropriation by youth that was, in essence, "a coherent body of shared images, ideas, and ideals." For the developing individual, an ideology provides "a coherent, if systematically simplified, over-all orientation in space and time, in means and ends" (Erikson 1968, 113); a "correspondence between the inner world of ideals and evils and the social world with its goals and dangers" (187); an "introduction into the ethos of the prevailing technology" (188); a basis for "fitting in" in a way that guards against social anxiety; and "a geographic-historical world image as a framework for the young individual's budding identity" (188). Erikson (1968) believed that adolescents' exploration and eventual commitment to an ideology, whether explicit or implicit, was essential for positive development during this period. In the absence of such commitment, he believed "youth suffers a *confusion of values* . . . which can be specifically dangerous to some but which on a large scale is surely dangerous to the fabric of society" (188).

Adolescence, with its characteristic changes in mind, body, and social settings, is a time when both nature and nurture provide unique opportunities to expose young people to worldviews that provide wisdom and guidance with respect to existential questions about the nature of ultimate reality, identity, divinity, life and death, and so on. Indeed, such concerns are what adolescence is all about, as young people begin exercising their ability to think beyond the real to the ideal, beyond the actual toward the possible, beyond the now to any number of imagined futures. Such ultimate concerns require adolescents to step beyond the actual to the possible in their thinking and imaginations, something that is intrinsic to this period of cognitive development, and also require vibrant institutions and cultural ideologies that can address such issues in age-appropriate and historically vital ways. Whether such ideologies exist for adolescents today, ideologies that eschew violence and materialism in favor of community and spiritual concerns, is a pressing question in many parts of the world.

Worldview beliefs important to consider when studying the spiritual domain of identity refer to beliefs about "ultimate concerns" (Emmons 1999). These include beliefs about the nature of God, human nature, the creation of the universe, time, death, truth and the possibility for spiritual salvation, among others (see Table 1; Arnett 2008; Koltko-Rivera 2004). Smith and Denton (2005) described the prevailing religious worldview among adolescents in the U.S. today—the vast majority of whom self-identify as Christian—in terms of Moralistic Therapeutic Deism (MTD). They suggested that the MTD worldview among American adolescents is "deism" because the sacred ideal is a deity, an ultimate being called "God" (or, less so in America, "Allah" and "Yahweh") who creates the universe, orders it with divine moral laws, and then watches over human life on earth. Given the central role of God in such a worldview, questions such as "What, in the end, does God want for us and want for us to do?" and "What is the way to God and happiness?" arise. The MTD worldview is "moralistic" in that it teaches that living a good and happy life on earth requires that one be a good and moral person. Adolescents in the United States believe that God wants them to be happy and that the way to happiness is by being morally good and obeying the moral laws laid down in religious scriptures. Being morally good not only leads to happiness, but, in general, youth believe that "good people go to Heaven when they die" (Smith and Denton 2005, 163). Third, the MTD worldview is "therapeutic" in that it frames God as an ultimate and benevolent being who assists us in feeling good and happy about ourselves and our lives through grace and the scriptures. Finally, Smith and Denton (2005) suggested that the "God" of MTD is "not one who is particularly personally involved in one's affairs—especially affairs in which one would prefer not to have God involved" (164). This "distant God," they suggest, is selectively available for taking care of needs.

In sum, the primary religious worldview among U.S. adolescents, according to these authors, is a theistic perspective in which a benevolent and loving God offers grace and the path of moral virtue as the way to happiness. Smith and Denton (2005) are rather critical of this worldview, which they refer to as a "parasitic faith" with respect to the "historical religious traditions" (p. 166). Smith and Denton (2005) appear to view MTD as a worldview characteristic of a broader "privatization of faith" and advance of individualism and narcissism in American cultural and religious life that has been discussed by others (e.g., Bellah 1985). Whether one agrees with this interpretation or not, this example demonstrates the potential scientific merit of the notion of a worldview in that it is a viable empirical construct that has functional significance for both positive and problematic forms of adolescent development. MTD, to the extent it exists, reflects an example of a particular kind of cultural-historical spiritual worldview—a *theistic one* (see Table 1).

A second controversial worldview has been said to exist alongside the mainstream "churched worldview" in the United States since its founding. According to Fuller (2001), there has existed what can generically be called an "unchurched worldview" in American cultural life since colonial times. Adherents to such a worldview have identified themselves as "spiritual, but not necessarily religious." This tradition has existed as a minority perspective against a backdrop of dominant, mainstream Christian churches (Fox 1983). The unchurched tradition has manifested itself in various specific cultural movements since the founding of the United States, and these movements, though diverse in form, are all characterized by five common worldview themes, according to Fuller (2001).

First, adherents of the unchurched tradition tend to feel some sense of misgiving about or impatience with institutional religion (e.g., mainstream Christian churches). Second, a key source of this sense of misgiving is the theological notion of the human being as inherently sinful—a belief at the heart of Christian Fall/Redemption theology (Fox 1983). Adherents of the unchurched tradition tend to reject such an identification of self with inherent "sin" and instead posit that "divinity lies within." They tend to view the human condition, and the suffering so engendered in that condition not as a function of inherent human sinfulness, but rather of limited human awareness and unrealized human potential with respect to identification with one's "inner divinity" (Fuller 2001).

Third, adherents of the unchurched tradition are said to view ongoing efforts to "see divinity in all things" rather than an ongoing commitment to a religious creed as a key manifestation of their faith (Fuller 2001). In a manner of speaking, the "spiritual, not religious" individual is interested and invested in "seeing" rather than "believing." As a consequence, a high value is placed upon spiritual experimentation and experience over religious belief. The fourth characteristic of the "spiritual, not religious" person is a concern with the right, if not the moral responsibility, of the individual to inquire personally into the truth-value of socialized religious beliefs. Finally, adherents to the unchurched tradition exhibit intense curiosity with regard to the universe, human beings' relationships with the universe, and the possible existence of that which lies beyond its physical manifestations (e.g., the metaphysical). In sum, adherents of the "unchurched tradition" are impatient with mainstream Christian churches and their theological dogma concerning the inherent sinfulness of the human being and the externality of divinity; are less concerned with commitment to creed than with seeing divinity in all as a manifestation of faith; believe that individuals need to inquire into the truth-value of religious claims for themselves; and value exploration and experimentation. This particular worldview sounds much like that of the contemplative spiritual worldview presented in Table 1.

Silberman (2005) has proposed that another aspect of worldviews that is important to consider is the information they provide about world change and what are the legitimated means for realizing world change in the direction of one's ultimate concerns in this world (e.g., meditation, suicide bombing). Thus, images of the "ultimate kind of world" that individuals and groups envision as the most ideal, as well as the sanctioned means that these individuals and groups espouse for achieving world change in the direction of their cherished ideals, are essential elements of religious and spiritual worldviews and key for understanding their functional consequences for positive and problematic youth development. What kind of world, ultimately, do young people hold to be sacred and desirable? What are the means they are learning for bringing such a world into being? How do such worldviews shape individual identities and senses of life purpose; and how do these differentially predict generosity, universal compassion, and peace on the one hand, or violence, sectarianism, and terror on the other?

Spiritual Life Stories. Consistent with other aspects of identity formation, we propose that individuals who self-identify as "spiritual" also begin to construct a "spiritual life story" that reflects their personal investigations into ultimate, transpersonal concerns, as well as what their journey has been like in pursuit of their sacred ideal (e.g., their "spiritual journey"). Spiritual experiences, we assume, provide important anchors for these stories (e.g., initiation events, mystical experiences, suffering, etc.). Consistent with other work, we assume that spiritual narratives begin to achieve some form of coherence in adolescence and beyond (McAdams 1997).

Spiritual Practices. Another important window into individuals' spiritual identities concerns the kinds of spiritual practices they engage in on a regular basis. Spiritual practices can be defined as everyday, deliberate activities, engaged in solitude or in the company of others, in which individuals seek to explore and extend their relationship with some conception of the sacred or divinity (Wuthnow 2001). Thus, the purpose of spiritual practices is to enrich one's spiritual life and, more generally, one's life. Such practices may serve to enhance psychological well-being through a variety of mechanisms, including social support, meaning-making, and the effects of engagement in spiritual practices on basic self-regulatory processes (e.g., Urry and Poey 2008). From a sociocultural perspective, examining transformations in individuals' patterns of participation in spiritual practices over time in particular communities of practice is one way of conceptualizing "spiritual development" (e.g., Rogoff 2003; Wenger 1998).

Spirituality as the Highest Stages of Human Development

According to Wilbur (2006), a third way that the word *spiritual* is sometimes used today is to denote the highest stages of the social, moral, and ethical development (Gilligan 1982; Kohlberg 1976; Perry 1981; Selman 1980); faith development (Fowler 1981); and personal identity/worldview development (Cook-Greuter 2000; Erikson 1973; Gebser 1949; Loevinger 1966). These higher stages of development are rooted in both (a) systems-wide cognitive potentials for the creation of increasingly abstract, complex, and systemic (holistic) forms of mental representations; (b) potentials for alert, reflective, multiperspectival forms of awareness in the individual (Alexander and Langer 1990; Fischer and Bidell 2006; Kegan 1981; Piaget 1954; Wilbur 2000); and (c) highly supportive role models and developmentally appropriate social opportunities that scaffold the development of such potentials (Johnson and Boyatzis 2006; Lerner et al. 2006).

Given a mixture of self-effort, persistence, and consistent supportive environments (Fischer and Bidell 2006), it is posited that formal operational thought can be transcended and reorganized at a higher level of complexity and reflectivity. Conventional forms of social, moral, ethical, and faith development can, potentially, be transcended and reorganized around complex principles and embodied ways of being in the world; identity and views of the world can develop beyond the personal and fused forms and towards the transpersonal and "construct aware" stages; and one's empathic circle of care and contribution can extend beyond ego- and ethnocentric aims toward a concern with the whole world and all sentient beings. These higher stages of domain development, referred to as *spiritual*, all involve qualitative shifts in representational capacity and representational complexity. They also all involve intentional reflective ego-transcended awareness (e.g., Feldman 2008; Wilbur 2006). In sum, these "post-conventional" levels of development reflect a level of universality that is unique and relatively rare (Alexander and Langer 1990). Spirituality, from this perspective, is the attainment of the heights of human potential in the core domains of life (e.g., Maslow 1954). It is firmly grounded in this world and the potentials of the human being.

This final definition of spirituality relates to another key hallmark of spiritual identity development—transformations in the nature of the relationship between individuals' awareness and their mental constructs about self and world. Thus another meaning of spiritual identity development from the perspective of stage-structure theories is that, in the lower stages, I and ME are relatively fused. In the higher stages, they are relatively distinct; the very solidity of self itself is being transcended, and a new identity perspective characterized by more open, fluid, and "construct-aware" perspective of meaning is being established (Cook-Greuter 2000). The rare individual who attains this stable position of experience is said to be aware of the constructed nature of identity and experience itself through discursive thought, language, and automatic conceptual marking and evaluating of lived experience as "me." That is, the construction of experience by individuals' construction of self also becomes an insight and aspect of phenomenal awareness of the individual who has realized this level of awareness. Such "mindful" individuals are really being described as inhabiting a position in consciousness that transcends the relatively automatic conceptual construction of reality by identity. This is the ego-transcendent or construct-aware stage of development according to Cook-Greuter (2000) and Indian contemplative traditions more generally (Wilbur 2006). The *state* of ego-transcendence has become a *stable* meaning perspective. This meaning perspective is rarely attained, and, if it is to be attained, it usually requires individuals to undergo some specialized form of contemplative training (e.g., Lutz, Dunne, and Davidson 2007).

Similarly, another meaning of spiritual identity development from the perspective of stage-structure theories with regard to worldview beliefs is that, in the lower stages, individuals espouse a mythopoetic worldview wherein symbol and referent are fused. At this stage of awareness and understanding, there is no distinction between meaning and existence, perception and reality, thoughts and thinker (Cassirer 1955; Wilbur 2006). Spiritual identity development is marked by a growing awareness on the part of individuals that their mythopoetic, religious worldviews are but one "means of expression which, though they reveal a determinate meaning, must necessarily remain inadequate to it, which 'point' to this meaning but never wholly exhaust it" (Cassirer 1955, 239).

DISTINGUISHING RELIGIOUS IDENTITY FROM SPIRITUAL IDENTITY

How can one differentiate religion from spirituality in human development (Hill and Pargament 2003; Hill et al. 2000; Zinnbauer and Pargament 1999)? We propose that religious and spiritual identities are *distinct but often overlap-*

ping domains of identity development (cf. Ho and Ho 2007; Templeton and Eccles 2006). As discussed above, these domains are each characterized by distinctive states of awareness, mental representations and narratives, embodied behaviors, and social relationships that undergo transformations across the human lifespan. We conceptualize religious identities as primarily *cultural and collective* in nature, whereas we conceptualize spiritual identities as primarily *transcultural and contemplative* in nature.

We agree with others who have conceptualized the core of a *religious identity* as a personal identification of oneself with a social group characterized by a particular cultural-historical-religious tradition (Ashmore, Deaux, and McLaughlin-Volpe 2004; Templeton and Eccles 2006). Individuals who claim membership in a particular religious tradition share in common with other group members collective sacred worldviews and their associated "beliefs, practices, rituals, and symbols designed (a) to facilitate closeness to the sacred or transcendent (God, higher power, or ultimate truth/reality) and (b) to foster an understanding of one's relationship and responsibility to others in living together in a community" (Koenig, McCullough, and Larson 2001, 18). Self-identification with a particular religious group; the meaning of that identification to the person in terms of his or her representations of self, world, life purpose, and the (prescribed) good life; the centrality of the identification to a person's overall sense of identity; shared religious practices and the nature and number of social bonds with group members are all key substantive aspects of a *collective religious identity*. Functionally, collective religious identities fulfill individuals' basic needs for social belonging, esteem, self-understanding, meaning and purpose, transcendence, and contribution to something greater than the self through organized cultural forms (cf. Fowler 1981).

According to results of the National Survey of Youth and Religion (Smith and Denton 2005), the vast majority of U.S. adolescents ages 13–17 self-identify with one religion (81 percent). Most religious identifications are as *Christian* (75 percent); mainly *Protestant* (52 percent) and *Catholic* (23 percent). In addition, 2.5 percent self-identify religiously as *Mormon*, 1.5 percent as *Jewish*, 0.5 percent as *Muslim,* and another 1–2 percent identify with other religions (e.g., Jehovah's Witnesses, Hindus, Buddhists, Eastern Orthodox Christian, Unitarian Universalist, etc.). Approximately 3 percent of adolescents self-identify with two different religions. The rest, reflective of a substantial minority of adolescents (16 percent) did not report any collective religious identity. These youth were labeled *non-religious* (Smith and Denton 2005, 31). Among adolescents categorized as *non-religious*, most went on to further self-identified themselves as "just not religious" (10 percent), "atheist" (1.5 percent), or "agnostic" (1.5 percent). The remaining 3 percent of the 16 percent "non-religious youth" seemed uncertain about their religious identity (Smith and Denton 2005). This study, together with others on the religious self-identifications of American adults ages 18 years and above (Kosmin, Mayer, and Keysar 2001), suggests that, despite living in a predominantly Christian country, approximately 1 in 6 individuals ages 13 years and above in the United States do not identify with any particularly religious tradition. Among the U.S. adolescents raised in a religion who became nonreligious, the main reason was, by far, "intellectual skepticism and disbelief" (Smith and Denton 2005).

Cross-national studies have shown that religious identifications are associated with a particular set of guiding life values. Across twenty-one samples, Saroglou, Delpierre, and Dernelle (2004) found that individuals who self-identified as more religious than others were also more likely to say that *conformity* and *tradition* were important guiding principles in their lives and less likely to say that *self-direction* and *novelty-seeking,* or *power and achievement,* were important "guiding principles" in their lives. In sum, this study showed that religiosity was generally associated with the same values across the three monotheistic traditions: those who were more religious tended to favor values that promoted the maintenance of tradition and the social order and tended to disfavor values that promoted autonomous self-direction, change, novelty seeking, and self-enhancement. These results accord with other studies showing that higher religiosity is associated with higher authoritarianism, lack of openness to experience, greater risk aversion, greater need for closure, higher impulse control, and more negative attitudes towards sexuality. Those who were more religious also tended to express concern for the welfare of others, but they did not necessarily espouse a universal outlook in which all human beings were recognized as equals. As the authors note, religiosity may be associated with benevolence primarily toward in-group members. Seeing if these value orientations are true among religious adolescents across the world and how these results might compare to the value correlates of self-identifying as "spiritual" are two potential topics for future research.

SPIRITUALITY AND POSITIVE YOUTH DEVELOPMENT

Significant attention was devoted to the moral and religious development of adolescents in the early part of twentieth century (e.g., Hall 1904). An enduring legacy of this earlier work, work in which adolescence was defined as a period of normative "storm and stress," has been an emphasis on institutional religiosity in relation to amelioration of developmental risk during adolescence (e.g., sexuality, substance use, delinquency) rather than on spirituality and aspects of positive youth development (see Donelson 1999; Regnerus 2003). Thus, one of the key areas in need of future research is on how the inner spiritual life is or is not reflected in indicators of positive youth development and thriving, such as a sense of self-confidence and life purpose, healthy connections with other people, competence, psychological health, ethics of generosity and compassion, and acts that contribute to the welfare of others during adolescence and early adulthood (Benson, Roehlkepartain, and Rude 2003; Lerner, 2004; Sherrod and Spiewak 2008). To a certain extent, it seems as if relations between spirituality and positive youth development should be strong given that the main formulations of PYD in psychology parallel the moral ends that religious-spiritual institutions have traditionally tried to affect in humans historically: ethical behavior, the realization of the sacred ideal, and the use of the knowledge, power, love, and wisdom so gained for the betterment of the world. The chapters in this book are testaments to the interesting kinds of relations between spirituality and positive youth development that are being uncovered as this field develops. In the next section, we next highlight a few key topics in this regard.

FUTURE DIRECTIONS

One key direction for future research on self and identity processes in spirituality and positive youth development involves an examination of the kinds of people, opportunities, and social settings that nurture healthy and authentic forms of spirituality and spirituality ↔ positive development relations (see King 2008; Nasir 2008). As in any other domain of development, we believe spiritual development is a scaffolded performance, dependent on the assistance of more expert others (e.g., Rogoff 2003). Furthermore, adolescence, with its characteristic changes in thinking and feeling, is a prime time for young people to be exposed to, and engaged in, dialogue around ideas and philosophies bearing on ultimate existential questions of identity, purpose, and meaning.

Given that religion and spirituality are key facets of ethnicity, race, and culture (Slonim 1991), a second direction for future research concerns the intersectionality between young people's developing ethnic/racial, cultural, and religious and spiritual identities in shaping patterns of positive or problematic youth development. Virtually no research has examined the intersectionality among such identities with adolescents (e.g., Juang and Syed 2008). New research in this area would enhance our understanding of the roles that religion and spirituality can play in the positive development of ethnically, racially, and culturally diverse youth (e.g., Nicolas and DeSilva, 2008).

For instance, African-Americans tend to be among the most religious individuals in the world. For African-American adults, indicators of religious identity such as valuing religion and attending services is associated with many positive outcomes such as lower blood pressure, better health through religious coping, increased social support, increased life satisfaction, decreased alcohol consumption, and lower depressive symptoms (Steffen, Hinderliter, Blumenthal, and Sherwood 2001). This strong religiousness among African Americans has been linked to the history of the African-American church, the one institutional setting that afforded African Americans a communal place of gathering even during the days of slavery. Jagers (1997) suggested that the church serves as a source of emotional, spiritual, and instrumental support for African-American parents and children. In many cases, the church serves as an extended family, where members refer to each other as "brother" and "sister." The church family becomes a source of social support and sometimes financial support as well (e.g., Taylor, Chatters, and Levin 2004).

Despite the centrality of the church in African-American history, "almost no research focuses specifically on Black adolescents" with respect to religion and development today (Taylor, Chatters, and Levin 2004, 46). What is known is that African-American youth place more importance on religion than their European-American peers (Donahue and Benson 1995; Wallace, Forman, Caldwell, and Willis 2003). Furthermore, the few studies that do focus on the role of religion in the lives of African-American youth focus on its influence in deterring deviant behavior. For instance, religiousness has been found to be related with decreased delinquency in studies of African-American youth living in at-risk communities (Johnson, Jang, Li, and Larson 2000). Researchers suggest relationships developed through church involvement provide African-American youth with positive role models that function to deter deviant behaviors (Cook 2000). Thus, more studies of the intersection and functional significance of ethnic/racial/cultural, and religious/spiritual identities in the positive development of diverse youth are needed.

Similarly, we believe that religion and spirituality play a key role in the development of immigrants and their families (see Abo-Zena et al. 2008; Jensen 2008; Juang and Syed 2008; Roeser et al. 2008). How might religious institutions provide a "context of reception" for newcomers to the United States? How might spiritual beliefs support immigrants in their efforts to assimilate and bridge to the mainstream of American cultural and economic life? How might being an ethnic-minority immigrant who is also a member of the religious majority of the country affect youth development? Future studies examining such issues would increase our understanding of the development of immigrant youth in the United States today.

A third area for future research concerns the development of religious and spiritual worldviews during adolescence and early adulthood and their functional significance for well-being and life choices (e.g., Arnett 2008). How do young people's knowledge and understanding (or lack of knowledge and understanding) of the world's religions shape their own worldviews? How can we help all young people achieve a deep appreciation for the plurality of religions and wisdom traditions that characterize different facets of humanity today and, in doing so, promote greater mutual understanding and civil society? Which religious and spiritual worldviews are differentially associated with positive and problematic forms of human development (e.g., Feldman 2008; Silberman 2005)?

Finally, more research on the catalysts of spiritual development is needed. For instance, one key factor thought to affect the emergence of a reflective and intentional approach to living, in which happiness and a more satisfying life are sought, is the experience of suffering and an inability to address it sufficiently through prevailing identity commitments and worldviews (Corbett 2000). From this perspective, spiritual development is said to be triggered when "traditional religious beliefs and images from childhood no longer offer comfort from suffering or provide adequate reasons for injustices in the world" (Templeton and Eccles 2006, 255). The resultant loss of meaning and desire for livable solutions to questions of ultimate meaning (e.g., the existence of suffering) can catalyze new existential questioning, exploration, and seeking. Thus understanding how life events may trigger spiritual doubts, identity explorations, and ongoing commitments is another important topic of inquiry.

SUMMARY

The role of spirituality in informing and shaping the identity explorations and commitments of adolescents is only beginning to be explored in the developmental sciences. For many adolescents, however, it appears that religion and spirituality represent important sources of self-images, role models, ideals, and ideologies that can influence the course of their identity development. Thus, in this chapter, we devoted ourselves to conceptualizing *spirituality* from the perspective of human identity and human identity development. We proposed that spirituality can be fruitfully conceptualized as a particular domain of identity development, one characterized by specific substantive contents and functions having to do with ultimate transcultural concerns and spiritual ideals, with imagined possibilities rather than actualities, and with ways of being in the world. In contrast to religious identities, which we see as collective and cultural in nature, we defined spirituality identities as more contemplative and transcultural in nature in that they are developed around basic existential issues common to all human beings. Functionally, we posited that spiritual identity serves as a motivation and meaning system that energizes and directs individuals' efforts to realize a more purposeful, more meaningful, and more satisfying life and to identify with greater and greater subgroups of the human family. Developmentally, we posited that individuals grow in the spiritual domain much in the same way as they might in any other—through the processes of role modeling, socialization, and self-socialization. Development in this identity domain is like development in any other domain we proposed—it involves transformations in states of awareness; in psychological representations of self and world; and in the relations between awareness and one's representational constructs (cf. Epstein 1990). Although we view spiritual development as an important subdomain of individuals' overall psychosocial identity development, we nonetheless acknowledge that some individuals develop this aspect of their identities, while many others do not. We hope our conceptualization of spirituality in this chapter, as a unique subdomain of identity development during adolescence and beyond, can inform future efforts at understanding the links between spirituality and positive development during these critical stages in the life course.

NOTES

1. James' use of the term *spiritual* was, in most cases, equivalent to the contemporary term *psychological*.

REFERENCES

Alexander, C., and E. Langer, eds. 1990. *Higher stages of human development*. New York: Oxford University Press.

Abo-Zena, M. M., R. W. Roeser, L. Juang, S. Issac, and D. Du. 2008. On the relations of ethnic identity, religious/spiritual identity, and psychological well-being among immigrant adolescents. Paper presented as part of a symposium, *On the role of spirituality and religion in the lives of immigrant youth and their families,* at the biennial meeting of the Society for Research on Adolescence, Chicago, IL.

Allport, G. W., and J. M. Ross. 1967. Personal religious orientation and prejudice. *Journal of Personality and Social Psychology* 5: 432–43.

Arnett, J. 2008. From "worm food" to "infinite bliss": Emerging adults' views of life after death. In R. M. Lerner, R. W. Roeser, and E. Phelps, eds., *Positive youth development and spirituality: From theory to research,* 231–43. West Conshohocken, PA: Templeton Foundation Press.

Ashmore, R.D., K. Deaux, and T. McLaughlin-Volpe. 2004. An organizing framework for collective identity: Articulation and significance of multidimensionality. *Psychological Bulletin* 130: 80–114.

Batson, C. D., P. Schoenrade, and W. L. Ventis. 1993. *Religion and the individual: A social-psychological perspective.* New York: Oxford University Press.

Bellah, R. 1985. *Habits of the heart.* Los Angeles: University of California Press.

Bem, D. J. 1982. Self-perception theory. In L. Berkowitz, ed., *Advances in experimental social psychology* Vol. 6, 1–62. New York: Academic Press.

Benson, P. L., M. J. Donahue, and J. A. Erickson. 1989. Adolescence and religion: A review of literature from 1970 to 1986. *Research in the Social Scientific Study of Religion* 1: 153–81.

Benson, P. L., E. C. Roehlkepartain, and S. P. Rude. 2003. Spiritual development in childhood and adolescence: Toward a field of inquiry. *Applied Developmental Science* 73: 204–12.

Bodenhausen, G. V. 1992. Identity and cooperative social behavior: Pseudospeciation or human integration? In A. Combs, ed., *Cooperation: Beyond the age of competition,* 12–23. Philadelphia: Gordon and Breach.

Case, R. 1991. Stages in the development of the young child's first sense of self. *Developmental Review* 11: 210–30.

Cassirer, E. 1955. *The philosophy of symbolic form.* New Haven, CT: Yale University Press.

Cook, K. V. 2000. "You have to have somebody watching your back, and if that's God, then that's mighty big": The church's role in the resilience of inner-city youth. *Adolescence* 35: 717–30.

Cook-Greuter, S. R. 2000. Mature ego development: A gateway to ego transcendence? *Journal of Adult Development* 7: 227–40.

Corbett, L. 2000. A depth psychological approach to the sacred. In D. P. Slattery and L. Corbett, eds., *Depth psychology: Meditations in the field,* 73–86. Carpenteria, CA: Pacifica Graduate Institute.

Coté, J. E., and C. G. Levin. 2002. *Identity formation, agency, and culture: A social psychological synthesis.* Mahwah, NJ: Erlbaum.

D'Aquili, E., and A. B. Newberg. 1999. *The mystical mind: Probing the biology of religious experience.* Minneapolis: Fortress Press.

Dalai Lama, HH. 2001. *An open heart: Practicing compassion in everyday life.* New York: Little, Brown and Company.

Damasio, A. 1999. *The feeling of what happens: Body and emotion in the making of consciousness.* New York: Harcourt, Brace, and Co.

Damon, W., and D. Hart. 1988. *Self-understanding in childhood and adolescence.* New York: Cambridge University Press.

Damon, W., J. Menon, and K. C. Bronk. 2003. The development of purpose during adolescence. *Applied Developmental Science* 73: 119–28.

Dennett, D. C. 2006. *Breaking the spell: Religion as a natural phenomenon.* New York: Viking.

Donahue, M. J., and P. L. Benson. 1995. Religion and the well-being of adolescents. *Journal of Social Issues* 51: 145–60.

Donelson, E. 1999. Psychology of religion and adolescents in the United States: Past to present. *Journal of Adolescence* 22: 187–204.

Elkind, D. 1999. Religious development in adolescence. *Journal of Adolescence* 22: 291–95.

Emmons, R. A. 1999. Religion in the psychology of personality: An introduction. *Journal of Personality* 67: 873–88.

Emmons, R.A., and R. F. Paloutzian. 2003. The psychology of religion. *Annual Review of Psychology* 54: 377–402.

Epstein, S. 1990. Cognitive-experiential self-theory. In L. A. Pervin, ed., *Handbook of personality: Theory and research*, 165–92. New York: Guilford Press.

Erikson, E. H. 1950. *Childhood and society*. New York: Norton.

———. 1968. *Identity, youth, and crisis*. New York: Norton.

———. 1970. Reflections on the dissent of contemporary youth. *International Journal of Psychoanalysis* 51: 11–22.

———. 1973. *Dimensions of a New Identity*. New York: W.W. Norton and Company.

———. 1985. Pseudospeciation in the Nuclear Age. *Political Psychology* 6: 213–17.

Fehr, E., and U. Fischbacher. 2003. The nature of human altruism. *Nature* 425: 785–91.

Feldman, D. H. 2008. The role of developmental change in spiritual development. In R. M. Lerner, R. W. Roeser, and E. Phelps, eds., *Positive youth development and spirituality: From theory to research*, 167–96. West Conshohocken, PA: Templeton Foundation Press.

Fields, G. S. 2001. *Religious therapeutics: Body and health in Yoga, Ayurveda, and Tantra*. Albany: State University of New York Press.

Fischer, K. W., and T. R. Bidell. 2006. Dynamic development of action and thought. In R. M. Lerner and W. Damon, eds., *Handbook of child psychology: Theoretical models of human development*, 6th ed., 1:313–399. Hoboken, NJ: John Wiley and Sons.

Fowler, J. 1981. Stages of faith: The psychology of human development and the quest for meaning. San Francisco: Harper San Francisco.

Fox, M. 1983. *Original blessing*. Santa Fe, NM: Bear and Co.

Fuller, R. C. 2001. *Spiritual, but not religious: Understanding unchurched America*. Oxford: Oxford University Press.

Furrow, J. L., P. E. King, and K. White. 2004. Religion and positive youth development: Identity, meaning and prosocial concerns. *Applied Developmental Science* 8: 17–26.

Gebser, J. 1949. *The ever-present origin*. Athens: University of Ohio Press.

Gilligan, C. 1982. *In a different voice: Psychological theory and women's development*. Cambridge, MA: Harvard University Press.

Greenberg, A. 2007. OMG! How Generation Y is redefining faith in the iPod era. Downloaded from http://www.greenbergresearch.com/articles/1218/1829_rebootpoll.pdf (March 11, 2008).

Haidt, J. 2003. Elevation and the positive psychology of morality. In C. L. M Keyes and J. Haidt, eds., *Flourishing: Positive psychology and the well-lived life*, 275–89. Washington, D.C.: American Psychological Association.

Hall, G. S. 1904. *Adolescence: Its psychology and its relations to physiology, anthropology, sociology, sex, crime, religion, and education*. 2 vols. New York: Appleton.

Hanh, T. N. 1995. *Living Buddha, living Christ*. New York: Riverhead Books.

Harris, S. 2004. The end of faith: Religion, terror, and the future of reason. New York: W.W. Norton and Company.

Harter, S. 2006 The self. In N. Eisenberg, ed., Social, emotional, and personality development, 505–70. Vol. 3: Handbook of child psychology. 6th ed. Hoboken, NJ: Wiley.

Hay, D. 2007. *Something there: The biology of the human spirit*. Philadelphia: John Templeton Foundation Press.

Higher Education Research Institute. 2007. *The spiritual lives of college students. Executive Summary: A National Study of College Students' Search for Meaning and Purpose*. University of California, Los Angeles. Downloaded from http://spirituality.ucla.edu/spirituality/reports/FINAL_EXEC_SUMMARY.pdf (December 12, 2007).

Hill, P. C., and K. I. Pargament. 2003. Advances in the conceptualization and measurement of religion and spirituality: Implications for physical and mental health research. *American Psychologist* 58: 64–74.

Hill, P. C., K. I. Pargament, R. W. Hood, M. E. McCullough, J. P. Swyers, D. B. Larson, and B. J. Zinnbauer. 2000. Conceptualizing religion and spirituality: Points of commonality, points of departure. *Journal for the Theory of Social Behavior* 30: 51–77.

Ho, D. Y. F., and R. T. H. Ho. 2007. Measuring spirituality and spiritual emptiness: Toward ecumenicity and transcultural applicability. *Review of General Psychology* 11: 62–74.

Jagers, R. 1997. Afrocultural integrity and the social development of African American children: Some conceptual, empirical, and practical considerations. *Journal of Prevention and Intervention in the Community* 16: 7–31.

James, W. 1890. *The principles of psychology.* New York: Holt.

———. 1902. *The varieties of religious experience: A study in human nature.* New York: Longmans, Green. Repr. in *William James: Writings 1902–1910.* New York: Library of America, 1987.

Jensen, L. A. 2008. Immigrant civic engagement and religion: The paradoxical roles of religious motives and organizations. In R. M. Lerner, R. W. Roeser, and E. Phelps, eds., *Positive youth development and spirituality: From theory to research,* 247–61. West Conshohocken, PA: Templeton Foundation Press.

Johnson, B. R., S. J. Jang, S. D. Li, and D. Larson, 2000. The "invisible institution" and black youth crime: The church as an agency of local social control. *Journal of Youth and Adolescence* 29: 479–98.

Johnson, C. 2008. The spirit of spiritual development. In R. M. Lerner, R. W. Roeser, and E. Phelps, eds., *Positive youth development and spirituality: From theory to research,* 25–41. West Conshohocken, PA: Templeton Foundation Press.

Johnson, C. N., and C. J. Boyatzis. 2006. Cognitive-cultural foundations of spiritual development. In E. C. Roehlkepartain, P. E. King, L. M. Wagener, and P. L. Benson, eds., *Handbook of spiritual development in childhood and adolescence,* 211–23. Thousand Oaks, CA: Sage.

Juang, L., and L. M. Syed. 2008. Ethnic identity and spirituality. In R. M. Lerner, R. W. Roeser, and E. Phelps, eds., *Positive youth development and spirituality: From theory to research,* 262–84. West Conshohocken, PA: Templeton Foundation Press.

Kasser, T., and R. M. Ryan. 1993. A dark side of the American dream: Correlates of financial success as a central life aspiration. *Journal of Personality and Social Psychology* 65: 410–22.

Kegan, R. 1982. *The evolving self: Problem and process in human development.* Cambridge, MA: Harvard University Press.

King, P. E. 2008. Spirituality as fertile ground for positive youth development. In R. M. Lerner, R. W. Roeser, and E. Phelps, eds., *Positive youth development and spirituality: From theory to research,* 55–73. West Conshohocken, PA: Templeton Foundation Press.

Koenig, H. G., M. E. McCullough, and D. B. Larson. 2001. *Handbook of religion and health.* New York: Oxford University Press.

Kohlberg, L. 1976. Moral stages and moralization: The cognitive approach. In T. Lickona, ed., *Moral development and behavior: Theory, research, and social issues,* 31–53. New York: Holt, Rinehart and Winston.

Koltko-Rivera, M. E. 2004. The psychology of worldviews. *Review of General Psychology* 8: 3–58.

Kosmin, B.A., E. Mayer, and A. Keysar. 2001. *American religious identification survey.* New York: The Graduate Center of the City of New York.

Lerner, R.M. 2004. Liberty: Thriving and civic engagement among America's youth. Thousand Oaks, CA: Sage Publications.

Lerner, R. M., J. V. Lerner, J. Almerigi, C. Theokas, E. Phelps, S. Gestsdottir, et al. 2005. Positive youth development, participation in community youth development programs, and community contributions of fifth-grade adolescents: Findings from the first wave of the 4-H Study of Positive Youth Development. *Journal of Early Adolescence* 251: 17–71.

Lieberman, M. D., R. Gaunt, D. T. Gilbert, and Y. Trope. 2002. Reflection and reflexion: A social cognitive neuroscience approach to attributional inference. In M. Zanna, ed., *Advances in experimental social psychology*, 199–249. New York: Academic Press.

Loevinger, J. 1966. The meaning and measurement of ego development. *American Psychologist* 21: 195–206.

———. 1978. *Ego development*. San Francisco: Jossey-Bass.

Lutz, A., J. P. Dunne, and R. J. Davidson. 2007. Meditation and the neuroscience of consciousness. In P. Zelazo, M. Moscovitch, and E. Thompson, eds., *Cambridge handbook of consciousness*, 499–554. New York: Cambridge University Press.

Markstrom-Adams, C., G. Hofstra, and K. Dougher. 1994. The ego virtue of fidelity: A case for the study of religion and identity formation in adolescence. *Journal of Youth and Adolescence* 23: 453–69.

Markus, H., and S. Kitayama. 1991. Culture and the self: Implications for cognition, emotion, and motivation. *Psychological Review* 98: 224–53.

Maslow, A. 1954. *Motivation and personality*. New York: Harper and Row.

McAdams, D. P. 1997. *The stories we live by: Personal myths and the making of self*. New York: Guilford.

Nasir, N. 2008. Considering context, culture, and development in the relationship between spirituality and positive youth development. In R. M. Lerner, R. W. Roeser, and E. Phelps, eds., *Positive youth development and spirituality: From theory to research*, 285–304. West Conshohocken, PA: Templeton Foundation Press.

Nicolas, G., and A. DeSilva. 2008. Spirituality research with ethnically diverse youth. In R. M. Lerner, R. W. Roeser, and E. Phelps, eds., *Positive youth development and spirituality: From theory to research*, 305–21. West Conshohocken, PA: Templeton Foundation Press.

Oman, D., and C. E. Thoresen. 2003. Spiritual modeling: A key to spiritual and religious growth? *International Journal for the Psychology of Religion* 13: 149–65.

Oser, F. K., W. G. Scarlett, and A. Bucher. 2006. Religious and spiritual development throughout the lifespan. In R.M. Lerner, ed., *Theoretical models of human development*, 942–98. Vol. 1: *Handbook of child psychology*, 6th ed. New York: John Wiley.

Pargament, K. I. 2007. *Spiritually integrated psychotherapy: Understanding and addressing the sacred*. New York: Guilford.

Peck, S. C., J. S. Eccles, and O. Malanchuk. 2006. *Stability and change in identity profiles and pathways: Integrating qualitative and quantitative approaches to the study of identity complexity*. Unpublished manuscript. University of Michigan.

Piaget, J. 1954. The construction of reality in the child, trans. M. Cook. New York: Basic Books.

Perry, W. G., 1981. Cognitive and ethical growth: The making of meaning. In Arthur W. Chickering, ed., *The modern American college*, 76–116. San Francisco: Jossey-Bass.

Posner, M. I., and M. K. Rothbart. 2000. Developing mechanisms of self-regulation. *Development and Psychopathology* 12: 427–41.

Regnerus, M. D. 2003. Religion and positive adolescent outcomes: A review of research and theory. *Review of Religious Research* 44: 394–413.

Roehlkepartain, E. C., P. E. King, L. M. Wagener, and P. L. Benson, eds. 2006. *Handbook of religious and spiritual development in childhood and adolescence*. Thousand Oaks, CA: Sage.

Roeser, R.W. 2005. An introduction to Hindu India's contemplative spiritual views on human motivation, selfhood, and development. In M. L. Maehr and S.A. Karabenick, eds., *Religion and motivation*, 297–345. Vol. 14: *Advances in motivation and achievement*. New York: Elsevier.

Roeser, R.W., and S. Peck. 2007. *An education in awareness: Self-identity, motivation and self-regulated learning in contemplative perspective.* Unpublished manuscript.

Roeser, R. W., S. C. Peck, and N. S. Nasir. 2006. Self and identity processes in school motivation, learning, and achievement. In P. A. Alexander, P. R. Pintrich, and P. H. Winne, eds. *Handbook of educational psychology*, 2nd ed., 391–424. Mahwah, NJ: Lawrence Erlbaum.

Roeser, R. W., R.M. Lerner, L. A. Jensen and A. Alberts. 2008. Exploring the role of spirituality and religious involvement in patterns of social contribution among immigrant youth. Paper presented as part of a symposium, *On the role of spirituality and religion in the lives of immigrant youth and their families* at the biennial meeting of the Society for Research on Adolescence, Chicago.

Rogoff, B. 2003. *The cultural nature of human development.* Oxford: Oxford University Press.

Ryan, R. M., and E. L. Deci. 2006. Self-regulation and the problem of human autonomy: Does psychology need choice, self-determination, and will? *Journal of Personality* 74: 1557–86.

Saroglou, V., V. Delpierre, and R. Dernelle. 2004. Values and religiosity: A meta-analysis of studies using Schwartz's model. *Personality and Individual Differences* 37: 721–34.

Selman, R. L. 1980. *The growth of interpersonal understanding: Developmental and clinical analysis.* New York: Academic Press.

Sheldon, K. M. 2004. *Optimal human being: An integrated, multilevel perspective.* Mahwah, NJ: Lawrence Erlbaum Associates.

Sherrod, L., and G. S. Spiewak. 2008. Possible interrelationships between civic engagement, spirituality/religiosity, and positive youth development. In R. M. Lerner, R. W. Roeser, and E. Phelps, eds., *Positive youth development and spirituality: From theory to research*, 322–38. West Conshohocken, PA: Templeton Foundation Press.

Shiffrin, R. M., and W. Schneider. 1977. Controlled and automatic human information processing: II. Perceptual learning, automatic attending, and a general theory. *Psychological Review* 84: 127–90.

Shiota, M. N., B. Campos, and D. Keltner. 2003. The faces of positive emotion: Prototype displays of awe, amusement, and pride. *Annals of the New York Academy of Sciences* 1000: 296–99.

Silberman, I. 2005. Religion as a meaning system: Implications for the new millennium. *Journal of Social Issues* 61: 641–63.

Slonim, M. 1991. *Children, culture, and ethnicity.* New York: Garland.

Smith, C., and M. L. Denton. 2005. *Soul searching: The religious and spiritual lives of American Teenagers.* Oxford: Oxford University Press.

Smith, H. 1992. *Forgotten truth: The common vision of the world's religions.* San Francisco: Harper.

Steffen, P. R., A. L. Hinderliter, J. A. Blumenthal, and A. Sherwood. 2001. Religious coping, ethnicity, and ambulatory blood pressure. *Psychosomatic Medicine* 63: 523–30.

Tart, C. 1975. *States of consciousness.* New York: Dutton.

Taylor, R. J., L. M. Chatters, and J. S. Levin. 2004. *Religion in the lives of African Americans: Social, psychological and health perspectives.* Newbury Park, CA: Sage.

Templeton, J. L., and J. S. Eccles. 2006. The relation between spiritual development and identity processes. In E. C. Roehlkepartain, P. E. King, L. M. Wagener, and P. L. Benson, eds., *The handbook of spiritual development in childhood and adolescence*, 252–65. Thousand Oaks, CA: Sage Publications.

———. 2008. Spirituality, "expanding circle morality," and positive youth development. In R. M. Lerner, R. W. Roeser, and E. Phelps, eds., *Positive youth development and spirituality:*

From theory to research, 197–209. West Conshohocken, PA: Templeton Foundation Press.

Tutu, D. 1999. *No future without forgiveness.* New York: Doubleday.

Urry, H. L., and A. P. Poey. 2008. How religious/spiritual practices contribute to well-being: The role of emotion regulation. In R. M. Lerner, R. W. Roeser, and E. Phelps, eds., *Positive youth development and spirituality: From theory to research,* 145–63. West Conshohocken, PA: Templeton Foundation Press.

Wallace, B., and S. Shapiro. 2006. Mental balance and well-being: Building bridges between Buddhism and western psychology. *American Psychologist* 61: 690–701.

Wallace, J. M., Jr., T. A. Forman, C. H. Caldwell, and D. S. Willis. 2003. Religion and U.S. secondary school students: Current patterns, recent trends, and sociodemographic correlates. *Youth and Society* 35: 98–125.

Weaver, A., K. Pargament, J. Flannelly, and J. E. Oppenheimer. 2006. Trends in the scientific study of religion, spirituality, and health: 1965–2000. *Journal of Religion and Health* 45: 208–14.

Wenger, E. 1998. *Communities of practice: Learning, meaning, and identity.* New York: Cambridge University Press.

Wilbur, K. 2000. *Integral psychology: Consciousness, spirit, psychology, therapy..* Boston, Shambhala Publications.

———2006. *Integral spirituality: A startling new role for religion in the modern and postmodern world.* Boston: Shambhala Publications.

Wuthnow, R. 2001. Spirituality and spiritual practices. In R. Fenn, ed., *The Blackwell companion to sociology of religion,* 306–20. Malden, MA: Blackwell Publishers Ltd.

Youniss, J., J. A. McLellan, and M. Yates. 1999. Religion, community service, and identity in American youth, *Journal of Adolescence* 22: 243–53.

Zinnbauer, B., and K. I. Pargament. 1999. Emerging meanings of religiousness and spirituality: Problems and prospects. *Journal of Personality* 67: 889–919.

PART 2

*Biological Contributions
to the Spirituality–PYD
Relation*

6

Genetics of Faith

An Oxymoron Worth Examining

Elena L. Grigorenko

If you have ever tried to survey friends, family, or strangers with the question "Where does your faith come from?" you have probably never gotten the answer that faith is something with which we are born.[1] Typically in such surveys, especially if carried out in the Westernized world, answers range from "I was brought up [insert religious affiliation] . . ." to "It came to me through my suffering." Why are people uncomfortable with the idea that their faith might be rooted in their genes?

There are many plausible reasons for such discomfort. First, in Westernized views of the world, the distinction between the body and the soul is fundamental. Implicitly, genes are identified with the body, and faith is identified with the soul. Thus, intuitively, these concepts cannot be in causal relationships; therefore, the idea that genes might trigger faith, religiosity, or spirituality does not sound right.

Second, as per Heider's classic attribution theory (Heider 1958), people tend to characterize causal forces influencing someone's behavior as internal (i.e., predispositional) and external (i.e., contextual). Genetic factors, by definition, are determined by someone else (parents) and, therefore, are external and uncontrollable. People readily accept the assertion that they cannot control their genes, but what about their faith? Can they control their faith? From a Westernized point of view, yes. So, the idea that something that should be controllable (faith) is caused by something uncontrollable (genes) produces cognitive dissonance and discomfort.

A relevant aside is Nettler's (Nettler 1959) distinction of free will and determinism within the context of opposing versus justifying capital punishment. This distinction is tightly related to the concept of human dignity and integrity. The greater the degree of free will in one's behavior, the greater one's degree of responsibility (i.e., openness to punishment) and dignity; determinism of one's behavior by factors other than free will reduces culpability, but it also diminishes dignity. Relevant to the discussion here is the observation that a greater emphasis of free will was found to be associated with more conservatism, and a greater emphasis on determinism was associated with more liberalism (Tygart 1982). For example, it has been shown that the greater the degree to which people attributed the causes of homosexuality to genetics, the greater was the support for extending civil rights for gays and lesbians (Tygart 2000). It is possible that admitting the role of genetic factors in faith, religiosity, or spirituality is difficult because this admission is potentially associated with liberalism, whereas, generally, faith and religiosity are viewed as traditional and conservative values.

Finally, it is frequently mistakenly assumed that anything genetic is deterministic and unchangeable. Given how vivid, engaging, and fluid people's spiritual and religious experiences are, there is very little, if anything, immutable about them.

All these considerations create cognitive tension and dissonance when the phrase "genetics of faith" is used. Yet, if feelings of discomfort can be put aside, this seemingly contradictory association of genetics and faith can be examined. Correspondingly: (1) What is the evidence for the role of genes in faith, spirituality, religiosity, and related traits? (2) Assuming that this evidence exists, can anything be said about the magnitude of genetic effects? (3) Finally, what are the potential mechanisms that might engage genes in the formation of spirituality, religiosity, and related traits?

In this chapter, these questions are answered through reviews of heritability, molecular-genetic and population-genetic studies of faith, spirituality, religiosity, and related traits.

HOW "NATURAL" (READ *INNATE*) ARE FAITH AND RELIGION?

Religion has been viewed by some as a form of biological adaptation (Kirkpatrick 1999; Koenig and Bouchard 2006; Looy 2005). The argument is that religion is a complex form of mental representation that permits affecting others' subjective views of the world by implanting and repairing false beliefs and manipulating emotions and behaviors (Dennett 2006). As such, through these manipulations, it is adaptive for controlling the genetic fitness of oneself and one's kind (Bering and Johnson 2005; Cacioppo, Hawkley, Rickett, and Masi 2005). This is not a universally accepted point of view (Dawkins 1976; Wilson 2002), but it is an influential one.

Recently, this view has penetrated developmental psychology, a discipline with no established track record in studying faith, religiosity, and spirituality. Some recent publications suggest that religion is natural and possibly innate (Barrett 2000; Bloom 2007) and that faith, religiosity, and spirituality might be related to the fifth system (in addition to objects, actions, number, and space) of core knowledge, the system representing social partners (Spelke and Kinzler 2007). Interest in the etiology of these traits will likely continue to grow.

Entertaining the view that faith and religion could indeed be a product of human evolution might, at least partially, reduce the seemingly oxymoronic nature of the phrase "genetics of faith." Because biological evolution definitely involves genes, religion might somehow involve genes as well. What evidence for such involvement is available in the field today?

THE ROLE OF GENES IN FAITH AND RELIGION: RESEARCH EVIDENCE

Three lines of evidence are considered in this brief overview: evidence of the involvement of genetic factors obtained (1) through heritability studies of faith, religiosity, and spirituality; (2) from molecular-genetic studies; and (3) from population-genetic studies.

Heritability Studies

Familial resemblance has been investigated for a variety of facets of religiosity, including church affiliation, religious attitudes, and church attendance; the general assumption unifying this work was the sociocultural nature of the transmission mechanism (Dudley and Dudley 1986; Hoge, Petrillo, and Smith 1982; Myers 1996).

Yet, there is a rather small body of behavior-genetic research that includes measures of religiosity, spirituality, or related traits. These data oppose the conclusion of exclusive sociocultural transmission. For example, in an early behavior-genetic study, high-school-aged twins recruited through their participation in the National Merit Scholarship Test were asked questions about their religious beliefs (Loehlin and Nichols 1976). Although the results of this study are difficult to interpret because of measurement issues, this study is the first known attempt to investigate the dominant assumption that the transition of religiosity is cultural using behavior-genetic methodologies and twin data.

This early study was followed by a number of other twin studies. For example, the data on transmission of *religious affiliation* from a large sample of Australian twins suggested the presence of primarily environmental effects, although genetic effects were found to be relevant for female twins living apart (as cited in D'Onofrio, Eaves, Murrelle, Maes, and Spilka 1999). Another large sample of female twins from the Virginia Twin Registry also implicated the role of environmental factors in transmission of church affiliation (Kendler, Gardner, and Prescott 1997).

Data from both of these sizable Australian and British twin samples yielded modest (.22 to .35) but significant heritability estimates (Martin et al. 1986) of the patterns of transmission of *religious attitudes and values.* Another study of twins reared both together and apart corroborated these results and presented evidence for more substantial genetic effects (heritability estimates were reported to be approximately 50 percent; see Waller, Kojetin, Bouchard, Lykken, and Tellegen 1990).

Twin studies of *church attendance, personal devotion,* and *religious salience* have provided inconsistent evidence, with some indicating the presence and others the absence of genetic influences (D'Onofrio et al. 1999; Kendler, Gardner, and Prescott 1997; Kirk, Eaves, and Martin 1999; Truett, Eaves, Meyer, Heath, and Martin 1992). Of note is that, when present, the genetic factor appears more relevant to the religiosity of females than of males (Kirk, Eaves, and Martin 1999; Truett et al. 1992).

The trait of *spirituality,* defined as a derivative of *religious well-being* and *existential well-being,* was examined in a relatively small sample of twins from the U.S. Vietnam Era Twin Registry (Tsuang, Williams, Simpson, and Lyons 2002). Heritability of spirituality was estimated at 23 percent, whereas its components demonstrated higher levels of genetic influence (.37 for religious and .36 for existential well-being).

There have also been studies of other traits associated with religiosity and spirituality. For example, the heritability of *morality* was explored in the context of a U.S. longitudinal twin study (McGuire et al. 1999). Morality was measured across two time points, when twins, on average, were 13.6 and 16.2 years of age. The heritability estimates were not statistically significant at either time point, and the majority of the variance in morality was attributed to nonshared environmental effects.

Measures of *self-transcendence* and *psychological well-being* were administered to a sample of twins older than 50 in Australia (Kirk, Eaves, and Martin 1999). Heritability estimates were found to be statistically significant at approximately 40 percent for *self-transcendence*.

In summary, there is only a handful of relevant heritability studies. Overall, the results of these studies are inconsistent, yet, when considered as a body of literature, suggest the relevance of genetic factors as a source of variability in individual differences in faith, religiosity, spirituality, and related traits.

Molecular-Genetic Studies

Signaling in the brain is carried out in a variety of ways, one of which involves neurotransmitters. Among the many different types of neurotransmitters is a large class of monoamine neurotransmitters. Monoamines include, most notably, the biochemicals known as catecholamines (dopamine [DA], norepinephrine [NE, also referred to as noradrenaline], and epinephrine [Epi], also referred to as adrenaline) and serotonin (5-HT). Monoamines are transported in or out of a cell by specific proteins called transport proteins. In addition to these "dedicated" transporters (e.g., dopamine and serotonin transporters), there are vesicular monoamine transporters (VMATs). VMATs bring monoamines from the cytoplasm to the synaptic vesicles.[2] These less-specific monoamine transporters are able to transport all the different monoamines and, thus, influence the type, amount, and rate of sequestering of monoamines in vesicles.

Monoamine transmitters have been implicated in a wide range of typical and atypical behaviors and states. Correspondingly, genetic variation in many monoamine genes has been related to variation in behavior. Allelic variants in these genes that result in their altered expression have been associated with vulnerabilities to addictions, ADHD, sleep disorders, depression, schizophrenia, and other complex human traits (Uhl et al. 2000). Thus, monoamine-related genes have been considered as candidate genes for a variety of complex human traits, including personality traits.

The gene *VMAT2* entered the genetic landscape of religiosity and spirituality via a discussion by Dean Hamer in his book *The God Gene: How Faith Is Hardwired in Our Genes* (Hamer 2004). The book reports on a study in which Hamer looked for genetic associations between a number of candidate genes (notably, genes involved with the turnover of monoamines in the brain) and the trait of *self-transcendence*. Specifically, the variation in a single nucleotide polymorphism (SNP) appears to differentiate those who report high (Cytosine) self-transcendence and those who report low (Adenine) self-transcendence. The book generated controversy in the press (Kluger 2004; Oppenheimer 2004), but the scientific community has not responded with attempts at replication, in part because there has been no presentation of the *VMAT*2 findings in a peer-reviewed journal.

Thus, at this point, there is no robust evidence pointing to a particular gene or set of genes whose variation can be associated with individual differences in faith, religiosity, or spirituality.

Population-Genetic Studies

In any commentary on the links between genes and faith, population-genetic studies must be included in the discussion since, quite often, religious affiliations are equitable to ethnic or population belonging. It is important to note here that I put aside issues related to the "true" (if there is such a thing) identification of religious, ethnic, or racial backgrounds. I intentionally and selectively talk only about self-identification because this is the central concept for the formation and expression of one's identity in all its aspects, including faith, religiosity, and spirituality.

Genetics and genomics have instruments to address questions of ethnicity (e.g., who qualifies as Nenets in the Russian North, or as Yup'ik in Alaska), caste (e.g., who qualifies as a Kshatriya warrior or a Shudra worker in India), religious (e.g., who qualifies as Jewish or Muslim), and family identity (e.g., who qualifies as a member of the Romanoff czar family in Russia). It is important to mention that, genetically and culturally, identities tend to overlap, often because of shared geographical or historical circumstances. Yet, there are situations in which these identities contradict each other.

An illustration of one such contradiction comes from a small South African tribe called the Lemba (Parfitt and Egorova 2005). Although the Lemba, like most of their neighbors, speak Bantu, they are known for their specific religious practices, customs, and beliefs (Johnston 2003a). These beliefs and practices, which are transmitted orally, are remarkably similar to those of Judaism and are suggestive of Jewish ancestry. This anthropological link has been supported by genetic evidence: The Lemba carry a large percentage of genetic polymorphisms characteristic of non-Arab Semitic groups, several of which are on the Y chromosome, suggesting an ancestral link to the Jews in general and to the Bene Israel group and Jewish Cohanim (or priests) in particular (Spurdle and Jenkins 1996; Thomas et al. 2000; Thomas et al. 1998).

Another illustration comes from the "Black Seminole." The Black Seminole are descendants of former slaves who rebelled and engaged in the Seminole War alongside "true" Seminole Indians; today Black Seminole descendants still live in rural Indian communities in Oklahoma, Texas, the Bahamas, and northern Mexico. Relevant to this discussion is the situation of the Black Seminole in Oklahoma (where they are known as Freedmen). Since 1866, the U.S. government had recognized the Freedmen as members of the Seminole Nation of Oklahoma, but in 2000, "the Seminole Nation of Oklahoma passed a resolution that would effectively expel" a substantial number of the Black Seminole (Johnston 2003b, 262). The issue pertained to an amendment in the Seminole nation's constitution regarding the criteria for membership; specifically, to qualify now, a member had to "demonstrate" the presence of one-eighth Seminole Indian blood. The dispute created a controversy: The Seminole nation recognized blacks because of their shared cultural tradition, but then it stepped back to require a genetic verification of "belonging."

These two examples lead to an interesting question: should genetic ties predominate historical, linguistic, geographic, or cultural ties? People's ability to demonstrate descent from members of a recognized geographical, historical, or cultural alliance has always mattered. The fundamental question is how this alliance is defined and whether it can be claimed to be genetically homogeneous. In the case of the Freedmen Seminole, the alliance was originally historical and social-psychological. By living side-by-side with Seminole Indians, some Freedmen inevitably acquired genetic "Indian-ness," but it was never central to their self-identity; their history of camaraderie and inclusion was!

Similarly, does the fact that the Lemba carry variants of genes from Jewish Cohanim make them more faithful, religious, or spiritual than other African tribes? And, are those Lemba who carry those variants more spiritual than the Lemba who do not?

This brings us back to the assertion that religiosity and spirituality might be related to biological evolution. As Jeffrey Kluger, a writer for *Time*, said, "Far from being an evolutionary luxury then, the need for God may be a crucial trait stamped deeper and deeper into our genome with every passing generation. Humans who developed a spiritual sense thrived and bequeathed that trait to their offspring. Those who didn't risked dying out in chaos and killing" (Kluger 2004).

Summing up this brief overview of what is known about the genetics of faith, religiosity, and spirituality is relatively easy. At this point, there is intriguing but limited evidence suggesting the involvement of genetic factors in the development and manifestation of these complex human traits. This conclusion is accompanied by several subconclusions: (1) the field of relative studies is very small; (2) there is a significant need for behavioral work that might help identify traits of particular sensitivity to genetic impacts; and (3) there is a significant need for measurement work that will enable the separation of "true" variance in faith, religiosity, and spirituality from noise.

And yet, with all cautionary statements issued, even based solely on the limited evidence discussed earlier, the hypothesis of the importance of genetic factors in faith, religiosity, and spirituality cannot be rejected. Thus, supposing that the issue here is not the fact of genetic involvement but how it is realized, what are the potential genetic mechanisms in the formation of spirituality, religiosity, and related traits?

GENES AND ENVIRONMENT: THE INTERTWINED WHOLE

Years of so-called behavior-genetic research (i.e., research on similarities and dissimilarities between various combinations of relatives—twins, siblings, parent-offspring, and so forth), largely aimed at partitioning variance in various human traits into main effects of genes and environments and various high-order effects between them, made abundantly clear to behavioral scientists the importance of genetic forces for all aspects of human development. In fact, it is difficult to think of a human trait that reliably shows no genetic effects (i.e., whose heritability estimates are close to zero). Although heritability estimates vary substantially for variable traits, for the overwhelming majority of human traits examined, the impact of genes appears to be statistically significant and variable from non-negligible to substantial. Yet, this multi-year exercise in the quantification of heritabilities has not resulted (and was not intended to result!) in revelations concerning the genetic causal links or mechanisms underlying these estimates.

Substantial progress in molecular-genetic research, on the other hand, led to the completion of the Human Genome Project in 2003 (http://www.ornl.gov/sci/techresources/Human_Genome/home.shtml). Subsequent work has resulted in the discoveries of genes and gene variants and preliminary specification of the roughly 24,500 genes (Pennisi 2003) carried by humans. Because approximately 55 percent of mouse genes are expressed in the mouse brain (Sandberg et al. 2000), it is reasonable to assume that a comparable portion of human genes is expressed in the human brain.

Thus, on one hand, there is evidence that the majority of (if not all) identifiable and measurable human traits are formed under or in the presence of genetic influences; on the other hand, there are approximately 24,500 genes that realize these influences. And approximately 55 percent of these genes make us talk, reason, rumor, and believe—all the specifically human things that are associated with the human brain. An obvious task is to associate the traits and genes with each other.

However, this is exactly where the problem lies—years of research into establishing these associations have proven this task far from trivial. In fact, although there are many candidate genes in which variation has been associated with human behaviors, no particular associations have been firmly established and replication failures are common (Insel and Collins 2003).

There are many reasons for this "frustration" in the field. First, as explicated by a number of researchers, most complex human traits are governed by many genes of small effect (Insel and Collins 2003). Such genes are harder to find and, once found, remain elusive for replications. Second, there is a huge body of literature indicating that environments have an impact on genes. This body of literature is especially powerful in cancer research (Ting, McGarvey, and Baylin 2006), but there is an increasing number of examples of the impact on behavior by genes whose expression was altered as a result of a presumed environmental impact (Froyen, Bauters, Voet, and Marynen 2006; Schanen 2006). Third, although both genetic and environmental factors manifest main effects, the majority of variation in human behavior is likely attributable to both genes and environment (Rutter 2006).

In this context, the results of the studies presented here do not look so hopeless. Although there is too limited an amount of data to generate firm hypotheses, in thinking about possible hypotheses, it is often useful to explore advances in other subfields. So, what insights can this literature provide?

Many Networked Genes of Small Effect

The dominant assumption in the field of genetics and genomics of complex traits today is that, in the general population, there are genetic components to individual differences in these traits and that these genetic components are, most likely, representative of the effects of many genes, each of which is characterized by relatively low effect sizes (Hamer 2002). In fact, given how proteins act and how much protein-protein interaction, both sequential and concurrent, is required, for example, for a transmission of a signal in the brain, the current expectation that genetic networks and pathways rather than isolated specific genes will be implicated in complex behaviors is not surprising.

This dogma has many direct implications for research, some of which are relevant in the context of this discussion. First, it is no surprise that research replications are rather difficult to obtain: small effects are difficult to identify and even more difficult to replicate (Roberts et al. 1999). Second, it is possible that heterogeneity (i.e., variability in the genetic mechanisms that result in the same behavior trait) is characteristic of genetic networks contributing to the development and manifestation of complex traits, presenting a difficult challenge in the field (Colhoun, McKeigue, and Smith 2003; Sillanpää and Auranen 2004). Thus, the very meaning of replication might change—the point of replication would then be to reliably identify networks and pathways, not specific alleles or genes. And third, and most importantly, the dogma implies probabilistic, not deterministic, mechanisms of genetic influences. This means that the causation chain between genes and behaviors is open to alteration. Below I discuss two types of such alterations—through (1) so-called epigenetic effects and (2) the coaction of genes and environments.

How are these remarks relevant to genetic studies of spirituality and religiosity? First, a genetic study of faith would require a large and carefully structured sample of participants. As indicated earlier, evidence obtained from behavior-genetic studies indicates the importance of genetic effects for at least some behavioral components of spirituality and religiosity. These results indicate the particular phenotypes that might be most informative in designing relevant molecular studies. Second, given the landscape of findings obtained through studies of other complex human traits, it is reasonable to anticipate the involvement of many genes of small effect. Thus, certain designs capable of detecting the small effects of multiple genes should be implemented. Finally, whatever design is adopted, the consideration of environmental factors relevant to religiosity and spirituality will be critical.

Epigenetic Effects

Epigenetic effects are heritable but reversible genetic effects involved in the regulation of the genome that take place and can be transmitted across generations without changing the sequence of DNA. In other words, epigenetic effects are associated with heritable changes in phenotype that do not involve changes in genotype. These changes occur in response to environmental and/ or other genetic factors. There are multiple epigenetic mechanisms, two of

which are particularly interesting for studying complex traits: DNA methylation and acetylation (histone modification). These mechanisms can regulate
gene expression, "manipulating" gene function and permitting an organism to
respond to the environment without introducing changes to DNA sequence
(Jaenisch and Bird 2003; Li 2002). Epigenetic mechanisms might explain the
dissimilarity of identical twins, which, based on their DNA sequences, are
genetic clones. Yet, when methylation and acetylation profiles were studied in
identical twin pairs, twins demonstrated dissimilar epigenetic profiles (Fraga
et al. 2005; Mill et al. 2006) that could be related to differences in their phenotypes (Oates et al. 2006). Of interest also is the observation that epigenetic
differentiation increases with age (Fraga et al. 2005).

Although human studies on the connection between complex traits and
epigenetic effects are still scarce, there is rich animal literature that underscores
the importance of such environmental factors as diet (Van den Veyver 2002),
parenting environment (Champagne et al. 2006), and adverse fetal environment (Seckl and Meaney 2006) on gene expression.

The question, however, is whether and how these exciting findings might
be applicable to research on complex human traits in general and religiosity
and spirituality in particular. At this point, of course, the answer is unknown.
Yet, the ideas and approaches developed in studies of epigenetic mechanisms
in animal models and cancer research in humans are very intriguing and
promising. There are multiple possibilities here. Because so many different
biological events happen prenatally and at very early ages, it is possible that
epigenetic mechanisms not only regulate cell differentiation but also help lay
the foundation for differentiating the gene networks and pathways that underlie complex human behaviors. For example, prenatally, gamma-aminobutyric
acid (GABA) acts as a major excitatory neurotransmitter, but postnatally it
becomes a major inhibitory neurotransmitter, with the switch being carried
out through complex mechanisms at birth involving the maternal hormone
oxytocin (Tyzio et al. 2006). The details of this switch are unknown, but it is
possible that the expression of GABA-related genes is altered epigenetically. In
fact, reprogramming carried out before and after birth by epigenetic mechanisms is now viewed as a "normal developmental process" (Whitelaw and
Whitelaw 2006, R134). Similarly, epigenetic mechanisms might be involved in
the differentiation of innate core systems of knowledge (Spelke and Kinzler
2007). Particular life experiences might also modify gene expression; because
dramatic life experiences have repeatedly been associated with emergent

spirituality and religiosity, epigenetic changes arising in response to such life experiences might form a biological substrate for these complex human traits. Of note is that the human epigenetic profile undergoes changes with normal aging, so that the genome of older adults is hypermethylated compared with younger adults (Brena, Huang, and Plass 2006); might these changes also be associated with increased spirituality and religiosity in aging? At this point, these suggestions are just informed speculation. However, as Michael Rutter stated, one of the main conclusions of epigenetic studies in animals is that "part of the pattern of nature-nurture interplay involves the environmentally induced chemical changes responsible for methylation and acetylation influences of gene expression" (Rutter 2007, 14).

Coaction of Genes and Environments

Nearly everything in all living organisms (i.e., all organisms with a genome) is characterized by the coaction of genes and environment. Because genes do not exist in a vacuum but rather in an environment, these coactions are typically described through gene-environment interactions and covariation.

Gene-environment Interaction (GxE interaction). Conceptually, GxE interaction refers to the conditional effect on the genotype of the presence, direction, and magnitude of an environmental effect. This concept is linked to the huge individual variation observed in the (1) response of different genotypes to both risk (e.g., poverty) and protective (e.g., wealth) environmental factors and (2) in outcomes of the same genotypes in different environments. For example, a deficient copy of the *PHA* gene results in a form of mental retardation, phenyketonuria (PKU), but the severity of the behavioral phenotype can vary dramatically depending on the environmental intervention provided to the individual with the deficient copy of the gene.

Statistically, GxE interaction signifies the presence of a statistically significant interaction effect between measured genotype (i.e., a specific allele, for example, a particular polymorphism within a particular gene or a combination of alleles, that is, a particular haplotype) and measured environment (i.e., an environmental characteristic defined either through a categorical or continuous indicator, e.g., the presence/absence of maltreatment and SES index, respectively). Here, it is important to note the difference between statistical methodologies of the (1) "old school" behavior genetic/quantitative genetic approaches in studying different types of relatives, mostly twins, where neither genes nor environments are measured; and (2) newer molecular genetic/

environmental pathogen studies, where both genes and environment are measured. These methodologies are different in their assumptions and may or may
not coincide in their estimates of the significance of GxE interaction for specific traits (Moffitt, Caspi, and Rutter 2005).

Biologically, GxE interaction assumes the "convergence of environmental
and genotypic effects within the same neural substrate that allows for the possibility of gene-environment interaction" (Caspi and Moffitt 2006, 585).

Might there be a place for GxE interaction in the emergence of spirituality
and religiosity? Clearly, it is a testable hypothesis—such a study simply needs
to be carried out. Following the recommendation in the literature (Moffitt,
Caspi, and Rutter 2005), to design such a study, measurable candidate environmental and genetic factors should be identified. The literature provides some
evidence implicating specific environmental factors (Roehlkepartain, King,
Wagener, and Benson 2006) in the development and maturation of spirituality and religiosity. "Technical eligibility requirements" for these factors assume
the presence of variation in individual responses to these environmental influences; indeed, whatever environmental factor is considered, there are observed
individual differences in spirituality and religiosity outcomes. The true challenge here, however, is identifying an "optimal" measure or set of measures
that will characterize the environment by capturing the variance relevant to
the emergence of spirituality and religiosity. In addition, this optimal measure
should be sensitive to age, gender, and other relevant factors. It is highly recommended that the candidate factor be explicitly measured, using information
collected in the past or concurrently; self- or relative-reported retrospective
information might introduce specific biases. Finally, the "proximity" to the
individual of the candidate environmental indicator is also a factor; apparently,
environmental characteristics that encroach on the individual directly have a
higher likelihood of relevance for such studies.

Similarly, there is literature suggesting the possible relevance of monoamin-
ergic genes (of which there are many). Again, there are some "technical" requirements for candidacy of a "measured genetic factor": specifically, the candidate
allele or haplotype should have some functional relevance (e.g., be associated
with a specific form of protein). It is also recommended that the considered
allele/haplotype be rather common in the general population. And, because
genes do not affect behavior directly but do so through neural substrates, it is
important to consider neuroscientific data implicating specific substrates relevant to spirituality and religiosity, and to investigate the expression patterns in

these substrates to attempt to identify relevant genes. Finally, design and sampling considerations must be taken into account. Although many approaches might be relevant (Yang and Khoury 1997), carefully designed case-control (Clayton and McKeigue 2001) and prospective longitudinal cohort studies have been marked as having great potential to deliver on our understanding of GxE interactions (Collins 2004). In whatever design is adopted, however, the issue of power is important (van den Oord 1999).

Gene-environment covariation (correlation, rGE). Conceptually, rGE is related to an overlap in variance between genes and environments. These correlations are typically characterized as passive, active, or evocative (Plomin, DeFries, and Loehlin 1977). Passive correlation occurs when parental genotypes contribute to variations in rearing environments that, in turn, covary with the child's genotype (e.g., genetic predispositions for religiosity are passed on to the child, which then correlates with the family environment of religious tradition). Active correlation occurs when the child seeks environments driven by genes (e.g., a spiritual child seeks spiritually enriched environments). Evocative correlation occurs when the child's genotype covaries with reactions from other people (e.g., a spiritual child might evoke spiritual reactions from other people).

There are no explicit studies of rGE with regard to the development of religiosity and spirituality. However, there are some studies that provide indirect, but relevant, information. Specifically, it was reported that the spousal correlation for religious affiliation is high and estimated at .67 (Hur 2003). This suggests that, were religious affiliation influenced by genetic factors, both parents would pass their relevant genes to their children and, most likely, these genes would influence home environment and parenting as well, thus creating passive rGE. In fact, there is evidence that religious fathers tend to be more involved and develop higher quality relationships with their children (King 2003).

In addition, there are multiple studies indicating the influence of religion on initiation behaviors related to smoking (Timberlake et al. 2006) and alcohol consumption (Koopmans, Slutske, van Baal, and Boomsma 1999). Although the presence of the positive effects of religiosity leading to lower rates of initiation was interpreted by researchers as an indicator of GxE, in fact, if religiosity is influenced by genes, it is possible that these findings are indicative of the presence of active rGE. The carriers of "religiosity-associated" genetic variants might actively avoid risk environments (i.e., nightclubs) and seek protective environments (i.e., church groups), leading to decreased risk of initiation.

CONCLUSIONS

To conclude, let us revisit the questions that opened our discussion. Regarding the evidence for the role of genes in faith, spirituality, religiosity, and related traits, we can conclude that there is enough to challenge the previously accepted belief that faith, religiosity, and spirituality are exclusively influenced by culture and environment. Yet, it is impossible to say, at this point, what facets of these complex traits are influenced by genes and how. Any possible genetic effects, clearly, are not major. In short, it is unlikely that there is a gene "for" faith, spirituality, and religiosity. Most likely, there are many relevant genes, each of which contributes both individually and interactively to the formation of neural substrates that permit the emergence, development, and manifestations of these traits. Variation in these genes contributes to variation in these neural substrates, which, in turn, contributes to individual differences in these traits. A relevant metaphor is a symphony orchestra: Individual musicians contribute to the quality of the orchestra, but many other factors such as the conductor, the quality of the instruments, and the music itself determine the resulting performance. Finally, as for potential mechanisms that engage genes in the formation of spirituality, religiosity, and related traits, this chapter outlined a few, but there are potentially many others. There is so much to learn and discover in these mostly uncharted waters! And the more knowledge that becomes available, the less discomfort there will be with the statement that faith is influenced by genes; it is, indeed, influenced by them but only when they are "alive." And what makes them "alive" are environments, ranging from cell temperature in response to assault, to emotions of transcendence inspired by music or art (purely cultural phenomena created by humans who are shaped by their genes!).

ACKNOWLEDGMENTS

Preparation of this essay was supported by a grant from the American Psychological Foundation (PI: Grigorenko) and a grant from the Foundation for Child Development (PI: Grigorenko). We express our gratitude to Ms. Robyn Rissman and Ms. Mei Tan and Ms. Lyn Long for their editorial assistance.

NOTES

1. This chapter does not intend to provide a clear separation between terms such as *faith, religion, religiosity,* and *spirituality.* Thus, throughout the text, these terms are used interchangeably, depending on what was used by particular authors in their respective studies or essays. Broadly

speaking, these concepts overlap in their reference to a complex human trait of relating to self and others while guarded by a particular set of behavioral and emotional norms either adapted from a cultural institution (i.e., an organized religion) or self-developed (i.e., through "psychological mapping" of the individual and common good). Thus, although the three concepts are unique and specific, there is undoubtedly a semantic junction where they meet. That junction is relevant to this chapter.

2. There are two different types of VMATs: VMAT1, which is expressed by chromaffin cells, a type of neuroendocrine cell (i.e., cells receiving neuronal input and, in response, releasing hormone molecules), and VMAT2, found primarily in neurons and histaminergic cells (Wang et al., 1997).

REFERENCES

Barrett, J. L. 2000. Exploring the natural foundations of religion. *Trends in Cognitive Sciences* 4: 29–34.

Bering, J. M., and D. D. P. Johnson. 2005. "O Lord . . . You Perceive my Thoughts from Afar": Recursiveness and the evolution of supernatural agency. *Journal of Cognition and Culture* 5: 118–42.

Bloom, P. 2007. Religion is natural. *Developmental Science* 10: 147–51.

Brena, R. M., T. H. Huang, and C. Plass. 2006. Quantitative assessment of DNA methylation: Potential applications for disease diagnosis, classification, and prognosis in clinical settings. *Journal of Molecular Medicine* 84: 365–77.

Cacioppo, J. T., L. C. Hawkley, E. M. Rickett, and C. M. Masi. 2005. Sociality, spirituality, and meaning making: Chicago Health, Aging, and Social Relations Study. *Review of General Psychology* 9: 143–55.

Caspi, A., and T. E. Moffitt. 2006. Gene-environment interactions in psychiatry: Joining forces with neuroscience. *Nature Reviews Neuroscience* 7: 583–90.

Champagne, F. A., I. C. Weaver, J. Diorio, S. Dymov, M. Szyf, and M. J. Meaney. 2006. Maternal care associated with methylation of the estrogen receptor-alpha1b promoter and estrogen receptor-alpha expression in the medial preoptic area of female offspring. *Endocrinology* 147: 2909–15.

Clayton, D., and P. McKeigue. 2001. Epidemiological methods for studying genes and environmental factors in complex diseases. *Lancet* 358: 1356–60.

Colhoun, H. M., P. M. McKeigue, and G. D. Smith. 2003. Problems of reporting genetic associations with complex outcomes. *Lancet* 361: 865–72.

Collins, F. S. 2004. The case for a US prospective cohort study of genes and environment. *Nature* 429: 475–77.

D'Onofrio, B. M., L. J. Eaves, L. Murrelle, H. H. Maes, and B. Spilka. 1999. Understanding biological and social influences on religious affiliation, attitudes and behaviors: A behavior genetic perspective. *Journal of Personality* 67: 953–84.

Dawkins, R. 1976. *The selfish gene*. London: Granada Publishing.

Dennett, D. C. 2006. *Breaking the spell: Religion as a natural phenomenon*. New York: Viking.

Dudley, R. L., and M. D. Dudley. 1986. Transmission of religious values from parents to adolescents. *Review of Religious Research* 28: 3–15.

Fraga, M. F., E. Ballestar, M. F. Paz, S. Ropero, F. Setien, M. L. Ballestar et al. 2005. Epigenetic differences arise during the lifetime of monozygotic twins. *Proceedings of the National Academy of Sciences of the United States of America* 102: 10604–9.

Froyen, G., M. Bauters, T. Voet, and P. Marynen. 2006. X-linked mental retardation and epigenetics. *Journal of Cellular and Molecular Medicine* 10: 808–25.

Hamer, D. 2002. Rethinking behavior genetics. *Science* 298: 71–72.

———. 2004. *The God gene: How faith is hardwired in our genes*. New York: Doubleday.

Heider, F. 1958. *The psychology of interpersonal relationships*. New York: Wiley.

Hoge, D. R., G. H. Petrillo, and E. I. Smith. 1982. Transmission of religious and social values from parents to teenage children. *Journal of Marriage and the Family* 44: 569–80.

Hur, Y.-M. 2003. Assortative mating for personality traits, educational level, religious affiliation, height, weight, and body mass index in parents of a Korean twin sample. *Twin Research* 6: 467–70.

Insel, T. R., and F. S. Collins. 2003. Psychiatry in the genomics era. *American Journal of Psychiatry* 160: 616–20.

Jaenisch, R., and A. Bird. 2003. Epigenetic regulation of gene expression: How the genome integrates intrinsic and environmental signals. *Nature Genetics* 33: 245–54.

Johnston, J. 2003a. Case study: the Lemba. *Developing World Bioethics* 3: 109–11.

———. 2003b. Resisting a genetic identity: the black Seminoles and genetic tests of ancestry. *Journal of Law Medicine and Ethics* 31: 262–71.

Kendler, K. S., C. O. Gardner, and C. A. Prescott. 1997. Religion, psychopathology, and substance use and abuse: A multimeasure, genetic-epidemiologic study. *American Journal of Psychiatry* 154: 322–29.

King, V. 2003. The influence of religion on fathers' relationships with their children. *Journal of Marriage and Family* 65: 382–95.

Kirk, K. M., L. J. Eaves, and N. G. Martin. 1999. Self-transcendence as a measure of spirituality in a sample of older Australian twins. *Twin Research* 2: 81–87.

Kirkpatrick, L. A. 1999. Toward an evolutionary psychology of religion and personality. *Journal of Personality* 67: 921–52.

Kluger, J. 2004. Is God in our genes? *Time*. October 17.

Koenig, L. B., and T. J. J. Bouchard. 2006. Genetic and environmental influences on the traditional moral values triad—authoritarianism, conservatism, and religiousness—as assessed by quantitative behavior genetic methods. In P. McNamara, ed., *Where God and science meet: How brain and evolutionary studies alter our understanding of religion*, 47–76. Vol. 1: *Evolution, genes, and the religious brain*. Westport, CT: Praeger Publishers/Greenwood Publishing Group.

Koopmans, J. R., W. S. Slutske, G. C. M. van Baal, and D. I. Boomsma. 1999. The influence of religion on alcohol use initiation: Evidence for genotype X environment interaction. *Behavior Genetics* 29: 445–53.

Li, E. 2002. Chromatin modification and epigenetic reprogramming in mammalian development. *Nature Reviews Genetics* 3: 662–73.

Loehlin, J. C., and R. C. Nichols. 1976. *Heredity, environment, and personality: A study of 850 sets of twins*. Austin: University of Texas Press.

Looy, H. 2005. The body of faith: Genetic and evolutionary considerations. *Journal of Psychology and Christianity* 24: 113–21.

Martin, N. G., L. J. Eaves, A. C. Heath, R. Jardine, L. M. Feingold, and H. J. Eysenck. 1986. Transmission of social attitudes. *Proceedings of the National Academy of Sciences of the United States of America* 83: 4364–68.

McGuire, S., B. Manke, K. J. Saudino, D. Reiss, E. M. Hetherington, and R. Plomin. 1999. Perceived competence and self-worth during adolescence: A longitudinal behavioral genetic study. *Child Development* 70: 1283–96.

Mill, J., E. Dempster, A. Caspi, B. Williams, T. Moffitt, and I. Craig. 2006. Evidence for monozygotic twin (MZ) discordance in methylation level at two CpG sites in the promoter region of the catechol-O-methyltransferase (COMT) gene. *American Journal of Medical Genetics. Part B, Neuropsychiatric Genetics* 141: 421–25.

Moffitt, T. E., A. Caspi, and M. Rutter. 2005. Strategy for investigating interactions between measured genes and measured environments. *Archives of General Psychiatry* 62: 473–81.

Myers, S. M. 1996. An interactive model of religiosity inheritance: The importance of family context. *American Sociological Review* 61: 858–66.

Nettler, G. 1959. Cruelty, dignity and determinism. *American Sociological Review* 24: 375–84.

Oates, N. A., J. van Vliet, D. L. Duffy, H. Y. Kroes, N. G. Martin, D. I. Boomsma et al. 2006. Increased DNA methylation at the AXIN1 gene in a monozygotic twin from a pair discordant for a caudal duplication anomaly. *American Journal of Human Genetics* 79: 155–62.

Oppenheimer, M. 2004. Godspells. *Washington Post*. September 5.

Parfitt, T., and Y. Egorova. 2005. Genetics, history, and identity: The case of the Bene Israel and the Lemba. *Culture, Medicine & Psychiatry* 29: 193–224.

Pennisi, E. 2003. Gene counters struggle to get the right answer. *Science* 301: 1040–41.

Plomin, R., J. C. DeFries, and J. C. Loehlin. 1977. Genotype-environment interaction and correlation in the analysis of human behavior. *Psychological Bulletin* 84: 309–22.

Roberts, S. B., C. J. MacLean, M. C. Neale, L. J. Eaves, and K. S. Kendler. 1999. Replication of linkage studies of complex traits: an examination of variation in location estimates. *American Journal of Human Genetics* 65: 876–84.

Roehlkepartain, E. C., P. E. King, L. Wagener, and P. L. Benson, eds. 2006. *The handbook of spiritual development in childhood and adolescence*. Thousand Oaks, CA: Sage Publications.

Rutter, M. 2006. *Genes and behavior*. Malden, MA: Blackwell.

———. 2007. Gene-environment interdependence. *Developmental Science* 10: 12–18.

Sandberg, R., R. Yasuda, D. G. Pankratz, T. A. Carter, J. A. Del Rio, L. Wodicka et al. 2000. Regional and strain-specific gene expression mapping in the adult mouse brain. *Proceedings of the National Academy of Sciences of the United States of America* 97: 11038–43.

Schanen, N. C. 2006. Epigenetics of autism spectrum disorders. *Human Molecular Genetics* 15: R138–R150.

Seckl, J. R., and M. J. Meaney. 2006. Glucocorticoid "programming" and PTSD risk. *Annals of the New York Academy of Sciences* 1071: 351–78.

Sillanpää, M. J., and K. Auranen. 2004. Replication in genetic studies of complex traits. *Annals of Human Genetics* 68: 646–57.

Spelke, E. S., and K. D. Kinzler. 2007. Core knowledge. *Developmental Science* 10: 89–96.

Spurdle, A. B., and T. Jenkins. 1996. The origins of the Lemba "Black Jews" of southern Africa: Evidence from p12F2 and other Y-chromosome markers. *American Journal of Human Genetics* 59: 1126–33.

Thomas, M. G., T. Parfitt, D. A. Weiss, K. Skorecki, J. F. Wilson, M. le Roux et al. 2000. Y chromosomes traveling south: The Cohen modal haplotype and the origins of the Lemba—the "Black Jews of Southern Africa." *American Journal of Human Genetics* 66: 674–86.

Thomas, M. G., K. Skorecki, H. Ben-Ami, T. Parfitt, N. Bradman, and D. B. Goldstein. 1998. Origins of Old Testament priests. *Nature* 394: 138–40.

Timberlake, D. S., S. H. Rhee, B. C. Haberstick, C. Hopfer, M. Ehringer, J. M. Lessem et al. 2006. The moderating effects of religiosity on the genetic and environmental determinants of smoking initiation. *Nicotine and Tobacco Research* 8: 123–33.

Ting, A. H., K. M. McGarvey, and S. B. Baylin. 2006. The cancer epigenome—components and functional correlates. *Genes and Development* 20: 3215–31.

Truett, K. R., L. J. Eaves, J. M. Meyer, A. C. Heath, and N. G. Martin. 1992. Religion and education as mediators of attitudes: A multivariate analysis. *Behavior Genetics* 22: 43–62.

Tsuang, M. T., W. M. Williams, J. C. Simpson, and M. J. Lyons. 2002. Pilot study of spirituality and mental health in twins. *American Journal of Psychiatry* 159: 486–88.

Tygart, C. E. 1982. Effects of religiosity on public opinion about legal responsibility for mentally retarded felons. *American Journal of Mental Deficiency* 86: 459–64.

———. 2000. Genetic causation attribution and public support of gay rights. *International Journal of Public Opinion Research* 12: 259–75.

Tyzio, R., R. Cossart, I. Khalilov, M. Minlebaev, C. A. Hubner, A. Represa et al. 2006. Maternal oxytocin triggers a transient inhibitory switch in GABA signaling in the fetal brain during delivery. *Science* 314: 1788–92.

Uhl, G. R., S. Li, N. Takahashi, K. Itokawa, Z. Lin, M. Hazama et al. 2000. The VMAT2 gene in mice and humans: Amphetamine responses, locomotion, cardiac arrhythmias, aging, and vulnerability to dopaminergic toxins. *FASEB Journal* 14: 2459–65.

van den Oord, E. J. 1999. Method to detect genotype-environment interactions for quantitative trait loci in association studies. *American Journal of Epidemiology* 150: 1179–87.

Van den Veyver, I. B. 2002. Genetic effects of methylation diets. *Annual Review of Nutrition* 22: 255–82.

Waller, N. G., B. A. Kojetin, T. J. Bouchard, D. T. Lykken, and A. Tellegen. 1990. Genetic and environmental influences on religious interests, attitudes, and values: A study of twins reared apart and together. *Psychological Science* 1: 138–42.

Wang, Y. M., R. R. Gainetdinov, F. Fumagalli, F. Xu, S. R. Jones, C. B. Bock et al. 1997. Knockout of the vesicular monoamine transporter 2 gene results in neonatal death and supersensitivity to cocaine and amphetamine. *Neuron* 19: 1285–96.

Whitelaw, N. C., and E. Whitelaw. 2006. How lifetimes shape epigenotype within and across generations. *Human Molecular Genetics* 15: R131–R137.

Wilson, D. S. 2002. *Darwin's cathedral*. Chicago: University of Chicago Press.

Yang, Q., and M. J. Khoury. 1997. Evolving methods in genetic epidemiology. III. Gene-environment interaction in epidemiologic research. *Epidemiologic Reviews* 19: 33–43.

7

Cooperative Behavior in Adolescence

Economic Antecedents and Neural Underpinnings

Tomáš Paus, Simon Gächter, Chris Starmer, & Richard Wilkinson

Social interactions across generations and among peers represent powerful mechanisms influencing the development of cognitive and emotional skills and the well-being of youth. In this chapter, we discuss possible negative relationships between economic inequality and a key indicator of positive youth development, prosocial behavior (Lerner et al. 2005) and outline several experimental approaches for the study of neural underpinnings of cooperative behavior, conflict resolution, and resistance to peer influences. Enhancing understanding of these links may provide insights about the brain–behavior-context relations involved in a developmental systems approach to positive youth development (Lerner, Roeser, and Phelps 2008).

BACKGROUND

There is evidence that the health of an individual is closely related to the local (family) and global (nation) economic environment. *Individual income* and *income inequality* are two powerful predictors of life expectancy, albeit their relative contribution and independence are hotly debated (Lynch et al. 2001; Mackenbach 2002; Wilkinson and Pickett 2006). In a recent report, Pickett and Wilkinson (2007) used the UNICEF index of child well-being and the ratio of the total annual household income received by the richest 20 percent of the population to that received by the poorest 20 percent of the population (ranging between 3.4 in Japan, the most equal country, to 8.55 in the U.S., the most unequal) and found a significant negative relationship ($r=-0.67$) between child well-being and income inequality across 23 countries, all of which belong to the richest 50 counties in the world; no such relationship was observed between child well-being and average income (r=0.15).

Two opposing explanations for this correlation favor material and psychosocial pathways, respectively (Marmot and Wilkinson 2001; Poulton and Caspi 2005; Poulton et al. 2002). The psychosocial model argues that income inequality serves as a measure of *social inequality* that is seen, in turn, as "socially corrosive, leading to more violence, lower levels of trust, and lower social capital" (Wilkinson and Pickett 2006, 1778). Thus, income inequality may result in a relative shift from prosocial (affiliative) behavior to a more individualistic (dominance-based) behavior; note that such a shift would affect all socioeconomic groups (Wilkinson 2005). There is now evidence that many social and behavioral problems are more common in more unequal countries, and U.S. states, suggesting that inequality is associated with psychosocial differences between societies (Wilkinson and Pickett 2007). Here, we focus on cooperative behavior and conflict resolution as two examples of prosocial behavior.

Cooperative Behavior

Cooperative behavior has been studied extensively in the context of evolutionary biology, social psychology, and experimental economics. Both theoretical and experimental work in this area examines the interaction of "cooperators" and "noncooperators" (free riders) in various social-dilemma scenarios, such as the Prisoner's Dilemma game or the Public Good game (Axelrod 1984; Fehr and Gächter 2002). In these games, the payoffs resulting from mutual cooperation are higher than those from mutual free-riding. But individuals have incentives to be noncooperative because the payoff to an individual is highest if he/she free rides on the cooperativeness of others. Such games have been the focus of so much interest because they capture, in microcosm, the tension between self-interest and collective interest. In the Public Good game with punishment, group members have the possibility to punish other members by paying a fee to reduce the income of the punished group members. Numerous studies show that the possibility to punish increases cooperation (Fehr and Gächter 2002). Yet, different social groups achieve starkly different cooperation levels; there are strong differences in punishment of free-riders and even cooperators (Gächter, Herrmann, and Thöni 2005). Thus, the Public Good game with punishment gives us a simple yet effective tool to measure people's tolerance of free-riding and their attitude toward cooperators.

An important question is how cooperative behavior can emerge and persist in a mixed group of cooperators and free riders. Several strategies have been proposed (Nowak 2006), some of which emphasize the need for repeated interactions between the players. One example is the concept of "reciprocal altruism" (Trivers 1971). It has been suggested that the emergence and maintenance of reciprocal altruism in human and nonhuman animals depend on specific cognitive abilities, such as delayed gratification or punishment of cheaters (Stevens and Hauser 2004). Furthermore, it is likely that processes underlying nonverbal communication, such as the ability to extract social cues from faces and bodies, also play a significant role in shaping cooperative behavior.

Conflict Resolution

Conflict resolution is potentially critical for healthy functioning of social groups. In many primate species, aggression is often followed by "conflict resolution" or "reconciliation" (Aureli and de Waal 2000). De Waal (2000, 587) argues that "individuals try to 'undo' the social damage inflicted by aggression; hence, they will actively seek contact, specifically with former opponents." The ability to reconcile is a positive characteristic of some youths. Several studies are beginning to document different forms of reconciliatory behavior in children observed in naturalistic settings (Butovskaya and Kozintsev 1999; Ljundberg, Westlund, and Lindqvist Foresberg 1999). Overall, we see cooperative behavior and conflict resolution as part and parcel of positive youth development, a construct used as a descriptor of external (family, community) and internal (personality attributes) conditions that codetermine the probability of prosocial behavior of an individual and his or her efforts, therefore, to transcend self-interest and contribute to people and issues beyond the self (Lerner et al. 2005). Such transcendence is what Lerner and Phelps (2008) have termed *spirituality*.

Here, we take a developmental perspective to begin unraveling the complexities of the possible relationships between economic environment and the "social" brain. It is well known that early life experiences exert powerful effects on the development of cognitive and emotional processes and their neural substrate (De Kloet et al. 2005; Pollak 2005). But does the social environment shape neural systems supporting cooperative behavior in the same way as, for example, musical training influences the (structural and functional) development of auditory cortex (Gaser and Schlaug 2003)? Clearly, multiple factors may affect the hypothesized links between economic inequality, social behavior, and the brain. These factors may influence specific neural systems at different times. The fact that the human brain continues to mature well into the third decade (Blakemore and Choudhury 2006; Lenroot and Giedd 2006; Paus 2005) suggests that environmental influences are likely to play a role beyond infancy and early childhood.

This will be particularly the case for late-developing neural structures, such as those supporting decision making and cognitive control in social context.

In general, there are at least two ways in which experience of economic inequality can shape the brain: stress (Anderson and Armstead 1995) and imitation (Rizzolatti and Craighero 2004).

Early life *stress* represents one of several possible mechanisms mediating the effects of socioeconomic status (SES) on health in general and on psychopathology in particular (Lupien et al. 2000; Melchior et al. 2007). The relationship between socioeconomic status, stress, and health begins in early stages of life (Dohrenwend 1973; Meaney and Szyf 2005). Indeed, individuals with lower SES report greater exposure to stress, higher salivary levels of the stress hormone cortisol in adolescence, and a diminished sense of mental and physical health than individuals with higher SES (Lupien et al. 2000; Adler et al. 1994). In experimental animals, early-life stress has long-term consequences vis-à-vis brain anatomy and biochemistry, as well as behavior (Bronzino et al. 1996; Denenberg et al. 1991; Matthews, Wilkinson, and Robbins 1996; Meaney et al. 1991; Rots et al. 1996; Sutano et al. 1996). In nonhuman primates, stressful early rearing induced by variable foraging demands results in a number of behavioral, endocrine, and neural consequences observed in adolescence and adulthood. Of particular interest are differences in affiliative behavior (Andrews and Rosenblum 1994) and brain structure (Mathew et al. 2003) in monkeys reared by mothers living under unpredictable and predictable foraging conditions.

Imitation is a powerful mechanism facilitating learning by observation of parents and/or peers. As such, it may play an important role in transmitting social behavior within family or across peers. Human and nonhuman primates engage a number of cortical regions when observing conspecifics, or "actors": (1) regions in the temporal cortex involved in the processing of biological motion (Allison, Puce, and McCarthy 2000); and (2) regions in the frontoparietal cortex involved in the programming and execution of motor actions (Grosbras and Paus 2006; Rizzolatti and Craighero 2004). The former extract information from visual cues embedded in the actor's movements, whereas the latter may support computations used to infer the actor's intentions and/or to facilitate initiation, by the observer, of actions matching those of the actor.

Overall, powerful neurobiological mechanisms exist that may mediate the potential effects of economic inequality on social behavior. An experimental approach that bridges biological and social sciences may facilitate integration of existing knowledge across the respective disciplines and generate new knowledge by applying methods developed in one field to solve problems in another.

EXPERIMENTAL APPROACH

The ultimate goal of our work is to understand the neural mechanisms under-lying the possible impact of the individual's environment on his/her cognitive and emotional development. Economic inequality and social behavior provide but two example areas of research that are, we believe, directly relevant for positive youth development, the key topic of this book. In the text below, we will outline two cases from our current research that illustrate how functional and structural imaging of the adolescent brain can be used to uncover neural underpinnings of social behavior.

Adolescence: Cognition-Affect Interplay

During adolescence, high demands are placed not only on the executive sys-tems but also on the interplay between cognitive and affective processes. Such cognition-emotion interactions are particularly crucial in the context of peer-peer interactions and the processing of verbal and nonverbal cues. Previous research in cognitive neuroscience has uncovered several key building blocks that are relevant for processing social cues.

The cortex of the superior temporal sulcus (STS) contains a set of regions engaged during the processing of nonverbal cues such as those carried by eye and mouth movements (e.g., Puce et al. 1998), hand movements/actions (e.g., Decety et al. 1997; Grezes et al. 1999; Beauchamp et al. 2002), or body move-ments (e.g., Bonda et al. 1996). As suggested by Allison, Puce, and McCarthy (2000) feed-forward and feedback interactions between the STS and amygdala may be critical for the discrimination of various facial expressions and for the attentional enhancement of the neural response to socially salient stimuli. Consistent with such an "amplification" mechanism, Kilts (2003) observed sig-nificantly stronger neural response to dynamic, as compared with static, facial expressions of anger in both the STS and amygdala. We have also observed a strong blood oxygenation-level dependent (BOLD) response in amygdala not only while adult subjects viewed video clips of angry hand movements or angry faces but also during viewing of (dynamic) neutral facial expressions (Grosbras and Paus 2006).

Although the basic aspects of face perception are in place shortly after birth (Goren 1975), both the quantity and quality of face processing continues to increase all the way through adolescence (e.g., Carey 1992; Taylor et al. 1999; McGivern et al. 2002). Developmental fMRI studies of the processing of facial expressions are consistent with this pattern. For example, *happy, but not sad, faces* elicit significant BOLD response in the amygdala in adolescent subjects (Yang et al. 2003). Studies of *fearful facial expressions* suggest that an increase in the BOLD signal in amygdala can be detected in adolescents (Baird et al. 1999) but is relatively weak (Thomas et al. 2001).

Adolescence: Influence of Peers

It is likely that the interplay between cognitive and affective processes is being particularly tasked in social situations in which the right balance must be struck between the peer-based influences and the individual's goals. Adolescents differ in their sensitivity to peer pressure and in their ability to resist peer influences. It should be noted that, depending on context, peer pressure could have both positive and negative consequences. For example, peer influence may be necessary for learning new social skills, such as democratic decision making and negotiation and for correcting the mistakes, or balancing the idiosyncrasies, of parents (Sullivan 1953).

Resistance to peer influence can be assessed with a self-report questionnaire designed to minimize socially desirable responding (Steinberg and Monahan 2007). Scores on the Resistance-to-Peer-Influence (RPI) scale stay low during early adolescence and increase linearly from 14 years of age to reach adult levels at 18 years (Steinberg and Monahan 2007).

Neural Underpinnings of the Resistance to Peer Influences: Function

Which neural systems are engaged differentially in children or adolescents who differ in their resistance to peer influences? We have asked this question by examining neural activity in the following three systems.

First, the *action-observation* (or "bottom-up") network is considered by many to represent the neural substrate of imitation (Iacoboni 2005; Brass and Heyes 2005; Rizzolatti and Craighero 2004). It consists of frontal and parietal regions involved in the preparation and execution of actions. So-called "mirror neurons" found within the inferior premotor cortex and/or inferior frontal gyrus, as well as in the anterior inferior parietal lobe, are active both when subjects perform a specific action themselves and when they observe another individual performing the same action (Rizzolatti and Craighero 2004).

Second, the *biological-motion processing* network plays an important role in extracting socially relevant cues, such as those imparted by the movements of eyes or hands; as described above, neurons within the STS respond selectively to the presentation of dynamic bodies, body parts, or faces.

Third, the *executive* (or "top-down") network supports a number of cognitive processes underlying decision making, working memory, and suppressing alternative programs interfering with planned actions (Duncan and Owen 2000; Miller and Cohen 2001; Petrides 2005; Paus 2001). It consists of a set of regions in the lateral and medial prefrontal cortex (PFC).

Whether an adolescent follows the goals set by peers or those set by himself/ herself might depend on the interplay between the above three neural systems, namely the fronto-parietal network (bottom-up imitation of actions), the STS network (social cues), and the PFC network (top-down regulation of actions).

FIGURE 1. Inter-regional correlations in fMRI signal during the observation of angry hand movements.

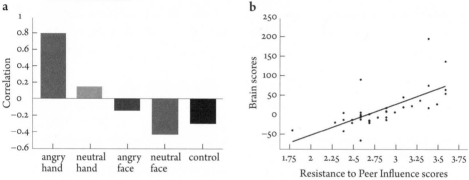

a, Latent Variable 1 (LV1) identified a combination of brain regions that, as a whole, correlated with the Resistance-to-Peer-Influence (RPI) scores. Note that high correlations are observed only for fMRI signal measured during the observation of angry hand movements.

b, Brains scores (weighted sum of all voxels in an image for each subject, using the weights derived from the brain LV1) derived from the fMRI signal measured during angry hand movements plotted as a function of RPI.

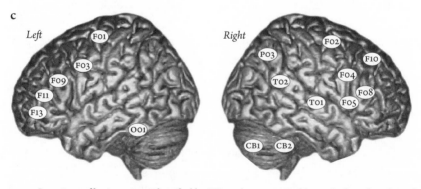

c, Locations of brain regions identified by LV1; only regions visible on the lateral surface of the left and right hemispheres are shown.

d

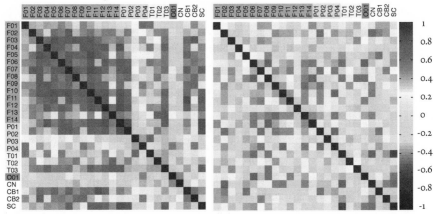

d, Correlation matrices depicting interregional correlations of fMRI signal measured during the observation of angry hand movements, as revealed by LV1, in subjects with high (left) and low (right) Resistance to Peer Influence. The high and low RPI subgroups correspond to the subjects with RPI scores above and below the group median, respectively.

e

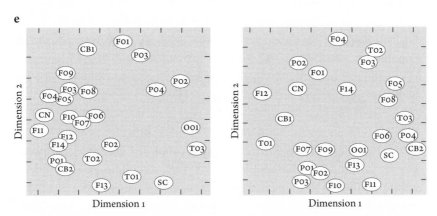

e, Multidimensional scaling (MDS) representations of the interregional correlations of the 26-D matrix depicted above; in the MDS 2-D plots, strongly correlated regions are placed close together. Note, for example, the close grouping of premotor (F03 and F04) and prefrontal (F08 and F09) fronto-cortical regions. F01, Premotor cortex, dorsal, left; F02, Premotor cortex, dorsal, right; F03, Premotor cortex, ventral, left; F04, Premotor cortex, ventral, right; F05, Frontal operculum, right; F06, Cingulate motor area, left; F07, Insula, anterior, left; F08, Prefrontal cortex, ventro-lateral, right; F09, Prefrontal cortex, dorso-lateral, left; F10, Prefrontal cortex, dorso-lateral, right; F11, Prefrontal cortex, ventro-lateral, left; F12, Anterior cingulate cortex, right; F13, Orbito-frontal cortex, lateral, left; F14, Prefrontal cortex, medial; P01, Posterior cingulate cortex; P02, Precuneus, left; P03, Parietal cortex, dorso-lateral, right; P04, Parietal cortex, dorso-medial, right; T01, Superior Temporal Sulcus, middle, right; T02, Superior Temporal Sulcus, posterior, right; T03, Hippocampus, right; O01, Fusiform gyrus, left; CN, Caudate nucleus, right; CB1, Cerebellum, right; CB2, Cerebellum, right; SC, Superior Colliculus, right.

To answer our question, we asked 10-year-old children to watch brief video clips containing face or hand/arm actions, executed in neutral or angry ways, while measuring changes in fMRI signal. Outside of the scanner, we administered the RPI questionnaire. We found that the children with high versus low RPI scores showed stronger interregional correlations in brain activity across the three networks while watching angry hand actions (Grosbras et al. 2007; Fig. 1). The pattern of interregional correlations identified by this method included both (1) regions involved in action observation: the fronto-parietal as well as temporo-occipital systems and (2) regions in the prefrontal cortex.

Note that we did not give the children any specific instruction regarding the video clips; there was no explicit task. The children simply watched the video clips as they would watch their peers in a situation where no clear goals have been formulated in advance. And yet, a number of prefrontal regions showed coordinated changes in the fMRI signal that correlated with those in the other two neural systems involved in action observation. Typically, prefrontal cortex is engaged when the subject performs an explicit task requiring, for example, manipulation of information in working memory, inhibition of prepotent responses, and/or suppression of interference, or planning and decision making (Petrides 2005). No such demands were explicitly made in this study. It is tempting to speculate that the brains of the children with high resistance to peer influence engaged *automatically* "executive" processes when challenged with relatively complex and socially relevant stimuli.

Neural Underpinnings of the Resistance to Peer Influences: Structure

Given the evidence for experience-driven structural plasticity (e.g., Draganski et al. 2004), is it possible that adolescents with high versus low resistance to peer influence differ not only in the degree of functional connectivity, as described above, but also in some morphological features? This might be the case for two very different reasons: first, early developmental events; and second, a repeated functional engagement of a given neural system. Although we cannot differentiate between the two scenarios, here we make an assumption that individuals who differ in their resistance to peer influence also differ in the probability of engaging the above neural networks. Over time, such a repeated and coordinated engagement would translate into structural changes. We have examined this possibility in a large sample of healthy adolescents (n=295, 12 to 18 years of age) and found that interregional correlations in *cortical thickness* in the same cortical regions revealed by the above fMRI study were higher in adolescents with high versus low resistance to peer influence (Paus et al. 2007, Fig. 2). Based on these results, we suggested that individuals with certain personality and cognitive characteristics, compatible with high resistance to peer influences, are more likely to engage relevant neural networks whenever challenged with relatively complex and socially relevant stimuli. These networks include cortical regions activated during action observation and cognitive/executive control. Over time, such a coordinated functional engagement is likely to shape morphological properties of these regions so that they become structurally alike.

FIGURE 2. Interregional correlations in cortical thickness.

a

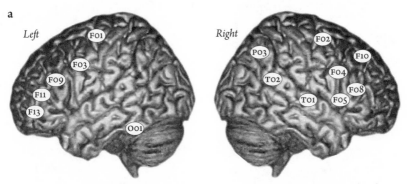

a, Locations of 22 brain regions identified in a previous functional MRI study of action observation (Grosbras et al. 2007); only regions visible on the lateral surface of the left and right hemispheres are shown.

b

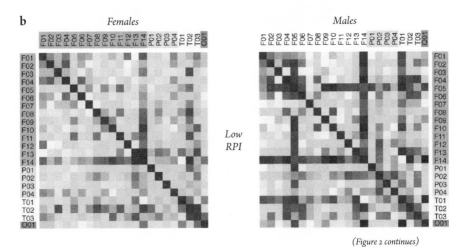

(Figure 2 continues)

(Figure 2 continued)

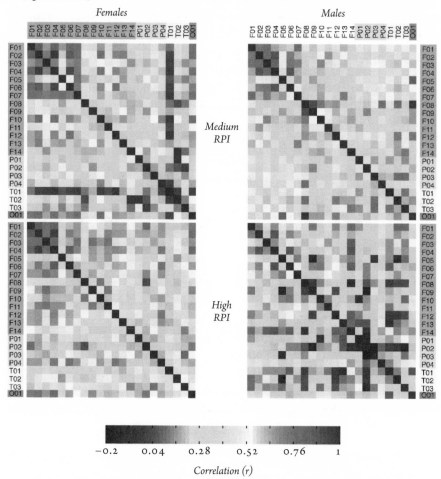

b, Correlation matrices depicting interregional pair-wise correlations (strength of the correlation indicated by gray tone) in cortical thickness across the twenty-two cortical regions in female (left column) and male (right column) adolescents with low (top row), medium (middle row), and high (bottom row) age-adjusted scores of Resistance to Peer Influence. Fo1, Premotor cortex, dorsal, left; Fo2, Premotor cortex, dorsal, right; Fo3, Premotor cortex, ventral, left; Fo4, Premotor cortex, ventral, right; Fo5, Frontal operculum, right; Fo6, Cingulate motor area, left; Fo7, Insula, anterior, left; Fo8, Prefrontal cortex, ventro-lateral, right; Fo9, Prefrontal cortex, dorso-lateral, left; F10, Prefrontal cortex, dorso-lateral, right; F11, Prefrontal cortex, ventro-lateral, left; F12, Anterior cingulate cortex, right; F13, Orbito-frontal cortex, lateral, left; F14, Prefrontal cortex, medial; Po1, Posterior cingulate cortex; Po2, Precuneus, left; Po3, Parietal cortex, dorso-lateral, right; Po4, Parietal cortex, dorso-medial, right; To1, Superior Temporal Sulcus, middle, right; To2, Superior Temporal Sulcus, posterior, right; To3, Hippocampus, right; Oo1, Fusiform gyrus, left.

Overall, several neural systems support the individual's ability to extract socially relevant information from people's faces and bodies. As this ability continues to develop during childhood and adolescence, so do the relevant neural systems. Importantly, the degree of the coordinated neural response to such stimuli across a set of cortical regions not only differs across individuals but also appears to predict certain behavioral characteristics, such as resistance to peer influence. It is also possible that repeated engagement of a particular neural system leads to subtle changes in brain structure.

Increasing Positive Youth Development?

It is quite likely that neural circuits are malleable to change vis-à-vis both their function and structure. New knowledge about the maturation and plasticity of the adolescent brain may provide additional, but not sole, information for the development of effective interventions. Quantification of intervention-related changes in brain structure and function may also provide useful information about neural mechanisms underlying the induced change in behavior. Knowing which neural systems have been modified may, in turn, provide additional insights relevant for fine-tuning of the initial strategy.

To illustrate a possible way forward regarding the design and evaluation of interventions targeting complex behavior during adolescence, let us conclude this review by describing a hypothetical evidence-based intervention aimed at increasing positive youth development.

Current theories of youth development emphasize that positive or healthy psychological and social functioning involves not only the absence of risk/problem behaviors but also the presence of characteristics of successful and adaptive behavior. In the "positive youth development" (PYD) perspective (Lerner et al. 2005), the presence of psychosocially and physically healthy change is indexed not by decreases in negative behaviors (e.g., substance abuse, bullying, or depression) but by increases in attributes indicative of thriving, such as the "Five Cs" of PYD: competence, confidence, connection, character, and caring (Lerner et al. 2005). These "Cs" have been found to be the outcomes of functionally and developmentally appropriate (adaptive) *interactions between the young person and his or her social ecology*; importantly, these attributes appear to be inversely related to indicators of both internalizing and externalizing risk/problem behaviors (Gestsdottir and Lerner 2007). In our previous research, we found that prenatal exposure to maternal cigarette smoking is associated with lower PYD scores in the adolescent offspring (Toro et al. 2007).

We also found that adolescents with low versus high resistance to peer influence have lower PYD scores (Paus et al. 2007). Could one design an intervention aimed at facilitating positive youth development? One way to approach this issue is to base the intervention directly on the PYD construct; for example, the PYD model emphasizes the role of contextual factors such as parental warmth and involvement in school or the presence of role models in the community. Another strategy could involve a "brain-based" approach and target cognitive and affective processes identified via imaging studies as distinguishing individuals with high versus low PYD scores. Note that this approach ignores the possible *causes*, environmental or genetic, that may have resulted in a more-or-less positive youth development of the adolescent up to this point. Using the latter approach, we could—for example—target cognitive and affective abilities necessary for extracting relevant social cues from faces and bodies of peers and to regulate decision making and action control in emotionally charged environments.

This approach identifies the so-called "hot cognition" as a critical aspect of adolescent cognitive development (Steinberg 2005). Given our findings on the neural underpinnings of the resistance to peer influence (see above) and the association between this behavioral characteristic and positive youth development, we may use changes in function and structure of the relevant network as an intermediate outcome measure (akin to endophenotypes used in genetic studies). In other words, we would predict that a successful intervention would promote functional connectivity in the network identified in the previous study and, depending on the length of the intervention, also increase interregional similarity of cortical thickness in the same network.

CONCLUSIONS

Recent advances in cognitive neuroscience, experimental economics, and social epidemiology opened up new opportunities for addressing important issues of the relationship between economic environment, prosocial behavior, and positive youth development. The use of functional and structural neuroimaging in this context provides a powerful tool for the quantification of the effects of environment on the individual's brain. But we need to be cautious in the interpretation of "brain images": we should not confuse a manifestation with a cause. Observing a difference, for example, in the size of structure X between two groups of individuals simply points to a possible neural mechanism mediating the effect of environment (or genes) on a given behavior; it is not the "cause" of this behavior. Neuroimaging-based assessment should be treated in the same way, and at the same level, as any other quantitative phenotype describing cognitive, emotional, endocrine, or physiological characteristics of an individual.

REFERENCES

Adler, N. E., T. Boyce, M. A. Chesney et al. 1994. Socioeconomic status and health: The challenge of the gradient. *American Psychologist* 49, no. 1: 15–24.

Allison, T., A. Puce, and G. McCarthy. 2000. Social perception from visual cues: Role of the STS region. *Trends in Cognitive Sciences* 4: 267–78.

Anderson, N. B., and C. A. Armstead. 1995. Toward understanding the association of socioeconomic status and health: A new challenge for the biopsychosocial approach. *Psychosomatic Medicine* 57: 213–25.

Andrews, M. W., and L. A. Rosenblum. 1994. The development of affiliative and agonistic social patterns in differentially reared monkeys. *Child Development* 65: 1398–1404.

Aureli, F., and F. B. M. de Waal. 2000. *Natural conflict resolution.* Berkeley: University of California Press.

Axelrod, R. 1984. *The evolution of cooperation.* New York: Basic Books.

Baird, A. A., S. A. Gruber, D. A. Fein, L. C. Maas, R. J. Steingard, P. F. Renshaw, B. M. Cohen, and D. A. Yurgelun-Todd. 1999. Functional magnetic resonance imaging of facial affect recognition in children and adolescents. *Journal of American Academy of Child and Adolescent Psychiatry* 38: 195–99.

Beauchamp, M. S., K. E. Lee, J. V. Haxby, and A. Martin et al. 2002. Parallel visual motion processing streams for manipulable objects and human movements. *Neuron* 34: 149–59.

Blakemore, S. J., and S. Choudhury. 2006. Development of the adolescent brain: Implications for executive domain and social cognition. *Journal of Child Psychology and Psychiatry* 47: 296–312.

Bonda, E., M. Petrides, D. Ostry et al. 1996. Specific involvement of human parietal systems and the amygdala in the perception of biological motion. *Journal of Neuroscience* 16: 3737–44.

Brass, M., and C. Heyes. 2005. Imitation: is cognitive neuroscience solving the correspondence problem? *Trends in Cognitive Sciences* 9:489–95.

Bronzino, J. D., et al. 1996. Neonatal isolation alters LTP in freely moving juvenile rats: sex differences. *Brain Research Bulletin* 41: 175–83.

Butovskaya, M. L., and A. G. Kozintsev. 1999. Aggression, friendship, and reconciliation in Russian primary school children. *Aggressive Behavior* 25: 125–39.

Carey, S. 1992. Becoming a face expert. *Philosophical Transactions of the Royal Society of London, Series B, Biological Sciences* 335: 95–102.

De Kloet, E. R., R. M. Sibug, F. M. Helmerhorst, and M. V. Schmidt. 2005. Stress, genes and the mechanism of programming the brain for later life. *Neuroscience & Biobehavorial Reviews* 29: 271–81.

De Waal, F. B. 2000. Primates—a natural heritage of conflict resolution. *Science* 289: 586–90.

Decety, J., J. Grezes, N. Costes et al. 1997. Brain activity during observation of actions. Influence of action content and subject's strategy. *Brain* 120: 1763–77.

Denenberg, V. H., R. H. Fitch, L. M. Schrott et al. 1991. Corpus callosum: Interactive effects of infantile handling and testosterone in the rat. *Behavioral Neuroscience* 105: 562–66.

Dohrenwend, B. P. 1973. Social status and stressful life events. *Journal of Personality and Social Psychology* 28: 225–35.

Draganski, B., C. Gaser, V. Busch, G. Schuierer, U. Bogdahn, and A. May. 2004. Neuroplasticity: changes in grey matter induced by training. *Nature* 427: 311–12.

Duncan, J., and A. M. Owen. 2000. Common regions of the human frontal lobe recruited by diverse cognitive demands. *Trends in Neuroscience* 23: 475–83.

Fehr, E., and S. Gächter. 2002. Altruistic punishment in humans. *Nature* 415: 137–40.

Fischbacher, U., S. Gächter, and E. Fehr. 2001. Are people conditionally cooperative? Evidence from a public goods experiment. *Economic Letters* 71: 397–404.

Gächter, S., B. Herrmann, and C. Thöni. 2005. Cross-cultural differences in norm enforcement. *Behavioral and Brain Sciences* 28: 822–23.

Gaser, C., and G. Schlaug. 2003. Brain structures differ between musicians and non-musicians. *Journal of Neuroscience* 23: 9240–45.

Gestsdottir, S., and R. M. Lerner. 2007. Intentional self-regulation and positive youth development in early adolescence: Findings from the 4-H Study of Positive Youth Development. *Developmental Psychology* 43: 508–21.

Goren, C. C. 1975. Visual following and pattern discrimination of face-like stimuli by newborn infants. *Pediatrics* 56: 544–49.

Grezes, J., N. Costes, and J. Decety, 1999. The effects of learning and intention on the neural network involved in the perception of meaningless actions. *Brain* 122: 1875–87.

Grosbras, M. H., K. Osswald, M. Jansen, R. Toro, A. R. McIntosh, L. Steinberg et al. 2007. Neural mechanisms of resistance to peer influence in early adolescence. *Journal of Neuroscience* 27: 8040–45.

Grosbras, M. H., and T. Paus. 2006. Brain networks involved in viewing angry hands or faces. *Cerebral Cortex* 16: 1087–96.

Herba, C., and M. Phillips. 2004. Development of facial expression recognition from childhood to adolescence: Behavioral and neurological perspectives. *Journal of Child Psychology and Psychiatry* 45: 1185–98.

Iacoboni, M. 2005. Neural mechanisms of imitation. *Current Opinion in Neurobiology* 15: 632–37.

Kilts, C. D. 2003. Dissociable neural pathways are involved in the recognition of emotion in static and dynamic facial expressions. *Neuroimage* 18: 156–68.

Knoch, D., A. Pascual-Leone, K. Meyer, V. Treyer, and E. Fehr. 2006. Diminishing reciprocal fairness by disrupting the right prefrontal cortex. *Science* 314: 829–32.

Lenroot, R. K., and J. N. Giedd. 2006. Brain development in children and adolescents: Insights from anatomical magnetic resonance imaging. *Neuroscience Biobehavioral Review* 30: 718–29.

Lerner, R. M., J. V. Lerner, J. Almerigi, C. Theokas, E. Phelps, S. Gestsdottir et al. 2005. Positive youth development, participation in community youth development programs, and community contributions of fifth grade adolescents: Findings from the first wave of the 4-H Study of Positive Youth Development. *Journal of Early Adolescence* 25, no. 1: 17–71.

Lerner, R. M., R. W. Roeser, and E. Phelps. 2008. Positive development, spirituality, and generosity in youth: An introduction to the issues. R. M. Lerner, R. W. Roeser, and E. Phelps, eds., *Positive Youth Development and Spirituality: From Theory to Research*, 3–22. West Conshohocken, PA: Templeton Foundation Press.

Ljundberg, T., K. Westlund, and A. J. Lindqvist Foresberg. 1999. Conflict resolution in 5 year old boys: Does post-conflict affiliative behavior have a reconciliatory role? *Animal Behavior* 58: 1007–16.

Lupien, S. J., S. King, M. J. Meaney, and B. S. McEwen. 2000. Child's stress hormone levels correlate with mother's socioeconomic status and depressive state. *Biological Psychiatry* 48: 976–80.

Lynch, J., G. Davey-Smith, M. Hillemeier, M. Shaw, T. Raghunathan, and G. Kaplan. 2001.

Income inequality, the psycho-social environment and health: Comparisons of wealthy nations. *Lancet* 358: 194–200.

Mackenbach, J. P. 2002. Income inequality and population health. *British Medical Journal* 324: 1–2.

Maguire, E. A., D. G. Gadian, I. S. Johnsrude, C. D. Good, J. Ashburner, R. S. J. Frackowiak et al. 2000. Navigation-related structural change in the hippocampi of taxi drivers. *Proceedings of the National Academy of Sciences, U S A* 97: 4398–4403.

Marmot, M., and R. G. Wilkinson. 2001. Psychosocial and material pathways in the relation between income and health: a response to Lynch et al. *British Medical Journal* 322: 1233–36.

Mathew, S. J., D. C. Shungo, X. Mao et al. 2003. A magnetic resonance spectroscopic imaging study of adult nonhuman primates exposed to early-life stressors. *Biological Psychiatry* 54: 727–35.

Matthews, K., L. S. Wilkinson, and T. W. Robbins. 1996. Repeated maternal separation of preweanling rats attenuates behavioral responses to primary and conditioned incentives in adulthood. *Physiology & Behavior* 59: 99–107.

McGivern, R. F., J. Andersen, D. Byrd et al. 2002. Cognitive efficiency on a match to sample task decreases at the onset of puberty in children. *Brain and Cognition* 50: 73-89.

Meaney, M. J., D. H. Aitken, S. Bhatnagar, and R. M. Sapolsky 1991. Postnatal handling attenuates certain neuroendocrine, anatomical, and cognitive dysfunctions associated with aging in female rats. *Neurobiology of Aging* 12, no. 1: 31–38.

Meaney, M. J., and M. Szyf. 2005. Maternal care as a model for experience-dependent chromatin plasticity? *Trends in Neuroscience* 28: 456–63.

Melchior, M., T. E. Moffitt, B. J. Milne et al. 2007. Why do children from socioeconomically disadvantaged families suffer from poor health when they reach adulthood? A Life-Course Study. *American Journal of Epidemiology* 166, no. 8: 966–74.

Miller, E. K., and J. D. Cohen. 2001. An integrative theory of prefrontal cortex function. *Annual Review of Neuroscience* 24: 167–202.

Nowak, M. A. 2006. Five rules for the evolution of cooperation. *Science* 314: 1560–63.

Paus, T. 2001. Primate anterior cingulate cortex: Where motor control, drive and cognition interface. *Nature Reviews Neuroscience* 2: 417–24.

———. 2005. Mapping brain maturation and cognitive development during adolescence. *Trends in Cognitive Science* 9: 60–68.

Paus, T., R. Toro, G. Leonard, J. Lerner, R. M. Lerner, M. Perron et al. 2007. Morphological properties of the action-observation cortical network in adolescents with low and high resistance to peer influence. *Social Neuroscience.* DOI: 10.1080/17470910701563558; First published September 20, 2007.

Pepler, D. J., and W. M. Craig. 1995. A peek behind the fence: Naturalistic observations of aggressive children with remote audio-visual recording. *Developmental Psychology* 31: 548–53.

Pepler, D. J., W. M. Craig, and W. L. Roberts. 1998. Observations of aggressive and non-aggressive children on the school playground. *Merrill-Palmer Quarterly* 44: 55–76.

Petrides, M. 2005. Lateral prefrontal cortex: architectonic and functional organization. *Philosophical Transactions of the Royal Society of London, Series B, Biological Sciences* 360: 781–95.

Pickett, K. E., and R. G. Wilkinson. 2007. Child well-being and income inequality: An ecological study of rich societies. *British Medical Journal.* Published electronically November 16, 2007, doi:10.1136/bmj.39377.580162.55.

Pollak, S. D. 2005. Early adversity and mechanisms of plasticity: integrating affective neuroscience with developmental approaches to psychopathology. *Development and Psychopathology* 17: 735–52.

Poulton, R., A. Caspi, B. J. Milne et al. 2002. Association between children's experience of socioeconomic disadvantage and adult health: a life-course study. *Lancet* 360: 1640–45.

Poulton, R., and A. Caspi. 2005. Commentary: How does socioeconomic disadvantage during childhood damage health in adulthood? Testing psychosocial pathways. *International Journal of Epidemiology* 34: 344–45.

Puce, A., T. Allison, S. Bentin et al. 1998. Temporal cortex activation in humans viewing eye and mouth movements. *Journal of Neuroscience* 18: 2188–99.

Rizzolatti, G., and L. Craighero. 2004. The mirror-neuron system. *Annual Review of Neuroscience* 27: 169–92.

Rots, N. Y., J. de Jong, J. O. Workel, S. Levine, A. R. Cools, and E. R. De Kloet. 1996. Neonatal maternally deprived rats have as adults elevated basal pituitary-adrenal activity and enhanced susceptibility to apomorphine. *Journal of Neuroendocrinology* 8: 501–6.

Séguin, J. R., and P. D. Zelazo. 2004. Executive function in early physical aggression. In R. E. Tremblay, W. W. Hartup, and J. Archer, eds., *Developmental origins of aggression*, 307–29. New York: Guilford.

Smeeding, T. 2000. Government programs and social outcomes: The United States in comparative perspective. Prepared for the Smolensky Conference: Poverty, the Distribution of Income and Public Policy. University of California-Berkeley. December 12–13.

Steinberg, L. 2005. Cognitive and affective development in adolescence. *Trends in Cognitive Science* 9: 69–74.

Steinberg, L., and K. Monahan. 2007. Age differences in resistance to peer influence. *Developmental Psychology*. 43: 1531–43.

Stevens, J. R., and M. D. Hauser. 2004. Why be nice? Psychological constraints on the evolution of cooperation. *Trends in Cognitive Science* 8: 60–65.

Sullivan, H. S. 1953. *The interpersonal theory of psychiatry*. New York: Norton.

Sutanto, W., P. Rosenfeld, E. R. de Kloet, and S. Levine 1996. Long-term effects of neonatal maternal deprivation and ACTH on hippocampal mineralocorticoid and glucocorticoid receptors. *Brain research. Developmental brain research* 92: 156–63.

Taylor, M. J., G. McCarthy, E. Saliba, and E. Degiovanni. 1999. ERP evidence of developmental changes in processing of faces. *Clinical Neurophysiology* 110: 910–15.

Thomas, K. M., W. C. Drevets, P. J. Whalen, C. H. Eccard, R. E. Dahl, N. D. Ryan, and B. J. Casey. 2001. Amygdala response to facial expressions in children and adults. *Biological Psychiatry* 49: 309–16.

Toro, R., G. Leonard, J. V. Lerner, R. M. Lerner, M. Perron, G. B. Pike et al. 2007. Prenatal exposure to maternal cigarette smoking and the adolescent cerebral cortex. Neuropsychopharmacology. Published electronically July 4, 2007, PMID: 17609681.

Trivers, R. L. 1971. The evolution of reciprocal altruism. *Quarterly Review of Biology* 46: 35–57.

Wilkinson, R. G. 2005. *The impact of inequality: How to make sick societies healthier*. London: Routledge.

Wilkinson, R. G., and K. E. Pickett. 2006. Income inequality and population health: A review and explanation of evidence. *Social Science and Medicine* 62: 1768–84.

———. 2007. The problems of relative deprivation: why some societies do better than others. *Social Science and Medicine* 65: 1965–78.

Yang, T. T., V. Menon, A. J. Reid, I. H. Gotlib, and A. L. Reiss. 2003. Amygdalar activation associated with happy facial expressions in adolescents: A 3-T functional MRI study. *Journal of the American Academy of Child & Adolescent Psychiatry* 42: 979–85.

8

How Religious/Spiritual Practices Contribute to Well-Being

The Role of Emotion Regulation

Heather L. Urry & Alan P. Poey

It is well documented that religious/spiritual (R/S) beliefs and practices are associated with higher levels of psychological well-being. For example, use of colloquial and meditative prayer is associated with higher levels of life satisfaction and happiness (Poloma and Pendleton 1991), and engaging in meditative practices like mindfulness meditation is associated with positive psychological outcomes (Wallace and Shapiro 2006). In addition, religiosity has been linked to lower levels of violent behavior in youth (Pearce, Little, and Perez 2003), and R/S practices have been associated with decreased participation in high-risk behaviors such as substance abuse (Cotton et al. 2006). Ellison and Levin (1998) have made the suggestion that the internalization of religious beliefs and morals can serve to guide and modify behavior in youth. The goal of this review is to determine whether emotion regulation might at least partially mediate associations between R/S practices and psychological well-being. Here we are guided by a simple heuristic, depicted in Figure 1.

In brief, the heuristic assumes that R/S practices (like prayer, meditation, rituals) impact psychological functions like attention and emotion, and these functions, in turn, have a direct influence on well-being. Bidirectional arrows between boxes in the figure represent the fact that brain structure and function both shape and are shaped by religious/spiritual practices, psychological functions, and well-being. Note that separate boxes for behaviors and brain

FIGURE 1. Depicted here is a heuristic representing theorized associations between religious/spiritual practices, psychological functions like emotion regulation, and well-being. Bidirectional arrows between these concepts and brain structure and function highlight the importance of understanding the neural level of analysis in these associations.

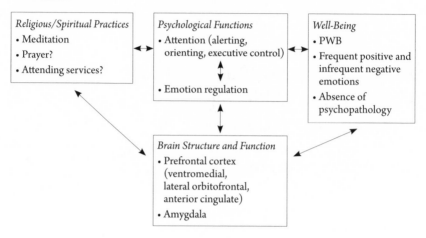

structure/function do not imply a dualist approach but rather highlight the different levels of analysis under consideration.

Here, we review evidence relating to each piece of this heuristic, which we use to frame the extant literature and suggest future study directions. We start with the psychological function of primary interest in this chapter, which is emotion regulation. After defining emotion regulation and describing ways of operationalizing it, we next discuss associations between emotion regulation and well-being. We then briefly review what is known about the neural correlates of emotion regulation and R/S practices, with a specific focus on meditation, since this practice has received quite a bit of attention by *both* psychology and neuroscience researchers.

WHAT IS EMOTION REGULATION?

Many emotion theorists suggest that emotions are adaptive responses to both challenge and opportunity (e.g., Tooby and Cosmides 1990) that are manifested in expressive, experiential, and physiological response tendencies (Bradley and Lang 2007). These response tendencies are thought to occur in a coordinated fashion such that the organism is prepared for action (Scherer 1984). Larsen (2000) writes: "Emotions provide information about the environment and the demands being placed on us by environmental events. Emotions signal what is going wrong or right, what is approaching us or going away from us in terms of threatening or rewarding objects in our external environment" (130). Despite their adaptive nature, however, we are often called upon to regulate our emotional responses in order to achieve our goals (Gross and Thompson 2007).

Emotion regulation has been defined in various ways and exhibits conceptual overlap with a number of related concepts, coping being a prime example. We adopt Thompson's (1994) definition in which emotion regulation reflects "the extrinsic and intrinsic processes responsible for monitoring, evaluating, and modifying emotional reactions, especially their intensive and temporal features, to accomplish one's goals" (27–28). Note that here emotion regulation means that emotion is the target of regulation rather than something that regulates other cognitive or interpersonal processes. In addition, differentiating emotion regulation from the broader domain of coping, both pleasant and unpleasant emotional reactions can be regulated, and the time course for this is on the order of seconds to minutes rather than the hours, days, weeks, etc., associated with the coping process.

The specific strategies that make up the ability to regulate emotion are numerous and varied, including such things as reappraisal, which refers to reconstruing the meaning of a situation to modify its emotional charge; acceptance; distraction by engaging in other activities; avoidance; suppression or exaggeration of expressive behavior; rumination; planning and problem-solving; seeking support; social isolation; seeking rewarding experiences; and behavioral strategies like consuming food, alcohol, or drugs, and exercising, sleeping, and relaxing. Emotion regulation strategies can be used at many different times during the emotion generative process, with a distinction between antecedent-focused (i.e., those that are used prior to the elicitation of an emotional response) and response-focused strategies (i.e., those that are used once the emotional response is under way) (Gross 1998; Gross and Thompson 2007). Importantly, different emotion regulation strategies have different cognitive, social, and physiological consequences. For example, suppressing facial signs of disgust results in higher sympathetic nervous system activation, whereas thinking about the emotion-provoking stimulus (e.g., an amputation film) in objective terms does not produce such increases (for a review of divergent consequences of suppressing outward behavior versus use of cognitive reappraisal, see Gross and Thompson 2007).

By way of a concrete example of emotion regulation, consider the behavior of a person at a job interview whose goal is to impress her interviewer. In this situation, the interviewee is apt to feel anxious about how she will be perceived; yet, rather than freely experiencing and expressing this anxiety, she will likely try to quell it. She might target her subjective experience of anxiety by using cognitive strategies (e.g., thinking about how she has succeeded in interview situations in the past, how the interview is not only about how she is perceived but also about figuring out whether the job is a good fit for her, or how there are worse things in life than failing to be offered this particular job). She might target her bodily arousal by taking long, slow breaths. In addition, she might make explicit attempts to hide the outward expression of her inward anxiety, perhaps by relaxing the muscles in her face and holding her arms in an open and natural manner rather than clasped tightly in her lap. In this example, then, the individual is making active attempts to decrease her anxiety with the aim of meeting her goal, which is to be seen as a good job candidate. Put quite simply, emotion regulation refers to the process of managing our emotions in the service of goal attainment. We turn now to the issue of how emotion regulation is measured.

SUBJECTIVE AND OBJECTIVE INDICATORS OF EMOTION REGULATION

The most common measurement strategy involves employing questionnaires designed to assess people's perceptions of their ability to regulate their emotional responses. Such questionnaires might be used as individual difference measures of how people think they manage their emotions in general or in response to specific discrete situations. Examples of individual difference measures include the Emotion Regulation Questionnaire (ERQ; Gross and John 2003), which concentrates on the use of cognitive reappraisal (e.g., "I control my emotions by changing the way I think about the situation I'm in") and expressive suppression (e.g., "I control my emotions by not expressing them") to deal with both negative and positive emotions in general. Another is the Cognitive Emotion Regulation Questionnaire (CERQ; Garnefski, Kraaij, and Spinhoven 2001), which asks respondents how much they use cognitive techniques to regulate their response to threatening or stressful life events, specifically positive refocusing, positive reappraisal, self-blame, other-blame, rumination, putting things into perspective, catastrophizing, acceptance, and planning. A third measure, the Measure of Affect Regulation Styles (MARS; Larsen and Prizmic 2004), includes instructions for state or trait assessments and is useful in measuring individual differences in emotion regulation.

Self-report questionnaires have the advantage of psychometric soundness as well as face validity for measuring personal perceptions of the emotion regulation strategies people believe they use. They are limited, though, to assessing those strategies one is aware of and willing to report using. Some strategies may be perceived as less socially acceptable and are therefore less likely to be endorsed, thus leading to biased estimates of emotion regulation. Other biases may arise when certain strategies are endorsed simply because they sound good, even if they are used infrequently or ineffectively. A person might report using reappraisal processes, for example, but a questionnaire can do little to measure reliably their rate of success in achieving the desired changes in emotional states. Moreover, the questionnaire items one endorses often require mental aggregation of the strategies used across multiple emotion-provoking events across time. It has been shown that such mental calisthenics are subject to systematic distortions such as the peak-end rule (Kahneman 1999), in which global retrospective evaluations are well predicted by an average of the most extreme and most recent affective values.

In addition to self-report questionnaire measures, researchers also use real-time, laboratory assessments of emotion regulation, operationalized as both observable and unobservable changes in emotional responses occurring through volitional attempts to regulate them. In this approach, the idea is to expose people to stimuli that provoke emotional responses (e.g., films, static pictures, script-driven imagery) and cue them to change their response on-line, by, for example, asking participants to make themselves actively feel more or less emotionally affected upon hearing an auditory instruction to "increase" or "decrease," respectively. The researchers then measure changes in expressive behavior, self-reported subjective experience, and physiology. Using these methods, researchers have shown that brief instruction in the use of cognitive reappraisal strategies to increase, maintain, or decrease one's response to negative pictures results in different ratings of subjective experience, as well as different patterns of startle eye-blink response and expressive behavior as measured using facial electromyography (EMG). Increasing negative affect results in larger startle magnitudes, greater corrugator muscle activity (good at capturing frowning), and higher ratings of subjective unpleasantness compared to maintaining negative affect, while decreasing negative affect results in smaller startle magnitudes, less corrugator activity, and lower ratings of subjective unpleasantness compared to maintaining negative affect (Dillon and Labar 2005; Jackson et al. 2000; Ochsner et al. 2004).

Fewer studies have evaluated the ability to regulate positive emotion.

Recent data suggest that startle eye-blink responses discriminate increasing, maintaining, and decreasing positive emotion: startle magnitudes are larger when increasing compared to maintaining positive emotion and smaller when decreasing compared to maintaining positive emotion (Dillon and Labar 2005). In addition, zygomatic muscle activity (good at capturing smiling) and subjective ratings have been shown to vary similarly in response to instructions to increase and decrease positive affect deliberately (Seo et al. 2005). As a whole, these findings suggest that eye-blink startle, corrugator and zygomatic EMG, and subjective ratings reliably index the ability to regulate both positive and negative emotion "on the fly."

Laboratory assessments of emotion regulation have the advantage of measuring emotion regulation success on-line, thus mitigating retrospective bias. In addition, when combining self-report and physiological indicators of emotional response, a more rounded, objective picture of emotion regulation success is obtained. However, there are downsides as well. For one, there are very likely multiple processes (e.g., cognitive, behavioral) that are engaged by tasks that ask people to use reappraisal to increase or decrease their emotional response. Such processes include reappraising stimulus significance, maintaining representations of the goal emotional state, monitoring the success of regulatory efforts, minimizing conflicts between the prepotent and desired emotional states, and gaze control, at least in visual tasks (see van Reekum et al. 2007). This means that it is unclear which process(es) is (are) responsible for producing the observed changes in emotional response. In addition, the test-retest reliability of laboratory assessments of emotion regulation has not been established, which means it is not certain that results reflect stable aspects of the individual. It is also a much more labor-intensive task to collect and process these data, and is perhaps impractical to do with very large samples. Moreover, very little is known about associations between laboratory estimates of emotion regulation and well-being in daily life. It is possible that the internally valid but artificial laboratory setting does not yield externally valid predictors of how we function, although there is evidence to suggest it does in some domains (see Urry et al. 2006).

EMOTION REGULATION AND RESILIENCE
IN THE WAKE OF ADVERSE EVENTS

Major negative life events can lead to problems of emotion. For example, some individuals experiencing or witnessing an event involving actual or threatened death or serious injury or other threat to one's physical integrity develop Post-Traumatic Stress Disorder (PTSD), which is characterized by recurrent intrusive thoughts of the traumatic event, heightened psychological and physiological reactivity to cues resembling the event, affective blunting and avoidance, irritability or outbursts of anger, and/or exaggerated startle responses. Whereas perturbations of emotion like these are a hallmark of many, if not most, forms of psychopathology, they are certainly not restricted to the realm of abnormal psychology. Death of a significant other, for instance, prompts severe sadness and anger, and there are many for whom the loss results in protracted anger, bitterness, recurrent distressing thoughts of the death, and painful emotions (Shear et al. 2005).

What the examples above fail to highlight is that most people who face adverse events never develop psychopathology or even display significant adjustment difficulties when bad things happen (Bonanno 2004). Self-report measures indicate that people who think of themselves as capable of bouncing back from adversity, which is the defining characteristic of resilience, report lower levels of negative affect and are more apt to interpret stressors as challenging rather than threatening (Tugade and Fredrickson 2004). Bereavement studies, for example, suggest that resilient individuals exhibit signs of greater emotional stability prior to the loss (Bonanno et al. 2002), are less negative when describing the meaning of the death (Gamino et al. 2004), have more positive appraisals (Stein et al. 1997), and are more likely to view the loss as an opportunity for growth (Tebes et al. 2004). These findings suggest that the ability to regulate emotional responses may dramatically alter the course of adjustment to adverse events (Emmons and McCullough 2003) and the recovery of subjective well-being. Responding effectively to such events is an important aspect of adjustment. However, there are many researchers who study positive aspects of well-being as a goal in and of itself.

BEING WELL (AND AVOIDING ILL)

There are two traditions of research on well-being in the psychological literature (Ryan and Deci 2001). Ryff's work (1989) on psychological well-being (PWB; Ryff 1989) exemplifies the eudaimonic tradition, in which well-being is defined in terms of the extent to which one has accomplished high levels of autonomy, environmental mastery, personal growth, positive relations with others, purpose in life, and self-acceptance. Hedonic well-being, on the other hand, is exemplified in the subjective well-being (SWB) literature (Diener

2000; Kahneman 1999) in which well-being is seen as the subjective sense of being satisfied with life overall and with important domains therein (e.g., work, family), as well as experiencing frequent pleasant and infrequent unpleasant emotions. Thus, hedonic well-being researchers highlight the subjective sense that life feels good, while eudaimonic well-being adherents emphasize that important life goals may or may not be accompanied by good feelings.

More recently, Duckworth, Steen, and Seligman (2005) have suggested that well-being should incorporate aspects of both traditions. They define happiness as comprising three things: pleasure (hedonic well-being), engagement (using one's strengths and talents to meet challenges), *and* meaning (belonging to and serving something larger than oneself, having a sense of purpose in life, i.e., eudaimonic well-being). In addition, they underscore the notion that well-being is more than the absence of unpleasantness: "Relieving the negatives, even in the rare event that we are completely successful, does not bring about 'happiness'; the skills of pleasure, engagement, and meaning are supplementary to the skills of fighting depression, anxiety, and anger" (Duckworth, Steen, and Seligman 2005, 640). However, depression, anxiety, and anger may be obstacles to achieving well-being and may reflect dysfunctional emotion regulation; thus, we include this in our review as well. In discussing relations between emotion regulation and well-being, we will report on results from both the eudaimonic and hedonic traditions, most of which have so far been limited to self-report assessments of these constructs.

EMOTION REGULATION AND WELL-BEING

Others have reviewed the association between emotion regulation and well-being or mental health (Gross and Munoz 1995; Kokkonen and Kinnunen 2006). Here, we describe the results of several studies using various methods to illustrate the phenomenon. Silk, Steinberg, and Morris (2003) used an experience sampling methodology with adolescents (ages 12 to 17) to sample negative events multiple times per day throughout the course of a week. Reported use of both disengagement from (e.g., denial) and involuntary engagement with (e.g., rumination) negative emotions were associated with poorer recovery from negative events and higher concurrent levels of depressive symptoms and problem behaviors. Shiota (2006) conducted a daily diary study in which undergraduate participants reported their most negative event and the manner in which they dealt with this event for a week. These strategies were coded into one of six coping categories including positive reappraisal, seeking positive sensory experiences, distraction, eating, problem-focused coping, and seeking social support. Correlational analyses showed that participants reporting more frequent use of positive reappraisal and seeking positive sensory experience and less frequent use of distraction techniques reported higher levels of positive affect, measured using the Positive and Negative Affect Schedule. More frequent use of social support was associated with lower levels of negative affect.

Using individual difference questionnaire measures, Garnefski, Kraaij, and van Etten (2005) conducted a correlational study in which they asked adolescents (ages 12 to 18) to complete the CERQ, an individual difference measure of emotion regulation, and a measure of internalizing (anxiety, withdrawal, and depression) and externalizing (aggression and delinquency) problems. Adolescents reporting lower use of positive reappraisal and greater use of self-blame, rumination, and catastrophizing reported higher levels of internalizing problems. Conversely, adolescents reporting greater use of positive refocusing, catastrophizing, and other-blame reported higher levels of externalizing problems. Also using the CERQ but with undergraduates, Martin and Dahlen (2005) demonstrated that lower levels of positive reappraisal and higher levels of self-blame, rumination, and catastrophizing were associated with higher concurrent levels of depression, anxiety, stress, and anger. Matsumoto and colleagues (2003), who were interested in understanding intercultural adjustment, demonstrated that self-reported emotion regulation predicts higher levels of satisfaction with life and lower levels of depression and anxiety among Japanese and non-Japanese immigrants to the United States. However, the measure of emotion regulation in this study, the Emotion Regulation scale from the Intercultural Adjustment Potential Scale (ICAPS), does not reflect emotion regulation as defined in this chapter because it mostly comprises items assessing emotional characteristics (e.g., "I feel happy most of the time," "I do not worry very much," and "I rarely feel anxious or fearful") as opposed to strategies used to deal with emotional responses.

Focusing on individuals facing a significant loss, Stein, Folkman, Trabasso, and Richards (1997) studied bereaved caregivers by quantitatively coding narrative interviews about the death for the types of appraisals being made two to four weeks later, and how those relate to later well-being, specifically depressive symptoms, positive morale and states of mind, and intrusive and avoidant thoughts. They found that higher proportions of positive appraisals were associated with higher levels of positive morale and states of mind and with lower levels of depression, both soon after the death and one year later. In another recent study, Kovacs and colleagues (2006) examined whether "contextual emotion-regulation therapy" (CERT), a ten-month treatment focused on training emotion regulation skills (including identifying effective strategies for diminishing distress and coping skills training), would alleviate childhood dysthymia. In an uncontrolled trial with twenty children (ages 7 to 12 years), treatment resulted in full or partial remission of dysthymia in approximately two-thirds of the sample. Methodological limitations prevent the conclusion that remediation of the disorder was brought on *specifically* by the emotion regulation training; however, the finding is consistent with this notion.

Studies of religious coping provide further evidence that emotion regulation may link R/S practices and subjective well-being. Positive reappraisal is often used in a religious or spiritual context to deal with highly negative life events. For example, Park (2005) studied 169 bereaved college students and found that religious positive reappraisal was associated with higher levels of subjective well-being and stress-related growth. According to Park (2005), "Religion can be involved in changing the appraised meaning of a stressful situation by (a) helping the individual to see the positive aspects that have come from the stressful situation, and (b) providing a means to make more benign reattributions" (712). Such reattributions, or reappraisals (e.g., "It must be part of God's plan" or "He's in a better place"), are prime examples of emotion regulation.

Together these studies suggest that emotion regulation, particularly the tendency to reappraise one's circumstances positively, is associated with higher levels of well-being (and lower levels of ill-being). We now turn to a brief review of what is known about the brain regions involved in emotion regulation. The neural level of analysis may be informative to the extent that brain areas involved in emotion regulation demonstrate overlap with the brain areas that are involved in or tuned by R/S practices.

BRAIN REGIONS INVOLVED IN EMOTION REGULATION

There have now been a fair number of studies focused on identifying the brain regions involved in emotion regulation using laboratory tasks like the ones described previously, but, so far, they have almost exclusively focused on using cognitive reappraisal. In our work, for example, participants are trained to change their emotional response by thinking differently about complex picture stimuli while brain activity is recorded. For example, participants are taught to imagine that the situation depicted gets worse or better or that the events depicted are more versus less personally relevant (Johnstone et al. 2007). Functional magnetic resonance imaging (fMRI) studies of the use of reappraisal to decrease negative emotion have implicated multiple areas of the prefrontal cortex (PFC) including lateral ventral, ventral and dorsal medial (including anterior cingulate), and dorsolateral regions (Beauregard, Levesque, and Bourgouin 2001; Bechara 2004; Eippert et al. 2007; Harenski and Hamann 2006; Johnstone et al. 2007; Kalisch et al. 2005; Kim and Hamann 2007; Levesque et al. 2003; Ochsner et al. 2002; Ochsner et al. 2004; Ohira et al. 2006; Phan et al. 2004; Phan et al. 2005; Urry et al. 2006; van Reekum et al. 2007). A few studies have suggested that these regions are more active with both attempts to increase and decrease felt emotion relative to a passive viewing control condition (Eippert et al. 2007; Ochsner et al. 2004; Urry et al. 2006). Such findings have been interpreted to indicate that various regions of PFC underlie the cognitive control processes necessary to regulate emotion deliberately, such as reappraising stimulus significance, maintaining representations of the goal emotional state, monitoring the success of regulatory efforts, and minimizing conflicts between the prepotent and desired emotional states.

Another potentially very interesting and important region is the amygdala, which is a small almond-shaped set of nuclei located bilaterally deep within the temporal lobe of the brain. The amygdala is important for acquiring defensive responses to previously innocuous stimuli and for detection of and visceral responses to potentially threatening information (for a review, see Davis and Whalen 2001). It has also been implicated in processing of positive stimuli (Hamann and Mao 2002; Hamann et al. 2002), suggesting that amygdala responses do not depend on threat alone. Interestingly, studies of deliberate emotion regulation have sometimes demonstrated that amygdala activation tracks changes in negative affect in line with instructions to regulate. Maintaining negative affect, for example, produces higher amygdala activation than passively viewing negative images (Schaefer et al. 2002), and decreasing negative affect often leads to lower amygdala activation compared to passive viewing of negative images (Ochsner et al. 2002). In studies that have examined both increasing and decreasing negative affect relative to a control condition, findings have been somewhat less consistent. Ochsner and colleagues (2004) found that amygdala activation was higher when increasing in one region of the amygdala and lower when decreasing in another region of the amygdala. However, other studies showed that amygdala activation was higher when increasing (although see van Reekum et al. 2007), but not lower when decreasing (Urry et al. 2006; Johnstone et al. 2007; Kim and Hamann 2007).

Regardless of the averaged findings comparing emotion regulation conditions within the amygdala, one thing that is apparent is that neither PFC nor the amygdalae operate in isolation. Instead, they exhibit bidirectional anatomical connections, thus providing the scaffolding that enables them to "talk" to one another (Ghashghaei, Hilgetag, and Barbasa 2007). We have shown, for example, that participants exhibiting higher levels of activation in ventromedial regions of PFC exhibit lower levels of amygdala activation when decreasing negative emotion (Urry et al. 2006) and that such inverse cross-talk does not work the same in people diagnosed with major depression (Johnstone et al. 2007).

All of these studies of the neural correlates of emotion regulation have assessed the brain regions involved in emotion regulation as it was actually taking place. There have been no studies assessing the possible long-term effects of daily emotion regulation practice on the brain. The closest approximation of such a linkage would be treatment studies in which people suffering from psychiatric disorders undergo psychotherapy, a large piece of which focuses on modifying maladaptive thoughts to improve emotions and behaviors. A recent review indicates that regions implicated in emotion regulation predict response to, or are influenced by, psychotherapy (Beauregard 2007). In the next section, we consider the literature addressing the question of how R/S practices affect the brain. R/S practices are numerous and include such activities as prayer, meditation, attending religious services, and engaging in rituals. We focus here on meditation, which has received quite a bit of attention in both psychology and neuroscience research circles. Of interest is discovering whether meditative states and/or traits are associated with brain regions involved in emotion regulation and attention.

MEDITATION: PSYCHOLOGICAL AND NEURAL CORRELATES

In a recent qualitative review, Cahn and Polich (2006) define meditation as "practices that self-regulate the body and mind, thereby affecting mental events by engaging a specific attentional set" (180). Two basic types of meditation include concentrative meditation, in which the individual focuses all attention on one sensory experience, for example, the breath or a word, and mindfulness (also known as receptive or opening-up) meditation, in which the individual focuses attention on all experiences in a nonjudgmental way. Kabat-Zinn (2003) defines mindfulness as follows: "An operational working definition of mindfulness is: the awareness that emerges through paying attention on purpose, in the present moment, and nonjudgmentally to the unfolding of experience moment by moment" (145). Research has focused on assessing effects of meditation in the moment (i.e., during the meditative state) as well as enduring effects of longer-term meditative practice.

Recent studies support the idea that meditation improves attention, specifically the ability to resist interference from distracting information. For example, Chan and Woollacott (2007) showed that concentrative and opening-up meditators showed lower interference in the Stroop Color-Word test, which they suggest assesses the executive control aspect of attention, compared to controls with no meditative experience. Jha, Krompinger, and Baime (2007) revealed effects on aspects of attention using the Attention Network Test ([ANT]; Fan et al. 2002), a task that provides scores for three aspects of attention, namely, alerting, orienting, and executive control. Three groups of participants completed the ANT twice; for two of the groups, this occurred before and after an intervention. For one group of previously nonmeditating participants, the intervention was an eight-week mindfulness-based stress reduction (MBSR) course in which both concentrative and receptive meditation were trained. For a second group comprising experienced meditators, the intervention was a one-month meditation retreat. Both intervention groups were compared to a control group who did not receive an intervention and had no meditation experience. At Time 1, the retreat group of experienced meditators showed enhanced executive control relative to the other two groups. At Time 2, the MBSR group exhibited stronger orienting responses compared to the control and retreat groups.

Meditative practice has also been shown to influence emotion. For example, recent work by Arch and Craske (2006) indicates that participants practicing fifteen minutes of focused breathing, a task that is often the first taught to novice meditators, reported more positive perceptions of neutral pictures presented afterwards compared to unfocused attention and worry groups. They also reported less emotional arousal to negative pictures compared to the worry group. Finally, a higher percentage of participants doing focused breathing was willing to view an entire series of highly negative images compared to the other two groups, who were more likely to discontinue the series. Assessing effects of repeated experience with mindfulness meditation on symptoms of depression and anxiety, Ramel and colleagues (2004) demonstrated that an MBSR course with veterans diagnosed with mood disorders exhibited decreases in rumination compared to a waitlist control group and that this decrease in rumination accounted for decreases in reported levels of depression and anxiety. A recent meta-analysis indicates that MBSR is associated with improvements in both mental and physical health outcomes (Grossman et al. 2004; but see Toneatto and Nguyen 2007).

The neural correlates of meditation have recently been reviewed by Cahn and Polich (2006). In brief, these authors review neuroimaging evidence suggesting that anterior cingulate cortex and dorsolateral PFC are involved in meditation. For example, Lazar and colleagues (2000) studied experienced meditators and demonstrated greater activation in, among other regions, dorsolateral PFC as well as in the amygdala when participants were meditating (focusing on their breath and covertly repeating mantras at inhalation and exhalation) compared to during a control condition in which they generated a random list of animals. In a later study, Lazar and colleagues (2005) demonstrated that the cortical mantle is thicker in regions of dorsolateral PFC in experienced meditators compared to control subjects. In a recent functional neuroimaging study, Creswell and colleagues (2007) hinted at possible effects of meditation on emotion processing in these very same regions. Participants completed a questionnaire that assessed mindfulness, which is "one's general tendency to be open and receptive to present moment experiences across cognitive, emotional, physical, interpersonal, and general life domains" (Creswell et al. 2007, 561). They also participated in an affect-labeling task, in which participants were shown images of various facial expressions of emotion and asked to choose the better of two labels for the emotion being portrayed. This condition was contrasted with a gender-labeling task. Results indicated that those individuals reporting greater trait mindfulness exhibited greater activation in dorsolateral, medial, ventromedial, and ventrolateral PFC but lower activation in the amygdalae. These findings are consistent with those reported above for emotion regulation, in which inverse associations were observed between ventromedial PFC and the amygdala when using cognitive reappraisal to decrease negative emotion. Brain studies of meditation thus provide suggestive evidence for the possibility that R/S practices like meditation may impact emotion regulation ability and thus well-being in general.

WHERE DO WE GO FROM HERE?

The literature reviewed in this chapter suggests that emotion regulation is associated with well-being and that, at least based on questionnaire measures, the nature of these associations depends in part on the emotion regulatory strategies that people employ. It further suggests that R/S practices, using meditation as the main illustrative example, are associated with emotion and attention regulation, and we have reason to believe that brain regions involved in emotion regulation overlap with regions implicated in meditation. We also know that other R/S practices like prayer and religious reappraisal have in numerous studies been associated with increased well-being. From these strands of evidence, we again propose that emotion regulation may provide one pathway by which R/S practices contribute to well-being.

From here, there are several potentially useful directions for future research to test this heuristic directly. Perhaps it goes without saying that it will be important for future studies to measure, in the same participants, the content and frequency of R/S practices, aspects of both hedonic and eudaimonic well-being, emotion regulation ability using self-report and laboratory assessments, and brain structure and function. In our current cross-sectional work, we have undertaken just such a study of older adolescents (18–20 years old) to address some key questions embedded in the model. For one, do R/S practices correlate positively with modulation of emotion when older adolescents use cognitive reappraisal to increase, decrease, or maintain their emotional response to vivid picture stimuli evoking pleasant and unpleasant feelings? How about when they simply take a deep breath, a common focus of attention in meditative practice? Second, is the laboratory assessment of regulation of emotion using cognitive reappraisal correlated with psychological well-being? What are the links between attention (alerting, orienting, executive control) and emotion regulation? Is emotion regulation a skill that falls within the larger domain of executive control, or are they overlapping but separable constructs? Is brain function during emotion regulation in regions like PFC and the amygdala associated with R/S practices and well-being, and does brain function, in fact, mediate these associations?

Cross-sectional correlational work, however, will not reveal the causal direction of the associations we uncover. For that, we will need to conduct longitudinal research in which we measure all of these constructs at multiple times within individuals to test theoretically specified causal models of temporal precedence. This longitudinal work will be most fruitful with a focus on adolescence because, and this will come as no surprise, adolescence is a time of incredible change in behavior, cognition, and social functioning, and thus also a time when the brain matures to a substantial degree. Paus (2005) reviewed research indicating that brain changes include increases in white matter volumes, which implies faster, more efficient transmission of signals, and changes in gray matter in frontal and parietal regions. Longitudinal work that incorporates the ideas described herein would enable us to determine whether such brain changes underlie the emergence of emotion regulation and whether there are important windows of opportunity during which R/S practices have their biggest impact. Finally, experimental work in which R/S practices are manipulated remains an important endeavor. Much of this work has been done in the area of meditation, but this should really be extended to other spiritual practices like prayer.

REFERENCES

Arch, J. J., and M. G. Craske. 2006. Mechanisms of mindfulness: Emotion regulation following a focused breathing induction. *Behaviour Research and Therapy.* 44: 1849–58.

Beauregard, M. 2007. Mind does really matter: Evidence from neuroimaging studies of emotional self-regulation, psychotherapy, and placebo effect. *Progress in Neurobiology* 81: 218–36.

Beauregard, M., J. Levesque, and P. Bourgouin. 2001. Neural correlates of conscious self-regulation of emotion. *Journal of Neuroscience* 21: art–RC165.

Bechara, A. 2004. Disturbances of emotion regulation after focal brain lesions. *International Review of Neurobiology* 62: 159–93.

Bonanno, G. A. 2004. Loss, trauma, and human resilience: Have we underestimated the human capacity to thrive after extremely aversive events? *American Psychologist.* 59: 20–28.

Bonanno, G. A., C. B. Wortman, D. R. Lehman, R. G. Tweed, M. Haring, J. Sonnega et al. 2002. Resilience to loss and chronic grief: a prospective study from preloss to 18-months postloss. *Journal of Personality and Social Psychology.* 83: 1150–64.

Cahn, B. R., and J. Polich. 2006. Meditation states and traits: EEG, ERP, and neuroimaging studies. *Psychological Bulletin* 132: 180–211.

Chan, D., and M. Woollacott. 2007. Effects of level of meditation experience on attentional focus: Is the efficiency of executive or orientation networks improved? *Journal of Alternative and Complementary Medicine* 13: 651–57.

Cotton, S., K. Zebracki, S. L. Rosenthal, J. Tsevat, and D. Drotar. 2006. Religion/spirituality and adolescent health outcomes: a review. *Journal of Adolescent Health* 38: 472–80.

Creswell, J. D., B. M. Way, N. I. Eisenberger, and M. D. Lieberman. 2007. Neural correlates of dispositional mindfulness during affect labeling. *Psychosomatic Medicine* 69: 560–65.

Davis, M., and P. J. Whalen. 2001. The amygdala: Vigilance and emotion. *Molecular Psychiatry* 6: 13–34.

Diener, E. 2000. Subjective well-being: The science of happiness and a proposal for a national index. *American Psychologist* 55: 34–43.

Dillon, D. G., and K. S. Labar. 2005. Startle modulation during conscious emotion regulation is arousal-dependent. *Behavioral Neuroscience* 119: 1118–24.

Duckworth, A. L., T. A. Steen, and M. E. P. Seligman. 2005. Positive psychology in clinical practice. *Annual Review of Clinical Psychology* 1: 629–51.

Eippert, F., R. Veit, N. Weiskopf, M. Erb, N. Birbaumer, and S. Anders. 2007. Regulation of emotional responses elicited by threat-related stimuli. *Human Brain Mapping* 28: 409–23.

Ellison, C. G., and J. S. Levin. 1998. The religion-health connection: Evidence, theory, and future directions. *Health Education and Behavior* 25: 700–720.

Emmons, R. A., and M. E. McCullough. 2003. Counting blessings versus burdens: An experimental investigation of gratitude and subjective well-being in daily life. *Journal of Personality and Social Psychology*. 84: 377–89.

Fan, J., B. D. McCandliss, T. Sommer, A. Raz, and M. I. Posner. 2002. Testing the efficiency and independence of attentional networks. *Journal of Cognitive Neuroscience*. 14: 340–47.

Gamino, L. A., K. W. Sewell, and W. Kenneth. 2004. Meaning constructs as predictors of bereavement adjustment: A report from the Scott and White grief study. *Death Studies* 28: 397–421.

Garnefski, N., V. Kraaij, and P. Spinhoven. 2001. Negative life events, cognitive emotion regulation and emotional problems. *Personality and Individual Differences* 30: 1311–27.

Garnefski, N., V. Kraaij, and M. van Etten. 2005. Specificity of relations between adolescents' cognitive emotion regulation strategies and internalizing and externalizing psychopathology. *Journal of Adolescence* 28: 619–31.

Ghashghaei, H., C. C. Hilgetag, and H. Barbasa. 2007. Sequence of information processing for emotions based on the anatomic dialogue between prefrontal cortex and amygdala. *NeuroImage* 34: 905–23.

Gross, J. J. 1998. Antecedent- and response-focused emotion regulation: Divergent consequences for experience, expression, and physiology. *Journal of Personality and Social Psychology*. 74: 224–37.

Gross, J. J., and O. P. John. 2003. Individual differences in two emotion regulation processes: Implications for affect, relationships, and well-being. *Journal of Personality and Social Psychology*. 85: 348–62.

Gross, J. J., and R. F. Munoz. 1995. Emotion regulation and mental-health. *Clinical Psychology-Science and Practice* 2: 151–64.

Gross, J. J., and R. A. Thompson. 2007. Emotion regulation: Conceptual foundations. In J. J. Gross, ed., *Handbook of Emotion Regulation*, 3–24. New York: Guilford Press.

Grossman, P., L. Niemann, S. Schmidt, and H. Walach. 2004. Mindfulness-based stress reduction and health benefits—A meta-analysis. *Journal of Psychosomatic Research* 57: 35–43.

Hamann, S., and H. Mao. 2002. Positive and negative emotional verbal stimuli elicit activity in the left amygdala. *NeuroReport* 13: 15–19.

Hamann, S. B., T. D. Ely, J. M. Hoffman, and C. D. Kilts. 2002. Ecstasy and agony: Activation of the human amygdala in positive and negative emotion. *Psychological Science* 13: 135–41.

Harenski, C. L., and S. Hamann. 2006. Neural correlates of regulating negative emotions related to moral violations. *NeuroImage* 30: 313–24.

Jackson, D. C., J. R. Malmstadt, C. L. Larson, and R. J. Davidson. 2000. Suppression and enhancement of emotional responses to unpleasant pictures. *Psychophysiology* 37: 515–22.

Jha, A. P., J. Krompinger, and M. J. Baime. 2007. Mindfulness training modifies subsystems of attention. *Cognitive Affective and Behavioral Neuroscience* 7: 109–19.

Johnstone, T., C. M. van Reekum, H. L. Urry, N. H. Kalin, and R. J. Davidson. 2007. Failure to regulate: Counterproductive recruitment of top-down prefrontal-subcortical circuitry in major depression. *Journal of Neuroscience*. 27: 8877–84.

Kabat-Zinn, J. 2003. Mindfulness-based interventions in context: Past, present, and future. *Clinical Psychology-Science and Practice* 10: 144–56.

Kahneman, D. 1999. Objective happiness. In D. Kahneman, E. Diener, and N. Schwarz, eds., *Well-being: Foundations of hedonic psychology*, 3–27. New York: Russell Sage Foundation Press.

Kalisch, R., K. Wiech, H. D. Critchley, B. Seymour, J. P. O'Doherty, D. A. Oakley et al. 2005. Anxiety reduction through detachment: Subjective, physiological, and neural effects. *Journal of Cognitive Neuroscience* 17: 874–83.

Kim, S. H., and S. Hamann. 2007. Neural correlates of positive and negative emotion regulation. *Journal of Cognitive Neuroscience* 19: 776–98.

Kokkonen, M., and M.-L. Kinnunen. 2006. Emotion regulation and well-being. In L. Pulkkinen, J. Kaprio, and R. J. Rose, eds. *Socioemotional development and health from adolescence to adulthood*, 197–208. Cambridge: Cambridge University Press.

Kovacs, M., J. Sherrill, C. J. George, M. Pollock, R. V. Tumuluru, and V. Ho. 2006. Contextual emotion-regulation therapy for childhood depression: Description and pilot testing of a new intervention. *Journal of the American Academy of Child and Adolescent Psychiatry* 45: 892–903.

Larsen, R. J. 2000. Toward a science of mood regulation. *Psychological Inquiry* 11: 129–41.

Larsen, R. J., and Z. Prizmic. 2004. Affect regulation. In R. F. Baumeister and K. D. Vohs, eds., *Handbook of self-regulation: Research, theory, and applications*, 40–61. New York: Guilford Press.

Lazar, S. W., G. Bush, R. L. Gollub, G. L. Fricchione, G. Khalsa, and H. Benson. 2000. Functional brain mapping of the relaxation response and meditation. *NeuroReport* 11: 1581–85.

Lazar, S. W., C. E. Kerr, R. H. Wasserman, J. R. Gray, D. N. Greve, M. T. Treadway et al. 2005. Meditation experience is associated with increased cortical thickness. *NeuroReport* 16: 1893–97.

Levesque, J., F. Eugene, Y. Joanette, V. Paquette, B. Mensour, G. Beaudoin et al. 2003. Neural circuitry underlying voluntary suppression of sadness. *Biological Psychiatry* 53: 502–10.

Martin, R. C., and E. R. Dahlen. 2005. Cognitive emotion regulation in the prediction of depression, anxiety, stress, and anger. *Personality and Individual Differences* 39: 1249–60.

Matsumoto, D., J. A. Leroux, M. Iwamoto, J. W. Choi, D. Rogers, H. Tatani et al. 2003. The robustness of the intercultural adjustment potential scale (ICAPS): The search for a universal psychological engine of adjustment. *International Journal of Intercultural Relations* 27: 543–62.

Ochsner, K. N., S. A. Bunge, J. J. Gross, and J. D. Gabrieli. 2002. Rethinking feelings: An FMRI study of the cognitive regulation of emotion. *Journal of Cognitive Neuroscience.* 14: 1215–29.

Ochsner, K. N., R. D. Ray, J. C. Cooper, E. R. Robertson, S. Chopra, J. D. Gabrieli et al. 2004. For better or for worse: Neural systems supporting the cognitive down- and up-regulation of negative emotion. *NeuroImage* 23: 483–99.

Ohira, H., M. Nomura, N. Ichikawa, T. Isowa, T. Iidaka, A. Sato et al. 2006. Association of neural and physiological responses during voluntary emotion suppression. *NeuroImage* 29: 721–33.

Park, C. L. 2005. Religion as a meaning-making framework in coping with life stress. *Journal of Social Issues* 61: 707–29.

Paus, T. 2005. Mapping brain maturation and cognitive development during adolescence. *Trends in Cognitive Sciences* 9: 60–68.

Pearce, M. J., T. D. Little, and J. E. Perez. 2003. Religiousness and depressive symptoms among adolescents. *Journal of Clinical Child and Adolescent Psychology* 32: 267–76.

Phan, K. L., D. A. Fitzgerald, P. J. Nathan, G. J. Moore, T. W. Uhde, and M. E. Tancer. 2004. Neural substrates for cognitive regulation of negative emotion. *Biological Psychiatry* 55: 57S.

———. 2005. Neural substrates for voluntary suppression of negative affect: A functional magnetic resonance imaging study. *Biological Psychiatry* 57: 210–19.

Poloma, M. M., and B. F. Pendleton. 1991. The effects of prayer and prayer experiences on measures of general well-being. *Journal of Psychology and Theology* 19: 71–83.

Ramel, W., P. R. Goldin, P. E. Carmona, and J. R. McQuaid. 2004. The effects of mindfulness meditation on cognitive processes and affect in patients with past depression. *Cognitive Therapy and Research* 28: 433–55.

Ryan, R. M., and E. L. Deci. 2001. On happiness and human potentials: A review of research on hedonic and eudaimonic well-being. *Annual Review of Psychology.* 52: 141–66.

Ryff, C. D. 1989. Happiness is everything, or is it—Explorations on the meaning of psychological well-being. *Journal of Personality and Social Psychology* 57: 1069–81.

Schaefer, S. M., D. C. Jackson, R. J. Davidson, G. K. Aguirre, D. Y. Kimberg, and S. L. Thompson-Schill. 2002. Modulation of amygdalar activity by the conscious regulation of negative emotion. *Journal of Cognitive Neuroscience.* 14: 913–21.

Scherer, K. R. 1984. Emotions—functions and components. *Cahiers de Psychologie Cognitive-Current Psychology of Cognition* 4: 9–39.

Seo, D., E. Bernat, M. Cadwallader, and C. J. Patrick. 2005. Temporal dynamics of emotional and attentional processing in an affective regulation task. Poster presented at Society for Psychophysiological Research.

Shear, K., E. Frank, P. R. Houck, and C. F. Reynolds III. 2005. Treatment of complicated grief: A randomized controlled trial. *JAMA* 293: 2601–8.

Shiota, M. N. 2006. Silver linings and candles in the dark: Differences among positive coping strategies in predicting subjective well-being. *Emotion* 6: 335–39.

Silk, J. S., L. Steinberg, and A. S. Morris. 2003. Adolescents' emotion regulation in daily life: Links to depressive symptoms and problem behavior. *Child Development* 74: 1869–80.

Stein, N., S. Folkman, T. Trabasso, and T. A. Richards. 1997. Appraisal and goal processes as predictors of psychological well-being in bereaved caregivers. *Journal of Personality and Social Psychology* 72: 872–84.

Tebes, J. K., J. T. Irish, M. J. P. Vasquez, and D. V. Perkins. 2004. Cognitive transformation as a marker of resilience. *Substance Use and Misuse* 39: 769–88.

Thompson, R. A. 1994. Emotion regulation: A theme in search of definition. *Monographs of the Society for Research in Child Development.* 59: 25–52.

Toneatto, T., and L. Nguyen. 2007. Does mindfulness meditation improve anxiety and mood symptoms? A review of the controlled research. *Canadian Journal of Psychiatry-Revue Canadienne de Psychiatrie* 52: 260–66.

Tooby, J., and L. Cosmides. 1990. The past explains the present: Emotional adaptations and the structure of ancestral environments. *Ethology and Sociobiology* 11: 375–424.

Tugade, M. M., and B. L. Fredrickson. 2004. Resilient individuals use positive emotions to bounce back from negative emotional experiences. *Journal of Personality and Social Psychology.* 86: 320–33.

Urry, H. L., C. M. van Reekum, T. Johnstone, N. H. Kalin, M. E. Thurow, H. S. Schaefer et al. 2006. Amygdala and ventromedial prefrontal cortex are inversely coupled during regulation of negative affect and predict the diurnal pattern of cortisol secretion among older adults. *Journal of Neuroscience.* 26: 4415–25.

van Reekum, C. M., T. Johnstone, H. L. Urry, M. E. Thurow, H. S. Schaefer, A. L. Alexander et al. 2007. Gaze fixations predict brain activation during the voluntary regulation of picture-induced negative affect. *NeuroImage* 36: 1041–55.

Wallace, B. A., and S. L. Shapiro. 2006. Mental balance and well-being: Building bridges between Buddhism and western psychology. *American Psychologist* 61: 690–701.

PART 3

*Individual Contributions
to the Spirituality–PYD
Relation*

9

The Role of Developmental Change in Spiritual Development

David Henry Feldman

This essay explores how an understanding of developmental change processes may contribute to the effort to understand and perhaps foster spiritual development, particularly among adolescents. The essay will begin with a discussion of the special, uniquely human processes of change that we describe as developmental. Next, the two topics—developmental change processes and spiritual development—will be brought more systematically together, with the hope that each topic will enrich and contribute to understanding the other. Finally, we will try to clarify distinctions between fixation of belief versus spiritual growth based on what is known about development.

Before turning to the main topics of this essay, it seems appropriate to provide the reader with some information about the author's experience in religious, spiritual, and cosmological matters. This is to inform the reader as to what theoretical and research as well as spiritual perspective on development is brought to bear on the present topic, a kind of scholarly full disclosure.

SPIRITUAL, RELIGIOUS EXPERIENCE

My early religious experience was in an observant, conservative, Eastern European-background Jewish family. We were members of the local synagogue in our small, western Pennsylvania town, attending services every week. I went to Sunday school, had a bar mitzvah, and was confirmed. I even taught Sunday school for a year during high school. My family celebrated all of the major Jewish holidays, often with members of our extended family, sometimes with other members of our synagogue. It is fair to say that our family's primary identification and affiliation was with the Jewish community of a few hundred people in our town of about ten thousand people. My older brother and I were expected to become good Jews and to follow our parents' example in matters of religion.

It did not work out that way. Although I dutifully did what I was expected to do, it never affected me spiritually. I found all formal religious observance wanting; my emotions ranged from bored to frustrated and even disoriented (I did not know then that I was hypoglycemic and could not tolerate the several-hours-long services, especially on major holidays). My experience was physically uncomfortable as well as spiritually lacking. To this day I still cannot remember when the major Jewish holidays appear during the year, tending to mix up and mislabel them.

For a brief period during high school, I considered converting to Catholicism for reasons that were never clear (other than that I was then in love with a very determined, very religious Catholic girl). And during my first year of college, I went through a very intense period (as many students do) of spiritual reexamination and reflection. I must have expressed some of these thoughts and feelings because, in my second year, I was elected chaplain of my fraternity, a post I held for only one term because it felt hypocritical to be leading prayers and seeking God's blessings when I did not believe in what I was doing. So, even though I lost a few privileges that came with being an officer, I resigned.

Once into adulthood, I rarely attended services or observed holidays, had little contact with formal religion of any sort. That tendency has continued more or less unabated, although, in recent years, I have been troubled by extreme forms of religious activity and the more violent manifestations of extreme belief that have been more frequent in most parts of the world. Although aware that violent and inhumane expressions of religiosity have been part of human experience for millennia, I feel that the current period seems particularly troubling and dangerous in its many manifestations, ranging from fanatical Jews killing the former prime minister of Israel to al-Qaeda and other terrorist groups that carry out their acts of destruction under the banner of one or another Muslim-inspired set of beliefs.

My main motivation in writing this chapter is to use whatever knowledge I may have acquired about positive developmental processes and positive forms of change to seek possible insights into religious, spiritual, and personal growth. As a scholar of development, I deeply believe that a better understanding of how positive change can be encouraged, and negative change discouraged, may be the best antidote to an increasingly poisonous human situation.

DEVELOPMENT

For more than thirty years, I have been preoccupied with the study of large-scale cognitive developmental change, the kind of change that is irreversible, wide-ranging, and profound. My interest is in some ways like that of William James' fascination with religious conversion. James said:

> Our ordinary alternations of character as we pass from one of our aims to another, are not commonly called transformations, because each of them is so rapidly succeeded by another in the reverse direction; but whenever one aim grows so stable as to expel definitively its previous rivals from the individual's life, we tend to speak of the phenomenon, and perhaps to wonder at it, as a "transformation." ... These alternations are the completest ways in which a self may be divided. (James 1902, 194)

To be sure, James's discussion was about religious conversion in all its manifestations, not only the cognitive. Indeed, psychology then was not divided into subspecialties as it is now. Still, the description of a major dividing point between what was true before and after the conversion experience is similar to the quality that has always interested me in cognitive development.

Development and conversion may be similar in being major, irreversible, and profound changes, but they are not necessarily similar in valence. When discussing development, there must be a demonstrated *positive* step toward a more advanced level (Scarlett 2005), not simply positive in how the individual undergoing the change sees it, but positive with respect to a set of criteria that have been established independent of the individual and of those who may have a vested interest in promoting the change. In conversion, for example, one might move from one destructive set of beliefs to another without gaining any deeper understanding or more profound insight into spiritual issues. In the film *The Manchurian Candidate*, manipulation and hypnotism were used to take control of a man's will, with the goal of using him as an assassin. Cults like Jim Jones' Jonestown sect, members of which took lethal doses of Kool-Aid en masse, would not meet criteria for being developmental, however much they involved beliefs based on spiritual themes.

Conversion, therefore, may or may not serve positive purposes (or, more likely, may be a mixture of both), but unless it leads to further transformations toward a known, higher, and more positive level, it would not qualify as a developmental change. To put it another way, development takes place within domains (Feldman 1994; Karmiloff-Smith 1992; Hirschfield and Gelman 1994; Wilber 1996). Specifying what the developmental levels of the relevant domain may be and what conditions foster movement from one to the next is a major challenge for a developmental analysis of spirituality.

Examples from my own specialty of cognitive development that share some of the power of conversion in William James' discussion might include Piaget's four major stages of cognitive development (Piaget 1971); each stage marks an irreversible divide between newly emerged qualities of the mind and those that have been eclipsed in the transition or transformation (as James would have put it). As Piaget taught us, the mind of the child (actually, the *minds* of the child) represent different logics, grounded in different principles of reasoning from the mind of the adolescent or adult. What Piaget called the "great mystery of the stages" (Piaget 1971, 9) was the amazing regularity with which the transformations of mind take place and the astonishing certainty that each mind will undergo the set, in order, as it moves from childhood to adulthood.

For me, the quest to understand better such profound positive transformations in mind led to the study of extreme examples that do not conform to Piaget's universal specifications in all respects, i.e., they are not inevitable and are not spontaneous, moving my study of profound cognitive developmental change a step closer to James'. Conversion is manifestly not a universal experience, and rarely does it occur without persistent catalyzing efforts by agents of the "revealed" new order. Camp meetings, evangelical preachers, calls to prayer, healers, and the like all aim to provide "crystallizing conditions" (Feldman 1974; Walters and Gardner 1986) to hasten a conversion. For my part, the kinds of great transformations that were interesting were, of course, developmental ones: child prodigies (Feldman 1986) and great creators (Feldman 2003; Gardner 1993; Gruber 1974).

The two kinds of individual developmental transformations that I have studied (mostly through case study methods) represent different kinds of developmental changes, one closer to conversion, the other closer to the topic we will be pursuing later in this essay, namely, spiritual development. For pure-case child prodigies, the succession of profound transitions in mastering a craft affirms and deepens a set of available structures of knowledge and processes of activity, as in conversion to a known form of belief. For great creators (on the scale of Picasso, Marie Curie, Einstein, and Beethoven), adequate structures and processes of activity of a domain are not available and must be created, so structures are *themselves* forever transformed along with the transformation of the individual or individuals responsible (Feldman 1974, 1980, 1989). The field of biology was forever transformed by the efforts of James Watson and Francis Crick to discover the molecular structure of DNA (Watson 1959). Picasso and Braque created a new form of visual art ("cubism"), changing their own and the Western world's understanding of representation and spawning a period of exploration and generalization built on their work (Gardner 1993).

My hope in studying extreme kinds of cognitive developmental transformations in the disciplines was that some of the fundamental processes of transformation, of irreversible, progressive change, might be revealed. I recognized that the kinds of universal changes that Piaget described were not isomorphic with the extreme, individual, and, in some cases, unique changes that I studied in child prodigies and high-register creative individuals (Gruber 1980; Feldman 1980; Morelock and Feldman 1999). Still, I thought that there might be fundamental processes of transformation and change in cognitive structures that would reveal themselves in the extreme cases, thereby helping to explain better why Piaget called the inevitable changes the great mystery of the stages (Piaget 1971, 9), i.e., how truly novel forms of thought can emerge from seemingly mundane mental processes and how total transformation of the mental apparatus can happen suddenly, irreversibly, and productively (Feldman 1994, 1995; Feldman and Fowler 1997).

Through these explorations, I learned things about both of my target topics, although, to be sure, both continue to be elusive. It would be fair to say that qualitative advances in thought of the sublime and of the mundane sorts are better understood than they have been but that both continue to require sustained efforts in theory, research, and analysis. The possibility that the two change processes share certain qualities seems still a viable assumption. Reciprocally, some of the critical differences that may distinguish between and among the various forms of major cognitive-developmental advances have also been better described (Feldman 1980; Gruber 1980).

The main difference between cognitive-developmental stage shifts of the universal sort and the kinds of advances in thought that irreversibly and permanently change the world is that the former do indeed apparently emerge from the exercise of ordinary and mundane cognitive processes, while the latter require additional qualities and processes. All children will predictably discover that objects continue to exist even when they are no longer part of the immediate visual scene or that the amount of juice in a glass is not changed when that juice is poured into a differently shaped glass. But only one person (or, arguably two; Gruber 1974) pushed the frontiers of knowledge to a new place by constructing a viable explanation for variation, selection, and survival among members of a species.

The main question for this essay is to what degree spiritual development is more like the common developmental change processes and common knowledge structures that all human minds use as they construct their understandings of the world, or, alternatively, more like the special, even unique, ensemble of structures and processes that made it possible for Darwin (or Georgia O'Keeffe or Gandhi or Mao) to transform their worlds in fundamental ways.

Of course, it will not be possible to answer this question fully. But we may be able to provide a partial or incomplete answer, and certainly we may point to some of the features of the various forms of large-scale cognitive-developmental advance that are involved as we continue to search for knowledge and understanding.

As a preliminary foray into the complex set of issues that must be addressed, let us stipulate that *all* of the major advances in knowledge and understanding that have been introduced thus far in the discussion are productively described as *developmental* in nature. Developmental in this context means that these forms of cognitive-developmental change share these qualities: *positive, qualitative, irreversible, sequential, large scale and/or pervasive, have profound emotional implications, and emerge through the exercise of intentional efforts on the part of the people who achieve them.*[1]

Each of the proposed attributes of a developmental change is necessary, and they all must be present for a change to be labeled *developmental*. As we will see, one or another missing attribute may help us distinguish between developmental changes and other kinds of changes that are pseudodevelopmental, nondevelopmental, or even *anti*developmental. The last of these, antidevelopmental change, may be important for distinguishing between different forms of religious and/or spiritual experiences. A brief review of each of the seven criteria follows.

Positive

Developmental changes, as defined here, are positive changes. Specifying what is meant by a positive change is, however, not straightforward. For our purposes, the sense in which positive is meant has to do primarily with expertise in recognized domains (Feldman 1994, 1995). If one is trying to become more accomplished as a violin player, there are levels of that activity that are generally recognized and affirmed through various practices. For example, an aspiring violinist may secure an audition at a prestigious music school. The invitation to an audition is based on a tape of performances, while the audition itself is judged in relation to standards set for admission to the school. The judgments are, of course, human and subject to criticism, second-guessing, and possibly correction; but the framework within which they take place is one in which expertise can be recognized and agreed upon (to the extent necessary) for decisions to be made based on performance.

Domains and activities vary in terms of how clearly marked their various levels may be or even the extent to which there is consensus that such levels exist. Yoga seems to be an activity, for example, that has genuine aversion to marked levels of expertise (Hao 2006), while Tae Kwon Do, a martial art, is carefully and transparently marked with colored stripes to recognized increased levels of expertise.

Major questions for an analysis of spiritual and/or religious development are the extent to which movement within the domain can be seen as positive, how well marked it is, and in what contexts judgments of levels of development or "expertise" may be made (Scarlett 2005).

Qualitative

Distinguishing a qualitative change from a more quantitative or modest or neutral change is not an easy task. Arguments over what constitutes a qualitative change have gone on for more than a century (e.g., Lerner 2002; Piaget 1962, 1970, 1971; Thorndike 1914, Vygotsky 1978), usually revolving around possible differences between "learning" and "development" (Kuhn 1995; Lerner 1995; Liben 1987). In a classic essay, Vygotsky (1978) reviewed the possible relationships between learning and development, with learning as critical to and necessary for development. For Piaget (1962), development is necessary for and critical to learning. It appeared that the two views were in direct conflict on the relationship of learning to development. But the two theorists defined learning and development differently, making it impossible to resolve their conflict.

Only by understanding that learning and development had different meanings in the two theoretical perspectives was it possible to begin to interpret their differences (Feldman and Fowler 1997). As it turns out, Piaget meant by the term *development* the emergence of new structures of thinking, while Vygotsky meant advances in the level of understanding of a specific domain (e.g. speech, chemistry, poetry). By *learning*, Piaget meant the accumulation of facts within the constraints of existing thought structures (e.g., adding vocabulary or solving arithmetic problems when concrete operational structures are operating), while Vygotsky meant successful efforts on the part of a teacher and student to transfer knowledge and understanding of a valued domain from one person to another (e.g., arithmetic procedures, chemical bonding rules, how to use alliteration properly).

Where one draws the line between "qualitative" changes and "nonqualitative" ones is not an altogether clear matter. In some respects, the field of developmental psychology's traditional emphasis on common or universal developmental changes has made the task more difficult (Feldman 1980, 1994, 1995). If a case can be made for both the universal kinds of changes that Piaget proposed as developmental and the more domain-specific kinds of changes that Vygotsky proposed (and, as we will see, others as well), we have shifted the task from defining what development *is* to a consideration of what is common and different among the various kinds of changes that are defined as developmental (Feldman 1995).

Examples are not a satisfactory way to make the distinction between qualitative and nonqualitative changes, but they are a good first step. One of the reasons that Piaget's theory has been so compelling for so long is that it provides powerful examples of what qualitative change means within its theoretical framework. When a child is unable to transcend her perception that two containers look different but hold the same amount; or if no marbles are added or taken away, the number of marbles remains constant, we witness a powerful cognitive structure that has been constructed and applied by the child. Being able to distinguish between perceptual appearance and logical necessity represents a qualitative advance for the growing child, a true developmental transformation (Bruinguier 1981).

Analogously, when Vygotsky describes a turning point in child thought between the use of a stick to retrieve an object on a high shelf versus a request to an adult to retrieve the object even though a stick is available, it requires little persuasion or argument to recognize that there has been a qualitative advance in the child's repertoire of responses to a frustrating situation. Other primates rarely learn from their conspecifics in this manner (Tomasello 1999).

A student in one of my classes recently brought up another example of a qualitative positive advance: blind Helen Keller first made the association between the tactile experience of feeling water pour over her hands and the word "water," a moment vividly described as the major turning point of her young life:

> We walked down the path to the well-house, attracted by the fragrance of the honey-suckle with which it was covered. Someone was drawing water and my teacher placed my hand under the spout. As the cool stream gushed over one hand, she spelled into the other the word water, first slowly, then rapidly. I stood still, my whole attention focused upon the motions of her fingers. Suddenly I felt a misty consciousness as of something forgotten—a thrill of returning thought; and somehow the mystery of language was revealed to me. . . .
>
> That living word awakened my soul, gave it light, hope, joy, set it free! (Keller 1903, 22)

Piaget referred to the sudden moment when a child realizes that he or she truly understands a new relationship or solution or basis for analysis of experience as a "taking of consciousness" (*prise de conscience*). This moment is the result of a protracted period of effort and is often accompanied by the child exclaiming his new status: "Ah, now I understand" as Piaget put it (quoted in Bringuier 1980, 45).[2]

And so when we speak of a qualitative advance as one of the criteria for a developmental change, we find that it is difficult to specify precisely just what a qualitative advance looks like or what are its key features. We know that it represents a change such that the emergent qualities can not be reduced to the pre-emergent qualities. When something new or novel has appeared, is sufficiently unlike what preceded it, and has sufficiently powerful advantages over previous structures, then the term *qualitative* seems appropriately placed upon it.

Irreversible

This criterion is more straightforward. If a change is irreversible, it means that the change is permanent or, if it is in a sequence of such changes, permanent until it is supplanted by a yet more advanced transformation. Because developmental changes are defined as being transformational, the irreversibility of a developmental change is assumed to be permanent but not necessarily complete. Irreversibility does not mean that every aspect of the transformed system will immediately function consistently at the new level; there may be fluctuations and reversions that follow even an irreversible developmental advance (Feldman 1994).

What makes the advance irreversible is that the newly achieved transformation will constrain activity in all situations, although specific behaviors might from time to time appear that reflect the earlier system but would not occur under the current one. For example, when a frustrated motorist curses at a stop light, the behavior would appear to be from a much-earlier period of development (probably emerging at ages 3 to 6), but that behavior is, nonetheless, under the control of a functioning system that allows the person to drive a car, behave responsibly in most situations, and reason about what route might be a good choice under a variety of scenarios.

Moving to a new developmental level does not mean losing all the behaviors from the previous (and all earlier) levels; it means having all behavior organized and carried out through the filters of the new, advanced system (Case 1984; Fischer 1980; Fischer and Pipp 1984). Cursing a stop light is understood to be a way of expressing frustration and of getting through the time while waiting for it to change. It is not understood to be communication with a living thing, nor is it believed to be under the actual control of the wishes of the driver. The actual behavior may not be distinguishable from the serious supplications of a 4-year-old toward the stop light, but, if a developmental advance has occurred, all such behavior would be motivated by a different set of goals, carried out under a different set of functions.

In Piaget's research, the presence of a well-consolidated developmental advance was tested by efforts to dissuade the child from his or her judgment through countersuggestion. Experimenters would try to convince the child that, for example, two containers of differing shape (but identical volume) really contain different amounts of liquid, even when the contents of one are poured into the other. Or, to make their arguments more compelling, they would pour the contents of one container into a third container before pouring it into the first one. For a child for whom conservation has become fully consolidated, i.e., fully a part of the child's functional repertoire, no amount of countersuggestion will change that child's judgment that the two containers contain an equal amount of liquid.

Piaget's term for this unshakeable state of mind is *logical necessity*. When, according to the system of logic that all minds share, a relationship holds across all apparent transformations, a conclusion transcends appearances and transcends all argument. If nothing is added or taken away, the amount stays the same. Although Piaget was, to be sure, not a student of spiritual development (at least, he was not a student of spiritual development after his early twenties), it may be worthwhile to consider the possibility of a kind of nonlogical necessity as a potential outcome of a spiritual search for a different kind of truth (Armstrong 2006).

The key quality, then, of an irreversible developmental shift is that, once achieved, there is no return to an earlier state of mind. This does not mean that, in a given situation (such as the stop light situation), behavior that was characteristic of an earlier period of development never occurs. If it does occur, it does so under the constraints and with the meaning of the behavior to the person dictated by the new system. Although we may curse the stop light with gusto, we do not seriously believe that our invectives will affect its behavior.

Sequential

At the heart of most large-scale developmental changes is the quality that each one can be placed in the context of a sequence of such changes. The usual assumption is that the set of changes forms an invariant sequence, i.e., there is a specified order and that each person must start at the first of the sequence of changes and move through them, in order, to the last. For sequences that can be shown to be universal, that all individuals will experience, it is further assumed that everyone will move from beginning to end of the sequence. Each person reaches the last stage in the process, having traversed all the major qualitative shifts that make up the set in order (Feldman 1994).

In more contemporary theorizing, sequences such as Piaget's stages of cognitive development have given way to less strict ones. In Carol Gilligan's (1981) work on girls' cognitive development, for example, perspectives tended to emerge on issues like abortion in sequential fashion, but not as traditional stages. The perspectives were described as less mature or less evolved and more mature and more highly evolved, each perspective in the sequence giving way to its later successor. It is in this less strict sense that we are using *sequence*.

Fritz Oser and his colleagues (e.g., Oser 1991) have studied a traditional, Piaget- and Kohlberg-based sequence of "stages of religious judgment" that were constructed much as earlier cognitive-development theories were. Based on interviews with children and adolescents in response to "religious dilemmas," a sequence of five stages of religious judgment was constructed. As with Kohlberg's stages of moral judgment and reasoning (e.g., Kohlberg, Levine, and Hewer 1983), participants from different religious backgrounds and different parts of the world were sampled, and the sequence of stages was found to appear in all groups.

The premise of Oser's work is that the stages of religious development are universal among all peoples, i.e., the same sequence of five stages will under-lie and provide the "deep structure" of religious reasoning around the world. More recent work has put less emphasis on the presumed universality of a single sequence of stages, moving, on the one hand, toward domain-specific sequences with varying "deep structures" (e.g., Keil 1990) or toward more vari-ation as a function of context (e.g., Cole 1996; Shweder and LeVine 1984).

My own work has placed sequences of developmental levels from *novice* to *master* (the labeling of the levels varies) at the center of my theoretical orienta-tion (Feldman 1994, 1995). While the idea of stages, of universal, system-wide transformations, may give way to more variable and varied kinds of systems, my view is that, especially for the field of cognitive development, it continues to be useful to use invariant sequence as a way to organize major change, important criticisms notwithstanding (Feldman 2004). It would also likely organize my first pass over the phenomena of spiritual development to look for evidence of a sequence of developmental levels that might mark its various forms, albeit sequences of the softer rather the harder variety. Some of the meditation disci-plines and yoga appear to have a sequence of levels that are achieved over the course of many years; these levels appear to be sequential, but the processes through which one moves from one level to another are varied, not rigidly sequential, and subject to interpretation within specific contexts (Hao 2006).

An important distinction can be made between universal sequences that all people experience and nonuniversal sequences that may be achieved by only a segment of the population. Oser's sequence was intended to be a univer-sal one, but, as I have discussed elsewhere (Feldman 1994), it is unlikely that domains like moral judgment and reasoning or stages of religious judgment meet the same criteria as Piaget's stages. And there are many who believe that even Piaget's stages do not meet the most demanding criteria for a universal sequence of stages.

"Nonuniversal" sequences share with universal ones the feature that the sequence from beginning or novice states to "master" or advanced states is invariant. They do not share the feature that they are inevitable, nor do they share the feature that everyone who embarks on the journey from novice to master will (or can) reach the end of the sequence. Therefore, two key mark-ers of universal sequences are not necessarily found in nonuniversal domains: they are not inevitable, and they do not guarantee that everyone will get to the most advanced level.

It seems likely that spiritual development in its many and varied forms may be a nonuniversal sequence or sequences. It is known that not all people seek spiritual growth and development (Fuller 2001), and it is known that only small numbers of seekers of enlightenment achieve the most advanced levels in the various faith traditions. The sequence assumption claims that it is possible to find a pattern of changes within each of the traditions that would conform to the requirements of an invariant sequence of developmental levels. It would not claim that choosing to pursue a particular form is inevitable, nor would it claim that all seekers would reach the most advanced levels of whatever domains they chose to pursue.

Large Scale or Pervasive

The key issue for this marker of a developmental change is to distinguish it from changes that are more local, smaller-scale changes. This is a challenging thing to do; some of the best minds in educational and developmental psychology have struggled with this issue for more than a century, most notably perhaps Lev Vygotsky (1978). Vygotsky argued that efforts to enhance development (large-scale change) begin with small-scale change and that the latter, cumulatively, makes possible and is the catalyst for the former. The many instances of practice of addition problems, carefully arranged so as to be optimally challenging, are smaller-scale, nondevelopmental accomplishments (*learning* in Vygotsky's terminology). An advance in the student's ability to understand addition as part of a system of operations that include subtraction, multiplication, and division is a developmental advance. It will not occur, Vygotsky argued, without the sustained, skilled efforts of a more experienced individual (typically, but not necessarily, a teacher).

The centerpiece of Vygotsky's developmental/educational psychology is the idea that relatively small-scale changes, organized under the tutelage of a skillful guide, represent the quintessential human catalyst for development, one of the qualities that makes human development uniquely human (Vygotsky 1978). In theory, the distinction is clear enough; in practice, it is very difficult to draw a line where "learning" ends and "development" begins—or to say what represents "large scale" versus "small scale" in any given domain at any level.

Even if Vygotsky was not right about the relationship between small-scale and large-scale change (i.e., the former causes the latter), he was right about the importance of being able to calibrate the magnitude of a change in knowledge and understanding, in skill and ability, within a given domain. However a developmental change is achieved, it is developmental in large part because it affects a substantial portion of the person's functioning. Just how to specify "substantial" is, of course, a challenge that needs to be met if this criterion is to be used systematically and productively.

One of the ways to detect a large-scale transformation is to assess the range of situations to which it can be applied. If the range is limited, the change may be better described as one of "learning," while, if broad, the change may be considered to be a "developmental" change. As with the specification of *large scale*, specifying what *narrow* and *broad* refer to is a considerable challenge. If these challenges can be met, however, there is a principled basis on which to distinguish learning from development.

A further refinement of this way of distinguishing developmental from more limited changes is to impose a requirement that the application of the more advanced reasoning extend across categories, as for example, when an advance in arithmetic can be applied in nonmathematical contexts. If a child has reached a point of mastery of the arithmetic operation of subtraction by being able to subtract most numbers from most other numbers, can the child extend this understanding to real objects, to continuous quantities, and to hypothetical situations? The greater the ability to do so, the more confident we may be that the advance in reasoning can be defended as developmental in nature.[3]

In Piaget's framework of stages, each of the four stages of cognitive development is defined as an essentially all-encompassing set of instruments of thought that can be applied to all situations, the widest possible transfer potential. In other developmental frameworks, the scope of the change is not necessarily so vast and is, therefore, less difficult to assess. For example, Carey (1985) studied conceptual change within certain domains such as biological knowledge, and was able to make a strong case for conceptual change within that domain. Keil (1990) was able to show developmental changes in the acquisition of language through a series of compelling studies in which, for example, a shift from surface characteristics to essential features was the basis for defining a word like *uncle*. In the former, an uncle is defined as an old man with a beard, while, in the latter, uncle is defined as the brother of one's mother or father. While it is possible to claim that a change such as this represents the accumulation of smaller amounts of learning, there is evidence consistent with the criterion of large-scale change if the reasoning extends to other words such as *aunt*, *grandmother*, and *cousin* and to other familial relationships, particularly if the application tends to follow rapidly from the first instance of the new form (e.g., Fischer and Pipp 1984).

Powerful Emotional Implications

The field of cognitive developmental change has recently begun to acknowledge that emotions are an important aspect of a major transformation. Even in Piaget's formulation, perhaps the most emotion-free account of cognitive development, two emotions, curiosity and discomfort (a reaction to cognitive disequilibrium), play important roles in motivating the child to construct more advanced structures. Curiosity drives the child to want to learn more about the world around her or him, while discomfort drives the child to try to resolve conflicts between mental structures that are incompatible with each other and/or between mental structures and what the world seems to require (Ginsburg and Opper 1988).

For Vygotsky, a need for affiliation, to be a member of the group, was the central motivating force in propelling the child to internalize the mediating systems being used by those with whom he came in contact. Belonging is the deep need that propels the process of cognitive development in Vygotsky's framework. In order to belong (to the family, the neighborhood, the culture, or history itself), it is necessary to share a set of understandings made possible by the special mediating systems present in those contexts. Language is, of course, the premier mediating system for Vygotsky, but perhaps less well understood is the importance of the particular form of language, even the local dialect, in forming bonds of association among members who share its use.

Vygotsky's express goal was to specify what makes human beings uniquely human. To be able to use language with the facility that human beings do is, of course, a major distinction, but Vygotsky went further. What makes human beings uniquely human is their need to form, preserve, and share the specific forms of speech and other mediating systems (e.g., religious beliefs, calculating techniques, musical expressions, and the like) that define their distinctive form of human society (Vygotsky 1978).

In our own work on cognitive developmental transitions within skill areas where expertise can be assessed (such as martial arts, sports, music, chess, and other domains), we have found that certain distinctive patterns of emotions accompany phases in the transition from one level of expertise to another. We have proposed a recursive six-phase pattern, with each phase of a cognitive-developmental shift marked by a set of emotions. For example, in the initial or novice phase, emotions like *interested, motivated, compliant, content*, and *optimistic* are commonly reported (participants are usually college students enrolled in a cognitive-development course; Feldman 1994).

In subsequent phases, emotions tend to shift from positive to negative valence, especially during the third phase, to which we have given the label *stagnation*. The phase on the skill side is marked by little change in the level of skills, a kind of static phase when little progress or increase tends to occur. The emotions that tend to accompany this phase are *bored, frustrated, impatient, pessimistic, unhappy*, and the like. This phase tends to be followed by a "break-through" phase in which one or more skills advance dramatically, moving forward in rapid-fire fashion. A distinctive set of emotions is usually reported to accompany this phase, notably *surprise, delight, excitement*, and *optimism*.

Students are often able to track retroactively their movement through levels of expertise based on identifying patterns of emotions recorded in journals they keep, and to reckon or cross-check their findings against either their own or their instructors' reports of changes in skills (Hao 2006). For the class assignment, students are required to engage in and pursue an activity where expertise is likely to change in the eight weeks or so that they are able to devote to the effort. The addition of emotions to the record has been a substantial advantage in the goal to chart and interpret change over time within their areas of activity.

In some of his last work on transitions in cognitive development, Piaget himself pointed to powerful emotions that accompany and follow a full consolidation of a major structure (Bringuier 1980; Piaget 1975). In the phenomenon that Piaget called *prise de conscience* (taking of consciousness), as mentioned earlier, Piaget reported that children often seem relieved, proud, satisfied, and confident when they had reached the end of a protracted construction effort. Not the "break-through" moment when something new actually appears, *taking of consciousness* refers to the point at which the new structure has become fully consolidated, fully integrated, and fully functional as the basis on which the child will operate.

For example, in his conversations with Claude Bringuier and elsewhere, Piaget describes the moment when a child fully comprehends that the number of objects does not change when they are rearranged: "Once one knows, one knows for ever and ever" (Piaget 1971, 5). Indeed, it was reading examples such as these that helped push me to include emotions in the work on transitions in levels of expertise; phases 1 and 6 of our recursive cycle closely resemble Piaget's description of the point at which a set of cognitive-developmental structures has been fully consolidated (Feldman 1994).

In another application of emotional markers inspired by Piaget's descriptions of children who are constructing their understandings of the world, I reinterpreted Piaget's stages so that some of their long-standing issues could be addressed, particularly the problem of the "immaculate transition" that has plagued Piaget's account of stage transitions for decades. In the revision, each of Piaget's four major stages is divided into two halves, with the marker for the shift from the earlier, construction or acquisition, phase to the later, extension or elaboration, phase being a kind of "taking of consciousness" or consolidation that Piaget described (Feldman 2004).

The systematic inclusion of emotion in the study of cognitive-developmental transitions is likely to be productive when it comes to the study of spiritual development. Since spiritual development is not the same as cognitive development, nor equivalent to emotional development nor sociocultural nor physiological (the usual categories in developmental psychology), it will require its own set of markers. But among those markers it is a good bet that emotions and emotional change will be among the more useful ones.

Intentional Efforts by Those Who Achieve New Levels

The question to what extent developmental changes are intentional is a venerable one. As far back as Plato, sources beyond the control of the individual have been assigned responsibility for bringing about major advances in the products of thought. For Plato, it was the gods; for behaviorists, it is the context; for nativists, it is the genes; and for dynamic-systems theorists, it is the routine functioning of the normal processes of thought. Lewis Thomas, the physician who wrote with such grace about the wonders of life, was an early dynamic-systems theorist (although not by that title). Thomas described the emergence of remarkable forms in this way:

It may be our biological function to build a certain kind of Hill. We have access to all the information of the biosphere, arriving as elementary units in the stream of solar photons. When we have learned how these are rearranged against randomness, to make, say, springtails, quantum mechanics, and the late quartets, we may have a clearer notion how to proceed. (Thomas 1974, 15)

Thomas adds later, "This is, when you think about it, really amazing. The whole dear notion of one's own Self—marvelous old free-willed, free-enterprising, autonomous, independent, isolated island of a Self—is a myth" (167).

In the field of creativity studies, emergence has been a theme pursued by several prominent scholars: Keith Sawyer (2003), Dean Keith Simonton (1999), and Robert Weisberg (1999), among them.

My own view is that the more profound kinds of advances, some of which I have described as developmental in this chapter, are marked by an unusually large amount of *intentionality* and/or *self-conscious pursuit of a goal*. For creative advances, the goal is often known to the individual or individuals pursuing it; for example, Charles Darwin knew perfectly well that he was trying to construct a viable, scientifically rigorous theory of evolution of species, and James Watson and Francis Crick knew perfectly well that their goal was to construct a viable and scientifically rigorous model of the genetic material. Indeed, in the latter case, Watson and Crick were one among several teams of scientists in a race to see which one would win the Nobel Prize (Watson 1968).

For the more universal developmental advances that have been the stock-in-trade of cognitive development for several decades, intentionality and self-conscious pursuit of a known goal are less clearly evident. It is unlikely that the 6-year-old child understands that she or he needs to construct an understanding of seriation, even as that seems to be precisely what the child is doing. As far as I know, children have not been interviewed to find out what they think they are doing when they endlessly arrange ordinal series of favorite flavors of ice cream, best friends, greatest athletes, or, as my daughter Betsy did at that age, what color marbles roll the farthest.

Chances are, though, that children through at least early adolescence are guided by goals that are part of their human heritage and are achieved through the interactions among their many relevant activities and their developing neurological system. There may be a rare 6- or 7-year-old who is inherently interested in logical relationships or abstract properties of symbol systems, but, for the most part, the logical capabilities that emerge during middle childhood do so relatively unselfconsciously.

For developmental advances between those that are all but inevitable and those that are unique and powerful enough to transform an established domain, the degree of intentionality and awareness of purposes and goals likely varies. Nonuniversal theory is a framework that tries to organize developmental domains on a continuum from universal developmental transformations (such as Piaget's stages) to unique ones (such as Georgia O'Keeffe's paintings), varying in a number of ways including how aware the individual involved is of what he or she is trying to accomplish (Feldman 1994). The "universal-to-unique" continuum of nonuniversal theory is shown in Figure 1.

As with the other markers for cognitive-developmental change, the quality of intentionality is not always easy to specify or assess. Most but not all of our evidence is from detailed case studies or accounts by the people involved

FIGURE 1. The universal to unique continuum. Reprinted with permission from Feldman 1980, 1994

/_____ /_____ /_____ /_____ /_____ /

Universal   Pancultural   Cultural   Disciplined   Idiosyncratic   Unique

(cf. Brandstadter 1998; Lerner 1982). A study of Charles Darwin's personal notebooks revealed that he was seeking a solution to the mystery of evolution from at least his middle-twenties, pursued it relentlessly if not with consistent success, and knew he had constructed a working model of the theory by his twenty-ninth year (Gruber 1974). Pablo Picasso and Georges Braque set out to deconstruct traditional forms of representation and self-consciously set the goal of constructing the new form that came to be called "cubism" (Gardner 1993). Braque and Picasso were very much aware that they had created something that did not previously exist, although, to be sure, it contained elements of existing ways of painting and required that the artists master those forms.

If a "Self" is a myth, as Lewis Thomas supposes, it is a very powerful and sustaining myth, and something like a self is critically involved, is likely a marker, for major developmental advances beyond those that are essentially guaranteed and universal ones (Brandstadter 1998). Even the child who struggles to learn to read in first grade has a goal in mind, a good set of criteria for what it will be like to achieve that goal, and a set of well-understood expectations that guide the effort. The argument put forward here, then, is that, in most if not all developmental advances, a strong and usually self-conscious intention drives and directs the process. When that process has achieved the change, it becomes a vital and valued part of the individual or individuals' sense of who they are and what they have done; and, in some cases, it becomes valued to the world (Feldman, Csikszentmihalyi, and Gardner 1994; Gardner 1998).

Spiritual development once again appears to be a special case, this time a special case of intentional change. I know of no instance of a major change in spiritual growth and/or conversion that is not in some sense intentional. Even in biblical accounts where God or a sign appears, it is when the person is seeking guidance, seeking affirmation, or seeking a way to resolve a dilemma or get help in a challenging situation. Those who seek salvation, who pray for a resolution to a dilemma, who beseech the Deity to heal their wounds are doing so intentionally and self-consciously, even if the help comes during a dream or at some unexpected moment.

The efforts that give rise to a spiritual change appear to be virtually always intentional, sought after, and pursued with diligence. Of course, precisely what is sought is often not clear, but that is true in other forms of developmental advance as well. While it was clear that the Wright brothers set out to invent a flying machine, it was not clear that Dr. Alexander Fleming (and codiscoverers E. B. Chain and Howard Florey) set out to invent penicillin. It was necessary that Fleming be generally well prepared to appreciate the new possibilities of a blue-green mold that grew on a staphylococcus plate culture, but not necessarily that he set out to invent an antibiotic. The field of antibiotics followed the unplanned discovery and did not even exist before the appearance of penicillin and the recognition of its bacteria-fighting ability, thus advancing its importance as a positive transformation in the medical treatment of infections.

Summary of Criteria for a Developmental Advance

To review briefly what we have discussed thus far, there appear to be several markers that apply, albeit neither uniformly nor in equal measure, to a set of positive changes that we have called *developmental*. These criteria may be used to help assess the degree to which a given instance of change may be confidently classified as developmental, which in this context means a major shift in thinking and reasoning.

The seven criteria are that the change must be *positive*, *qualitative*, substantial, and in some important ways a different form from what preceded it. The change must be *irreversible* such that, once transformed, there is no going back to an earlier framework or system; it must be in a mostly predicted and predictable *sequence* of such changes, starting at the first and ending at the last, in order; it must be of a *scale* that impacts a broad sector of behavior, either across domains or over the entire range of a major domain; it must be accompanied by a distinctive set of strong *emotions*; and, must be achieved as a consequence of efforts directed toward achieving the change, in some instances very clearly intended to solve a problem or accomplish a goal, in others less clearly motivated but *intentional* nonetheless.

SPIRITUAL DEVELOPMENT

Taking the criteria just discussed into account, what might we say about spiritual development as a developmental domain? What is spiritual development like as compared with other forms of developmental change, and how might positive spiritual development be encouraged given what we have learned about it?

We must first acknowledge that the realm of the spiritual is a distinctive area of human experience, related to but not the same as religious observance (Fuller 2001). It may or may not have the form of Oser's (1991) religious stages of judgment and may or may not have similar cognitive, social, cultural, physical, and emotional aspects to religious practice or to any other domain. Spiritual development may or may not be universal in the strict sense that Piaget's stages and Oser's stages of religious judgment are claimed to be. Considering each of the criteria we have discussed in relation to spiritual development may help reveal its similarities and differences from other domains within which development takes place.

Are spiritual-developmental changes positive and qualitative, or are they simply the uncovering of already understood truth? For a spiritual-developmental change to be shown to be *positive* and *qualitative*, it would need to display the kinds of markers that other such changes do. For example, a change would have to be of a different form of understanding, expressed through more mature and better explanations of the relevant areas of experience. These areas of experience would seem to be more of the intrapersonal, reflective kinds than those of religious judgment (Gardner 1983). Spiritual development might bring a significant change in one's understanding of one's responsibilities for those in need, while religious judgment might bring a significant change in why, for example, Muhammad's teachings represent an important enough change from those of Jesus to justify conversion. Each of the two areas of potential development is likely to inform and influence the other, but they do appear to be different forms of experience and of developmental change.

For spiritual development to be established as a developmental domain in the sense that we have defined it here, it would also need to exhibit a *sequence* of levels, or perspectives in the Carol Gilligan (1981) sense, that predictably tend to follow one another, each representing a positive forward step. These levels should hold across different religious traditions and different cultural contexts, although not as strictly perhaps as Oser's (1991) stages of religious judgment. Successive perspectives on the great questions of existence (e.g., "Why am I here?" "How should I lead my life?" What kind of God do I believe in?") are what Gardner (2002, film) called "existential intelligence" with its focus on issues of the meaning of life and the purpose of various kinds of actions.

A change in spiritual understanding, if it is to be considered a developmental change, must also be of a *large-scale* type. The change would have to apply across a wide range of areas of experience to qualify. Deciding to stop shoplifting may be the result of fear of being caught more than because it violates one of the principles of a good moral life, the former a local or small-scale change, the latter likely a large-scale one. To determine if a change is on a sufficiently large scale to be considered developmental is more of an art than a science, but the techniques for doing so very likely will involve evidence of application of broad principles across a variety of situations. And as with Piaget, it may be helpful to request from the individual the reasons that a change may be occurring through the use of countersuggestion and challenge (cf. Piaget 1971). This technique may also be useful for determining the degree to which a change is *irreversible,* enduring, and resistant to reversion to an earlier system.

The role that emotions might play in spiritual-developmental change is likely to vary at different points in the transition from one perspective on spirituality to the next. In our research on expertise in nonuniversal developmental domains, we have found that patterns of emotions are good indicators of where in a transition the individual is currently functioning (Feldman 1994; Hao 2006). The domains we have studied, to be sure, are different from spiritual development (mostly skill areas like music, chess, map drawing, juggling, yoga, and others), making our findings only indirectly relevant to what we might find in the spiritual domain. It seems certain, though, that the distinct sets of emotions are reflections of activity in the spiritual domain.

It might be expected, for example, that a sense of *uneasiness* or *discomfort* or *dissatisfaction* or a feeling that questions that used to be answerable no longer seem answerable might indicate the beginning of a transition from one level of spiritual understanding to another. Feelings of *relief, satisfaction, confidence, enthusiasm,* and *renewed energy* might be markers for the completion of a transition cycle toward a new level of spiritual understanding. Where in a transition one is may be reliably assessed using emotional markers in nonuniversal domains of the sort that we have studied. It seems reasonable to expect that, although the specific emotions and patterns of emotions may vary, research that aims to determine the emotional concomitants of changes in spiritual understandings could be very useful in advancing our knowledge about the processes of spiritual change.

Intentionality is likely to be a critical component of spiritual development, particularly at the more mature or advanced levels. There are compelling accounts of struggles, often over decades or even lifetimes, to reach deeper levels of understanding (e.g., Merton 1971; Gandhi 1951), to find more profound and encompassing truths. Unless one adheres to a belief that all understanding is given by entities or forces beyond one's control (and even then under certain interpretations), the drive toward more adequate understandings of the realm of the spiritual would appear to be one of its markers. The kinds of questions asked, the kinds of answers sought, the kinds of structures that form are no doubt distinctive to the domain, but an effort to study them as expressions of intentionality would appear to be a productive line of inquiry.

SPIRITUAL DEVELOPMENT VERSUS FIXED OR STATIC VIEWS

Are the criteria that help define developmental changes of use in distinguishing different forms of spiritual experience? In particular, can we use what we know about developmental change to help understand the difference between spiritual development and a fixed, or literalist, perspective?

Both kinds of change appear to be qualitative, particularly when conversion experiences are reported for those who embrace a fully formed framework. In the case of spiritual development, qualitative changes occur each time a new level of spiritual enlightenment or spiritual understanding is achieved. For the conversion kind of change, there is most often a single, life-changing transformation that leads to an unchanging and unchangeable set of beliefs and practices, beliefs and practices that represent the ideal of the framework. After the single great transformation that marks a conversion as occurring, any further changes are of a relatively minor sort.

On the criterion of large-scale, pervasive change, the fixed framework transformation would appear to be greater in scope and scale than the achievement of a new level of spiritual understanding, closer, in fact, to the traditional Piagetian notion of "structures as a whole" (Piaget 1971, 5). Spiritual-developmental change, while large scale, does not necessarily provide a set of constraints and controls that extend to all areas of experience. So, again, spiritual development, while almost certainly developmental, appears to be on a more limited scale than the embracing of a fixed and presumably immutable set of beliefs.

The sequential criterion is one where the two phenomena may differ more sharply than in degree. Other than whatever process leads to the fixation of belief within an unchanging or at least very stable system (Fodor 1976), there are by design no further major changes and no more advanced levels in a sequence in the fixed or revealed form of experience. The goal of literalist spirituality is to get to the stable state of firm belief, while the goal of spiritual development is to move toward evermore advanced levels of spiritual experience. Fixed-belief systems are intended to render further major spiritual changes unnecessary; they eschew reflection and critical analysis of the basic tenets of the system. Spiritual-development systems are intended to provide a framework within which one can move toward evermore enlightened spiritual understanding and practice through reflection, critical analysis, and insight.[4]

The emotions of the two kinds of spiritual activity are also likely to be different and may provide markers for what spiritual direction a person has taken. With respect to a conversion experience, we have been focusing on a single moment in what is intended to be a lifelong transformation, of course, and there may well be variations even in this one moment. Our goal is to find distinctive markers and indicators, ones that tend to occur for most people who undergo the experience in question. In our work with expertise, the emotions of transitions between levels of expertise have been striking in their consistency. Variations are often interpretable as deviations from the canonical, expected set (Feldman 1994; Hao 2006). While it may be unduly optimistic to think that a set of emotional markers could be used to distinguish between a tendency toward conversion versus development, an effort toward finding such markers may provide information early enough in the process to intervene.[5]

For the criterion of intentionality, the inherent ambiguity of the word makes it difficult to use it in differentiating between development and conversion. For development, the meaning seems clear enough; spiritual development is self-consciously chosen, a goal that can be pursued for a lifetime.

Conversion may also be marked by a kind of intentionality, but it would presumably be a different form. A fixed system presents a preexisting and fixed set of beliefs and practices, a set that promises to provide spiritual nourishment and, often, eternal salvation and happiness for adhering to the belief system. It would require consideration of the conditions under which people "choose" a commitment to fixed beliefs to sort out the degree of overlap between intentionality in the spiritual-development sense and the sense of a choice to convert to a fixed-belief system. The many efforts to provide such conditions in the form of prayer meetings, evangelical gatherings, church settings, faith campaigns, and the like, if successful, would seem to show understanding of how to create conditions under which a person would willingly choose to convert to the proffered system. These techniques might yield valuable information on the conditions that provide catalysts for conversion.

WHAT KIND OF DOMAIN IS SPIRITUAL DEVELOPMENT?

Spiritual development appears to fall into a region of developmental domains that are near universal, called pancultural in the universal-to-unique continuum (see Figure 1; also, Feldman 1994; Horowitz 1987). Along with speech, music, gesture, dance, and probably others, spiritual development is likely to occur in all children in all cultures. It is much less likely to occur in the absence of a culture, as, for example, in feral children or cases of intentional isolation from other people.

Within my theoretical framework, which marks domains in regions from universal to unique, spiritual development appears most likely to be found in what is called the "pancultural" region, requiring only that some other people and a viable culture and society be involved in the process (Feldman 1994). Only the few domains that have an even more powerful biological predisposition than spiritual development, such as the logical structures required for surviving in the physical world that Piaget described, are more universal. For these latter kinds of domains, virtually all children could, in principle, move to the highest level whether other people participated in the process and/or a set of social and cultural practices guide the process. But spiritual development is likely to be substantially directed by biological constraints and to exhibit especially sensitive periods to development like speech and attachment (Gopnik, Meltzoff, and Kuhl 1999; Keil 1990). As pancultural domains, evolution and biology essentially guarantee that development will proceed through developmental levels necessary for survival in human societies. Development beyond these guaranteed levels exists in various forms and practices but is not guaranteed. At each succeeding level, there are fewer and fewer people who have reached it; there are few successful poets relative to the number of people who are able to speak their native language fluently, for example.

The same is likely to be the case in spiritual development; there are few spiritual exemplars (Colby and Damon 1992) relative to the number of people who have acquired the spiritual understanding necessary to be able to function in their human communities. Reaching the more advanced levels of spiritual development requires years of training and discipline; Tibetan Buddhist lamas spend upwards of fifty years in systematic training (Thurman 1995), often beginning when they are still at the age when people are engaged in the more universal challenges of childhood.

If spiritual development proves to be a developmental domain and if it is shown to begin for human beings during their earliest years and proceed through a sequence of essentially guaranteed levels, then it should be possible to calibrate and provide indicators for where they are functioning within the domain. If it further proves to be true that there are particularly sensitive periods for spiritual development, as I believe there may be, then it would be valuable to know when they are occurring. These points may be the most-promising times to introduce opportunities for spiritual development, and, conversely, they may be periods when children are most susceptible to conversion to a fixed set of beliefs.

Although it is at best a guess, it appears to be the case that at least two periods of special sensitivity to spiritual issues and concerns appear in childhood, one from about the ages of 4 to about 9 (cf. Silverstein 1991), the other during middle adolescence from about 15 to 18. If so, knowing about the nature of such periods may be of use in designing activities and setting challenges for children eager to take them on.

CONCLUSION

The criteria discussed in this essay proposing to define development, and by extension define spiritual development, appear to be useful markers for detecting the presence of developmental change in the spiritual domain. They also may be useful in distinguishing between behavior that indicates the likelihood of spiritual development versus behavior that indicates the likelihood of conversion to a fixed set of beliefs. A period during which these indicators might be particularly useful is middle-to-late adolescence. Middle adolescence, along with the years from 4 to 9, appear to be particularly sensitive periods for engaging in and pursuing spiritual and religious goals (Silverstein 1991). During the earlier of the two sensitive periods, interest in spiritual and religious matters is part of the development of cognitive structures that will, eventually, enable the child's mind to form logical analytic structures. During the adolescent years, however, heightened interest in these same sorts of matters might be critical in crystallizing life decisions and long-term commitments to an identification with a particular religious and/or spiritual tradition.

If good markers and indicators of receptivity to and interest in spirituality and religiosity can be made available to those who parent, teach, guide, and encourage young people in their search for a life direction, then it should be possible to assess with reasonable confidence what kinds of activities and toward what goals a person is striving (Lerner 2007; Roeser 2006). If the purposes of those who guide adolescents are toward spiritual development, as I hope they will be, then using developmental criteria and markers for the various kinds of changes that may be under way in their charges should prove useful in these efforts. At the least, such markers can help in guiding the process of spiritual development, locating where that process may have been and where it might be going.

ACKNOWLEDGMENTS

Special thanks to my colleague George Scarlett, who read earlier drafts of this chapter and made helpful suggestions throughout. Thanks as well to colleagues Rob Roeser, Erin Phelps, and Rich Lerner, whose thoughtful suggestions and support were much appreciated during the preparation of the chapter.

NOTES

1. This last feature is likely to be controversial. The liveliest area of developmental theorizing these days is dynamic systems theory or DST (Lewis 2000). The presence of intentionality is essential, however, to the argument that I will be making later in the discussion concerning spiritual development. Most dynamic-systems theorists prefer explanations for major cognitive developmental changes that are random or at least unintentionally motivated processes and that, *by their very nature*, will from time to time produce major forward leaps or transformations (e.g., Lewis 2000; Lerner 2002; Hartleman and Molenaar 1999; Thelen and Smith 2006; van Geert 1992).

It should also be noted at this point that there is something of a hidden agenda not yet revealed in the discussion thus far. And note, of course, that I am bringing it up in an endnote, where only those who read the endnotes will learn of it. The hidden agenda is that I want to distinguish between and among not only the major cognitive-developmental changes so far introduced but also among all of the above and some *nondevelopmental* or even *antidevelopmental* changes that share certain features with their developmental counterparts. This last category may help distinguish between more negative and more positive forms of religious belief and practice.

2. Developmental changes are defined as positive, as advances in understanding, ability, or competence. There are negative changes that may be just as transformational, which leave the individual in as qualitatively different a state as positive ones, but these are not taken to be developmental. For example, when a person goes into a schizophrenic episode for the first time, when someone resolves to exact revenge for accumulated insults or sets buildings on fire because it is exciting, these are not likely to be developmental, although they are, to be sure, qualitative shifts. As with the other criteria for a developmental change, what is positive and what is negative may not be straightforward to determine.

3. For readers who are familiar with the educational-psychology literature, the criterion of "large-scale" change refers to the issue of "transfer of training" where, as Vygotsky put it, an ounce of training may yield a pound of advancement in performance (see, e.g., Thorndike 1914; Vygotsky 1978).

4. As my colleague George Scarlett has often reminded me, fundamentalist or literalist frameworks have frequently guided people to positive and worthwhile lives through their teachings and the requirement that their members adhere strictly to the rules of the system. The issue is not whether a particular system inevitably leads to good or bad behavior but that one type of system increases the likelihood of exploitation, deception, and corruption in the guise of a "revealed" set of truths, while the other reduces that likelihood.

5. Here I am, of course, revealing my bias toward development over conversion to a fixed set of beliefs. This bias, nonetheless, recognizes that there may be many positive things about living a life within a fixed, unchanging system. My own value system, though, is toward a life of seeking deeper understandings and more profound spiritual truths, without end. I am concerned that the risks associated with a set of fixed and exclusive beliefs far outweigh the possible benefits.

REFERENCES

Armstrong, K. 2006. *The great transformation: The beginning of our religious traditions*. New York: Knopf.

Baltes, P. B. 1987. Theoretical propositions of life-span developmental psychology: On the dynamics between growth and decline. *Developmental Psychology* 23, 611–26.

Brandstadter, J. 1998. Action perspectives on human development. In W. Damon, series ed., R. Lerner, vol. ed., *Handbook of child psychology*, 5th ed., vol. 1, 807–63. New York: John Wiley.

Bringuier, J.-C. 1980. *Conversations with Jean Piaget*. Chicago: University of Chicago Press.

Carey, S. 1985. *Conceptual change in childhood*. Cambridge, MA: MIT Press.

Case, R. 1984. The process of stage transition: A neo-Piagetian view. In R. Sternberg, ed., *Mechanisms of cognitive development*, 19–44. New York: W. H. Freeman and Company.

Colby, A., and W. Damon. 1992. *Some do care: Contemporary lives of moral commitment*. New York: The Free Press.

Cole, M. 1996. *Cultural psychology: A once and future discipline*. Cambridge, MA: Harvard University Press.

Feldman, D. H. 1974. Universal to unique: Toward a developmental view of creativity and education. In S. Rosner and L. E. Abt, eds., *Essays in creativity*, 45–85. Croton-on-Hudson, NY: North River Press.

———. 1980. *Beyond universals in cognitive development*. Norwood, NJ: Ablex Publishing Corporation.

———. 1986. *Nature's gambit: Child prodigies and the development of human potential*. New York: Basic Books.

———. 1989. Creativity: Proof that creativity occurs. In W. Damon, ed., *Child development today and tomorrow*, 240–60. San Francisco: Jossey-Bass Publishers.

———. 1994. *Beyond universals in cognitive development*. 2nd ed. Norwood, NJ: Ablex Publishing Co..

———. 1995. Learning and development in nonuniversal theory. *Human Development* 386: 315–21.

———. 2003. The creation of multiple intelligence theory: A study in high level thinking. In K. Sawyer and V. John-Steiner, eds., *Creativity and development*, 139–85. New York: Oxford University Press.

———. 2004. Piaget's stages: The unfinished symphony of cognitive development. *New Ideas in Psychology* 22: 175–231.

Feldman, D. H., and R. C. Fowler. 1997. The natures of developmental change: Piaget, Vygotsky and the transition process. *New Ideas in Psychology* 153: 195–210.

Feldman, D. H., M. Csikszentmihalyi, and H. Gardner. 1994. *Changing the world: A framework for the study of creativity*. Westport, CT: Praeger.

Fischer, K. 1980. A theory of cognitive development: The control and construction of hierarchies of skills. *Psychological Review* 87: 477–531.

Fischer, K., and S. Pipp. 1984. Processes of cognitive development: Optimal level and skill acquisition. In R. Sternberg, ed., *Mechanisms of cognitive development*, 45–80. New York: W. H. Freeman and Company.

Fodor, J. 1976. On the impossibility of acquiring "more powerful" structures. In M. Piattelli-Palmarini, ed., *Language and learning: The debate between Jean Piaget and Noam Chomsky*, 142–62. Cambridge, MA: Harvard University Press.

Fuller, R. C. 2001. *Spiritual but not religious: Understanding unchurched America*. Oxford: Oxford University Press.

Gandhi, M. 1951. *Non-violent resistance: Satyagraha*. New York: Shocken Books.

Gardner, H. 1988. Creative lives and creative works: A synthetic scientific approach. In R. Sternberg, ed., *Handbook of creativity*, 298–321. New York: Cambridge University Press.

———. 1993. *Creating minds*. New York: Basic Books.

———. 2002, film. *MI Millennium*. Los Angeles: Into the Classroom Media.

Gilligan, C. 1981. *In a different voice*. Cambridge, MA: Harvard University Press.

Ginsburg, H., and S. Opper. 1988. *Piaget's theory of intellectual development*. 3rd ed. Englewood Cliffs, NJ: Prentice-Hall.

Gopnik, A., A. Meltzoff, and P. Kuhl. 1999. *The scientist in the crib: What early learning tells us about the mind*. New York: Harper Collins.

Gruber, H. 1974. *Darwin on man: A psychological study of scientific creativity*. New York: E.P. Dutton.

———. 1980. Afterword. In D. H. Feldman, ed., *Beyond universals in cognitive development*, 175–80. Norwood, NJ: Ablex Publishing Corporation.

Gruber, H., and P. H. Barrett. 1974. *Darwin on man: A psychological study of scientific creativity*. New York: E.P. Dutton and Co.

Hao, H. 2006. *On the development of expertise: Transitions in novice yoga practitioners*. Ph.D. diss., Tufts University.

Hartleman, P., and P. Molenaar. 1999. Detection of developmental transitions. In G. Savelsbergh, H. van der Maas, and P. van Geert, eds., *Non-linear developmental processes*, 169–86. Amsterdam: KNAW.

Hirschfeld, L. A., and S. A. Gelman, eds. 1994. *Mapping the mind: Domain specificity in cognition and culture*. Cambridge: Cambridge University Press.

Horowitz, F. 1987. *Exploring developmental theories: Toward a structural/behavioral model of development*. Hillsdale, NJ: Lawrence Erlbaum Associates.

James, W. 1902. *Varieties of the religious experience*. New York: Longman, Green, and Co.

Karmiloff-Smith, A. K. 1992. *Beyond modularity: A developmental perspective on cognitive science*. Cambridge, MA: MIT Press.

Keil, F. 1990. Constraints on constraints: Surveying the epigenetic landscape. *Cognitive Science* 14: 135–68.

Keller, H. 1903. *The story of my life*. New York: Doubleday and Co.

Kohlberg, L., C. Levine, and A. Hewer. 1983. *Moral stages: A current formulation and a response to critics*. Vol. 10. Basel: S. Karger.

Kuhn, D. E. 1995. Special issue: Development and learning: Reconceptualizing the intersection. *Human Development* 386: 289–382.

Lerner, R. M. 1982. Children and adolescents as producers of their own development. *Developmental Review* 2: 342-70.

———. 1995. The place of learning within the human development system: A developmental contextual perspective. *Human Development* 38: 361–66.

———. 2002. *Concepts and theories of human development*. 3rd ed. Mahwah, NJ: Lawrence Erlbaum Associates.

———. 2007. *The good teen: Rescuing adolescence from the storm and stress years*. New York: Crown Publishers.

Lewis, M. 2000. The promise of dynamic systems approaches for an integrated account of human development. *Child Development* 711: 36–43.

Liben, L., ed. 1987. *Development and learning: Conflict or congruence?* Hillsdale, NJ: Lawrence Erlbaum Associates.

Merton, T. 1971. *The seven-storey mountain*. Garden City, NY: Image Books.

Morelock, M. J., and D. H. Feldman. 1999. Prodigies. In M. Runco and S. Pritzker, eds., *Encyclopedia of creativity*, 1901–10. San Diego: Academic Press.

Oser, F. 1991. The development of religious judgment. In F. Oser and W. G. Scarlett, eds., *Religious development in childhood and adolescence*, 5–26. San Francisco: Jossey-Bass.

Piaget, J. 1962. *Comments on Vygotsky's critical remarks on Language and thought in the child and Judgment and reasoning in the child*. Cambridge, MA: MIT Press.

———. 1964. Development and learning. In R. Ripple and V. Rockastle, eds., *Piaget rediscovered*, 7–19. Ithaca, NY: Cornell University Press.

——— 1970. Piaget's theory. In P. Mussen, ed., *Carmichael's manual of child psychology*, vol. 1: 703–32. New York: Wiley.

———. 1971. The theory of stages in cognitive development. In D. R. Green, M. P. Ford, and G. B. Flamer, eds., *Measurement and Piaget*, 1–11. New York: McGraw-Hill.

———. 1975. *The development of thought: Equilibration of cognitive structures*. New York: Viking.

Roeser, R. 2006. On the study of educational and occupational life-paths in psychology: Commentary on the special issue. *Educational Research and Evaluation* 12: 409–21.

Sawyer, K. 2003. Emergence in creativity and development. In K. Sawyer and V. John-Steiner, eds., *Creativity and development*, 12–60. New York: Oxford University Press.

Scarlett, W. G. 2005. Toward a developmental analysis of religious and spiritual development. In E. C. Roehlkepartain, P. King, L. Wagener, P.L. Benson, eds., *Handbook of spiritual development in childhood and adolescence*, 21–33. Thousand Oaks, CA: Sage Publications.

Shweder, R. A., and R. A. LeVine, eds. 1984. *Culture theory: Essays on mind, self, and emotion*. London: Cambridge University Press.

Silverstein, S. 1991. *Child spirit: Children's experience with God in school*. Santa Fe, NM: Bear and Co.

Simonton, D. K. 1999. Creativity from a historiometric perspective. In R. Sternberg, ed., *Handbook of creativity*, 116–33. New York: Cambridge University Press.

Thelen, E., and L. Smith. 2006. Dynamic systems theories. In W. Damon, series ed., R. M. Lerner, vol. ed., *Theoretical models of human development*, vol. 1, 258–312. Hoboken, NJ: John Wiley and Sons.

Thomas, L. 1974. *The lives of a cell*. Toronto: Bantam Books.

Thorndike, E. L. 1914. *The psychology of learning*. New York: Teachers College Press.

Thurman, R.A.F. 1995. *Essential Tibetan Buddhism*. San Francisco: Harper.

Tomasello, M. 1999. *The cultural origins of human cognition*. Cambridge, MA: Harvard University Press.

van Geert, P. 1992. A dynamic system model of cognitive and language growth. *Psychological Review* 98: 3–53.

Vygotsky, L. 1978. *Mind in society*. Cambridge, MA: Harvard University Press.

Walters, J.D., and H. Gardner. 1986. The crystallizing experience: Discovering an intellectual gift. In R. Sternberg, ed., *Conceptions of giftedness*, 306–31. New York: Cambridge University Press.

Watson, J. D. 1959. *The double helix*. New York: Atheneum Publishers.

Weisberg, R. 1999. Creativity and knowledge: A challenge to theories. In R. Sternberg, ed., *Handbook of creativity*, 226–50. New York: Cambridge University Press.

Wilber, K. 1996. *A brief history of everything*. Boston, MA: Shambhala Publications.

10

Spirituality, "Expanding Circle Morality," and Positive Youth Development

Janice L. Templeton & Jacquelynne S. Eccles

A key spiritual value and hallmark of positive youth development—our responsibility to care for others as well as ourselves—is conspicuously missing from most of the current research on religion and spirituality during adolescence. In this chapter, we attempt to remediate this gap in the literature by theorizing about the relation of adolescents' spirituality to positive aspects of their moral development. Specifically, we examine spirituality in relation to the development of an ethic of care and related actions aimed at making positive contributions to those beyond oneself. From our perspective, spirituality includes not only individuals' personal reflections on what to believe and how to behave but also their most cherished values, beliefs, and life purposes.

CIRCLES OF CARE, SPIRITUALITY, AND POSITIVE DEVELOPMENT

In this chapter, we explore why some young people come to value caring for others beyond the self and the in-groups as a core life purpose during adolescence. Specifically, we examine if religion or spirituality may play a role in the development of such prosocial ideals and the willpower needed to pursue such ideals across development. Does an individual's religion or spirituality serve as a source of meaning and motivation for acts of generosity, or can these serve as hindrance to the expansion of one's circle of empathic concern? To explore these issues, we describe a developmental phenomenon that we call "expanding circle morality" (ECM). ECM refers to the development of individuals' commitment to expanding their circle of caring continually to include greater and greater numbers of sentient beings. We propose that, when individuals develop a widening circle of empathic concern for greater numbers of people, animals, plants, and living things and when this value is turned into concrete actions that contribute to the welfare of others (e.g., Lerner 2004), a key hallmark of both spiritual and positive youth development has been realized. We begin by differentiating notions of spirituality and religiosity and then discuss potential relations between ECM, religiosity, and spiritual development during adolescence.

SPIRITUALITY AND RELIGIOSITY IN ADOLESCENT IDENTITY DEVELOPMENT

Previous research suggests that people differentiate between spirituality and religiosity. When adults were asked if they were spiritual and/or religious, most people endorsed both, and some, approximately 10–12 percent, reported they were spiritual but not religious (Shahabi et al. 2002; Zinnbauer et al. 1997). The same pattern holds true for adolescents, with most youth categorizing themselves as religious, but 10 percent of them endorsing the spiritual-but-not-religious category (Templeton and Eccles 2003). What factors contribute to the distinction that people make between spirituality and religiosity? We propose that a fruitful distinction between these terms can be made if one takes each to index a different kind of identity with particular functions in human development.

Templeton and Eccles (2006) defined religiousness as having a social-identity function and spirituality as having a personal-identity function. Religious identifications were conceptualized as collective identities reflecting membership in a particular religious group with which one shares common characteristics. This kind of identity serves the function of social belonging. In contrast, we defined a spiritual identity as a personal identity in that individuals tend to see their spiritual characteristics as unique to themselves rather than shared with a group. Furthermore, one's spiritual identity need not be associated directly with feelings of belonging to any particular religious group. Based on this distinction between religious identity as a collective identity, and spiritual identity as a personal identity, people could identify themselves as either religious or spiritual or both. For example, being Christian can mean being either religious or spiritual or both, depending upon the function of being Christian for the individual. If the identity function is primarily associated with being a member of the Christian community, then Christianity is serving primarily as a social or religious identity. On the other hand, if the identity functions not as a group membership but as characteristics unique to the individual and if it is based on thoughtful commitment to the tenets of this faith, then the person's identity is spiritual. If the person identified serves both of these functions, then we would say that the person is both religious and spiritual.

Qualitative methodology studies also suggest that people distinguish between religion and spirituality. Mattis (2000) asked African-American women to distinguish between religiousness and spirituality in day-to-day life. The women described spirituality as an "internalization and consistent expression of key values" and religiousness as "an individual's embrace of prescribed beliefs and ritual practices related to God" (Mattis 2000, 8). Religious values and practices were described as one way to achieve spirituality. Similarly, based on the spiritual life stories of thirty women, O'Brien (1996) defined spirituality as an "enduring and integrating core" providing the foundation to answer questions such as "Who am I?" "Where did I come from?" "What is the meaning of my life?" and "Where am I going?" (3).

These findings provide evidence that spirituality and religion mean something different to people, but these differences can be compatible and complementary. Commitment to a religious doctrine can fulfill not only social-belonging needs but also can foster the development of an internal psychological resource that provides strength and guidance in life and that is different from the benefits derived from social connection to a group. The facts that people distinguish between religion and spirituality and that they may endorse one or the other or both suggest it is important to understand these differences in order to capture all the important aspects of religion and spirituality in research on these topics.

Creating this conceptual distinction between religiousness and spirituality allows us to explore the personal psychological functions that spirituality, but not religiousness in the form of group membership, serves for the individual. This functional approach provides not only an investigative lens to stimulate inquiry into a variety of mechanisms to explain the link between these constructs and outcomes; it also circumvents the polarizing nature of definitions of religiousness and spirituality that make one better than the other. In addition, it provides different avenues for inquiry into the relation between religiousness, spirituality, and positive youth development.

SPIRITUALITY AND EXPANDING CIRCLE MORALITY

Now we turn our attention to the role of spirituality in people's lives. We believe that spirituality mirrors the individual's personal reflection about the role of spiritual beliefs in her or his life. Frankfurt (1988) proposed that philosophers have devoted their attention to questions of what to believe (epistemology) and how to behave (ethics) and have neglected the important human concern of what to care about. Extending Frankfurt's idea to spirituality, we put forward the idea that spirituality includes not only a reflection on what to believe and how to behave but that it is ultimately grounded in what we care about— our core beliefs and values—and that research on spirituality falls short when these values are not considered.

Frankfurt used the term *volitional necessity* to describe the choices people make to pursue what they care about because alternative courses are unthinkable regardless of the consequences. He proposed that volitional necessity is rooted in internalized, persistent core values (i.e., what we care about). Thus, unlike the addict struggling to control his addiction, a person constrained by volitional necessity feels he or she cannot take another course of action because of internal values rather than external dependencies. Blasi (2006) termed acting against one's core values *self-betrayal*. He argued that identity structures the will and that to act in a manner inconsistent with one's core identity betrays the self and is, therefore, unthinkable.

ECM Spiritual Beliefs

It is useful to explore core spiritual beliefs that fall into the category of what we care about and that could inspire volitional necessity. In particular, here, we are interested in a critical spiritual value, the responsibility to care for others, central to many religious and spiritual writings. In fact, most, if not all, religions hold in common some version of the "golden rule." The following excerpts are from *Oneness: Great Principles Shared by All Religions* (Moses 2002, 5–7):

"Do unto others as you would have them do unto you, for this is the law and the prophets."   *Christianity*

"What is hurtful to yourself do not to your fellow man. That is the whole of the Torah and the remainder is but commentary."   *Judaism*

"Do unto all men as you would they should unto you, and reject for others what you would reject for yourself."   *Islam*

"Hurt not others with that which pains yourself."   *Buddhism*

"Tzu Kung asked: 'Is there any one principle upon which one's whole life may proceed?' Confucius replied: 'Is not Reciprocity such a principle? What you do not yourself desire, do not put before others.'"   *Confucianism*

"This is the sum of all true righteousness—Treat others, as though wouldst thyself be treated. Do nothing to thy neighbor, which hereafter Thou wouldst not have thy neighbor do to thee."   *Hinduism*

"Treat others as thou wouldst be treated thyself."   *Sikhism*

"A man should wander about treating all creatures as he himself would be treated."   *Jainism*

"Regard your neighbor's loss as your own loss, even as though you were in their place."   *Taoism*

"Ascribe not to any soul that which thou wouldst not have ascribed to thee." *Bahá'í*

   In Christian doctrine, when asked which commandment is the most important, Jesus answered, "Love the Lord your God with all your heart, and with all your soul, and with all your mind, and with all your strength. The second is this, 'You shall love your neighbor as yourself.' There is no other commandment greater than these" (Mark 12:28–31). If we live the value of loving our neighbors (e.g., other countries, other religions, other species) as ourselves, would we be facing the social, political, and environmental issues of today? We believe the answer is no; thus, caring for others is an essential issue that needs to be addressed in the study of spirituality.

   We call this need to care for others *Expanding Circle Morality*. "Expanding Circle" (borrowed from Lecky [1870] and Singer [1981]) Morality (ECM) denotes a dynamic, lifelong, sustainable commitment to expanding our circle of caring continually to include all sentient beings as a core guiding principle in life (Templeton 2007). "Morality," in the sense of reducing harm to other sentient beings, rests on two types of responsibility: (1) our obligation and commitment to care for others and minimize harm and (2) accountability for our actions, both consistent and inconsistent, with our obligation.

ECM and Altruism

ECM is not the same as altruism. It is not "unselfish regard for or devotion to the welfare of others," as defined by the 1997 Merriam–Webster's dictionary, although individuals committed to ECM may exhibit altruistic-like behavior. Altruism, defined in the strict sense of disregard of self in service of others, is not sustainable. For example, if an individual gives all her money to a homeless person, she will not be able to care for herself. Instead, ECM operates as a life compass that encourages consideration of self as well as others in our thoughts, actions, and behaviors, while altruistic acts are byproducts of such an orientation to life. Thus, individuals may share some of their financial resources with a charity committed to helping the homeless while not sacrificing their own financial well-being.

Challenges to Expanding Circle Morality

Although most people endorse the belief that it is important to extend moral concern to others, the parameters of what constitutes "others" is controversial. The challenge is not in being friendly to your friends or loving those who love you or sharing with those close to you: the challenge is in loving, giving to, and caring for those different from you, including those who hate you, those whom you will receive nothing back from, and those who are not human but yet are sentient beings just like us. The magnitude of this challenge is heightened by both developmental and evolutionary influences that pull us toward narrow rather and expanding moral circles.

Developmental Influences

Categorizing the world around us is one of the most fundamental processes creating a boundary for our circle of moral concern. We begin categorizing the world around us at an early age. We categorize objects to understand them, and we categorize people to give meaning to our social environment (Gelman 2003). Children place themselves into gender, race, religion, and other social categories based on assigned or biologically determined characteristics. They also choose social categories based on the cultural opportunities available to them, such as being a soccer player or a ballet dancer. They learn the power of social and group-conformity norms and what it means to be included or excluded from a group (Dishion, McCord, and Poulin 1999). Children categorize themselves based on the concrete categories available in their social worlds, for example, the religion of their parents or primary caregivers. When the self is categorized as an in-group member, in-group bias usually results (Brewer 1991) and brings with it all the associated social issues (e.g., discrimination, prejudice, stereotyping).

Categorization is only an issue when it creates separateness and thus narrows rather than expands our circle of moral concern over the lifespan. According to self-categorization theory (Turner et al. 1989), categorization occurs at three different levels of inclusiveness—personal, social, and human. At the personal level, the self is perceived as different from other in-group members. At the social level, the self is seen as being like other in-group members but different from out-group members. At the human level, the self is like other human beings but different from nonhuman beings.

Evolutionary Influences

In addition to developmental influences, evolutionary mechanisms may also explain the pull toward a narrow moral circle. Henry and Wang (1998) explored the biological basis for self- versus species-preservation orientations to threat or stress responses. The self-preservation (i.e., flight or fight) system likely evolved because it helped species respond to life-or-death survival threats (Dickerson and Kemeny 2004). We believe the self-preservation system encourages self-categorization at the personal level of inclusion and thus promotes a narrow moral circle where only the individual matters.

Species-preservation behaviors (i.e., protective and caregiving behaviors) evolved in mammals giving birth and caring for young. Hormones such as gonadotropins, oxytocin, progesterone, and perhaps vasopressin encourage care for one's own kin (Henry and Wang 1998; Taylor et al. 2000), and hormones such as oxytocin and progesterone have been linked to affiliation and giving in humans (Depue and Morrone-Strupinksy 2005). Thus, our physiological/biological makeup may also contribute to limiting the size of our moral circles. In addition, David Sloan Wilson (Wilson and Csikszentmihalyi 2007) and other group-selection theorists argue that groups are sustained because of selection for group, rather than individual, characteristics.

Each of these evolutionary/biological and developmental systems predisposes us to narrow rather than expanding moral circles: self-preservation to a circle of one; species preservation and group selection to a circle including those genetically related to us or those with characteristics that benefit the survival of one group over other groups, respectively. Thus, inherent in the development of ECM spiritual beliefs is the need to override evolutionary and developmental propensities. This suggests that internalizing ECM as a core spiritual value is not only about personal development but also must include experiences that influence cognitive-developmental processes in ways that promote development of an ECM spiritual identity.

ECM Spiritual Development

We find Kohlberg's stage-like theory of moral development (Kolberg 1984) relevant to our understanding of the development of ECM. For our purposes here, we begin at the conventional morality level where good interpersonal relationships define morality. This morality means living up to the expectations of family and community and displaying "good" behaviors, i.e., those that are shared by one's community. Here, the circle of moral concern expands from self to the community of which one is a member. At Stage 4, the circle expands to include moral concern for society as a whole. Individuals at this stage adopt external scripts endorsed by that society (i.e., obeying laws and respecting authority to maintain the social order). In Stage 4, the boundary of the moral circle expands to include the society in which one perceives membership but is still limited to the members who also adopt the norms of that society. Therefore, endorsing the norms of one's society could also mean endorsing a very narrow moral circle, as demonstrated in Nazi Germany.

The circle of moral concern expands again as people begin thinking about what makes a good society in Stage 5 of postconventional morality. In this stage emerges the belief that a good society benefits all members of that society, even those in different social groups and those with different values, and that the basic rights of life and liberty should be extended to all members of the society via a democratic process. The next shift is toward universal principles, Stage 6, where equality among all members of society is the only acceptable outcome, and this value must be upheld even if the democratic process does not succeed. This stage is much like Frankfurt's concept of volitional necessity (Frankfurt 1988) described earlier, where not extending moral concern to all others is unthinkable. Thus, in Kohlberg's postconventional morality, the boundary of moral concern expands to all humans in a society.

Although Kohlberg's theory of moral development does not extend to non-human life, the stage descriptions are useful in describing changes consistent with ECM spiritual development. Thus, how do we reconcile the fact that neither Kohlberg's universal principle stage of moral development nor Expanding Circle Morality spirituality is predicted by the evolutionary and developmental propensities we described earlier? If we are biologically predisposed to care only for our genetic relations or tribal group, is there hope for expanding our circle of moral concern?

Expanding Our Circle of Moral Concern

We believe the answer is yes, but we must turn our attention to identifying essential characteristics of ECM spirituality as well as the experiences and contexts that influence cognitive-developmental processes to promote the development of an ECM spiritual identity. As we have already suggested, ECM spirituality is more than operating from an external script. For example, Colby and Damon's (1992) moral exemplars fit our description of people guided by ECM spiritual values. They found their moral exemplars lived by moral commitments more fully and personally than could be explained by simply following external rules. They argued the moral certainty of their moral exemplars was not rooted in dogmatism, but rather the exemplars demonstrated a lifelong commitment to learning about the world, finding new ways to act on their commitments, and openness to examining their ideas when challenged even though their core moral values were relatively stable throughout their life. For example, if they devoted their lives to an ideal like the equality of all people, only their approaches to social justice issues like poverty changed—their core values did not vary. In other words, they did not have a dogma that drove how they approached the issue.

If Colby and Damon's moral exemplars did not live their lives by external scripts, what were the characteristics of their developmental contexts that supported ECM spiritual development? Hints of the contextual experiences that may encourage us to expand our moral circle are found in the laboratory as well as in naturalistic settings. We first describe laboratory studies related to ECM spirituality, and then we turn our attention to social contexts that promote the development of ECM spirituality.

Laboratory Studies

Wohl and Branscombe (2005) found that increasing the level of social-category inclusiveness increased forgiveness and decreased collective-guilt assignment to historical and present-day perpetrator groups. Another laboratory approach stems from terror-management theory (Greenberg et al. 1986). Death awareness seems either to narrow people's moral circle actively or, at least, to make the boundaries of their moral circles more salient. Templeton (Templeton and Eccles 2006) used a typical terror-management procedure first to narrow participants' moral circles. The death-awareness manipulation was followed by a guided lovingkindness meditation, a spiritual practice that promotes compassion, to expand participants' circle of moral concern. Participants who were made aware of their own death and who then listened to a guided lovingkindness meditation judged moral transgressors less harshly than those who listened to a progressive-relaxation exercise following death awareness.

Another set of studies related to ECM spirituality revolves around a motivational framework developed by Learning as Leadership, Inc., and described by Crocker, Nuer, Olivier, and Cohen (2006). They define egosystem motivation as the desire to construct, maintain, protect, and enhance positive images of the self. They propose that individuals motivated by egosystem goals place their perceived needs above the needs of others. On the other hand, ecosystem motivation borrows the ecology term, *ecosystem*, as a metaphor to describe a human motivational state in which people see themselves as part of a larger interdependent system. Individuals who perceive themselves as embedded in a larger system consider the effects of their actions on others throughout the system and recognize that their impact on the system can also have repercussions for their own well-being. Compared to egosystem motivation, ecosystem-motivated individuals chose goals larger than themselves that benefit the self as well as others. Initial findings reveal that ecosystem motivations are associated with feelings of closeness and social support, as well as psychological well-being as defined by vitality, engagement in the present moment, self-esteem, and reduced anxiety and stress (Crocker et al. 2006). These laboratory studies provide preliminary evidence that an individual's moral circle can be expanded and that expanding one's circle of moral concern may benefit not only others but the individual as well.

Naturalistic Settings

If we agree that Expanding Circle Morality is critical to individual well-being, as well as social and environmental well-being, then what social and cultural contexts provide experiences that might support the development of an ECM spiritual identity among adolescents? As already suggested, religious institutions are the obvious contexts that can promote ECM values as well as provide experiences that allow youth to reflect on and internalize these values. Nonreligious groups can also provide experiences that promote ECM values. One such group is Learning as Leadership, Inc., mentioned earlier. The mission of the leadership-development organization is to provide tools and practices to help corporations as well as individuals shift from egosystem to ecosystem leadership.

In addition, there currently exists a wide variety of volunteer and service experiences that have the potential to promote ECM values (e.g., the Peace Corps, Habitat for Humanity, food banks, homeless shelters, and numerous other national and community organizations) by encouraging behaviors that contribute to the well-being of others. These programs provide an opportunity structure that promotes other-oriented behavior.

In one national volunteer-service program, the Teen Outreach Program (TOP), adolescents explore and evaluate their future life options and develop life skills and autonomy in a context featuring service to others, along with strong social ties to adult mentors (Philliber and Allen 1992; Allen et al. 1994, 1997, 1990). Although prevention of teen pregnancy and risky sexual activity are the primary goals of the program, neither of these outcomes is an explicit part of the programming. Participant volunteer activities range from working as aides in hospitals and nursing homes, to participating in walkathons, to tutoring peers. Classroom discussions, facilitated by adult mentors, help students learn from their service experiences. In addition to the individual benefits of reduced pregnancy rates, reduced school failure and school suspension were achieved in all evaluations (Philliber and Allen 1992; Allen et al. 1994, 1997, 1990). There are alternate explanations for the positive outcomes associated with this program, yet service to others appears to be a key component. For example, implementation quality of the TOP curriculum did not significantly influence program outcomes (Allen et al. 1997), suggesting that community service and mentoring are the most important aspects of the program. Service-learning programs and volunteer opportunities, sponsored by religious institutions or otherwise, offer naturalistic opportunities to study the link between other-oriented behavior and the development of an ECM spiritual identity.

CONCLUSIONS

We propose that the development of an ECM spiritual identity requires social contextual experiences that promote the merging of cognitive-developmental process with developmental issues related to changes in ECM spiritual identity. In his book examining the religions of the world, Huston Smith writes, "Normally, people will follow the path that rises from the plains of their own civilization; those who circle the mountain, trying to bring others around to their paths, are not climbing. . . . It is possible to climb life's mountain from any side, but when the top is reached the trails converge. At base, in the foothills of theology, ritual, and organizational structure, the religions are distinct. Differences in culture, history, geography and collective temperament all make for diverse starting points" (Smith and Smith 1991).

To use Smith's metaphor, we argue that ECM spirituality is at the mountaintop and that many contexts provide pathways up the mountain. Religious institutions that teach religious doctrine provide a start up the mountain, but doctrine alone is not sufficient. Adults who demonstrate ECM spiritual values come from a variety of religious and spiritual family backgrounds. They may have grown up in families actively engaged in their religion and religious community or in families engaged in community service and civic issues or in families who spent lots of time outdoors and were very active in conservation efforts. Whatever the pathway, identifying and supporting the common threads that contribute to development of individuals who extend moral concern to all sentient beings will make the world a better place for us all.

REFERENCES

Allen, J. P., G. Kuperminc, S. Philliber, and K. Herre. 1994. Programmatic prevention of adolescent problem behaviors: The role of autonomy, relatedness, and volunteer service in the teen outreach program. *American Journal of Community Psychology* 225: 617–38.

Allen, J. P., S. Philliber, S. Herrling, and G. P. Kuperminc. 1997. Preventing teen pregnancy and academic failure: Experimental evaluation of a developmentally based approach. *Child Development* 68, no. 4: 729–42.

Allen, J. P., S. Philliber, and N. Hoggson. 1990. School-based prevention of teen-age pregnancy and school dropout: Process evaluation of the national replication of the teen outreach program. *American Journal of Community Psychology* 18, no. 4: 505–24.

Blasi, A. 2006. Personal communication, April 16.

Brewer, M. B. 1991. The social self: On being the same and different at the same time. *Personality and Social Psychology Bulletin* 17, no. 5: 475–82.

Colby, A., and W. Damon. 1992. *Some do care: Contemporary lives of moral commitment.* New York: Free Press.

Crocker, J., J. Breines, P. Denning, R. K. Luhtanen, and L. Matison. 2006. *From egosystem to ecosystem: Self-oriented and other-oriented goals and well-being.* Unpublished manuscript.

Crocker, J., N. Nuer, M. A. Olivier, and S. Cohen. 2006. *Egosystem and ecosystem: Two motivational orientations for the self.* Unpublished manuscript.

Depue, R. A., and J. V. Morrone-Strupinsky. 2005. A neurobehavioral model of affiliative bonding: Implications for conceptualizing a human trait of affiliation. *Behavioral and Brain Sciences* 28, no. 3: 313–50.

Dickerson, S. S., and M. E. Kemeny. 2004. Acute stressors and cortisol responses: A theoretical integration and synthesis of laboratory research. *Psychological Bulletin* 130, no. 3: 355–91.

Dishion, T. J., J. McCord, and F. Poulin. 1999. When interventions harm: Peer groups and problem behavior. *American Psychologist* 54, no. 9: 755–64.

Frankfurt, H. G. 1988. *The importance of what we care about: Philosophical essays.* Cambridge [England]; New York: Cambridge University Press.

Gelman, S. A. 2003. *The essential child: Origins of essentialism in everyday thought.* Oxford: Oxford University Press.

Greenberg, J., T. T. Pyszczynski, and S. Solomon. 1986. The causes and consequences of a need for self-esteem: A terror management theory. In R. F. Baumeister, ed., *Public self and private self,* 189–212. New York: Springer-Verlag.

Henry, J. P., and S. Wang. 1998. Effects of early stress on adult affiliative behavior. *Psychoneuroendocrinology* 23, no. 8: 863–75.

Kohlberg, L. 1984. *The psychology of moral development: The nature and validity of moral stages.* 1st ed. San Francisco: Harper and Row.

Lecky, W. E. H. 1870. *History of European morals from Augustus to Charlemagne.* New York: D. Appleton and Co.

Lerner, R. M. 2004. *Liberty: Thriving and civic engagement among America's youth*. Thousand Oaks, CA: Sage Publications.

Mattis, J. S. 2000. African American women's definitions of spirituality and religiosity. *Journal of Black Psychology: Special Issue: African American culture and identity: Research directions for the new millennium* 26, no. 1: 2101–22.

Moses, J. 2002. *Oneness: Great principles shared by all religions*. New York: Ballantine Books, 5–7.

O'Brien, M. 1996. Spirituality and older women: Exploring meaning through story telling. *Journal of Religious Gerontology* 10, no. 1: 3–16.

Philliber, S., and J. P. Allen 1992. Life options and community service: Teen outreach program. In B. C. Miller, J. J. Card, R. L. Paikoff, & J. L. Peterson, eds., *Preventing adolescent pregnancy: Model programs and evaluations*, 139–55. Sage Focus Editions 140. Thousand Oaks, CA: Sage Publications.

Shahabi, L., L. H. Powell, M. A. Musick, K. I. Pargament, C. E. Thoresen, D. Williams et al. 2002. Correlates of self-perceptions of spirituality in American adults. *Annals of Behavioral Medicine: Special Issue: Spirituality, Religiousness, and Health: From Research to Clinical Practice* 24, no. 1: 59–68.

Singer, P. 1981. *The expanding circle: Ethics and sociobiology*. New York and Scarborough, Ontario: New American Library, Times Mirror.

Sloan Wilson, D., and M. Csikszentmihalyi. 2007. Health and the ecology of altruism. In S. G. Post, ed., *Altruism and health: Perspectives from empirical research*, 314–31. Oxford: Oxford University Press.

Smith, H., and H. Smith. 1991. *The world's religions*. San Francisco: Harper.

Taylor, S. E., L. C. Klein, B. P. Lewis, T. L. Gruenewald, R. A. R. Gurung, and J. A. Updegraff. 2000. Biobehavioral responses to stress in females: Tend-and-befriend, not fight-or-flight. *Psychological Review* 107, no. 3: 411–29.

Templeton, J. L. 2007. *Expanding circle morality: Believing that all life matters*. Doctoral dissertation, University of Michigan.

Templeton, J. L., and J. S. Eccles. 2003. Understanding spirituality in the lives of adolescents. Paper presented at the biennial meetings of the Society for Research on Child Development, Tampa, FL. April.

———. 2006. The relation between spiritual development and identity processes. In P. L. Benson, E. C. Roehlkepartain, P. E. King, and L. Wagener, eds., *The Handbook of Spiritual Development in Childhood and Adolescence*, 252–65. Thousand Oaks, CA: Sage Publications.

Turner, J. C., M. A. Hogg, P. J. Oakes, S. D. Reicher, and M. S. Wetherell. 1989. A self-categorization theory. *American Journal of Sociology* 94: 1514–16.

Wohl, M. J. A., and N. R. Branscombe. 2005. Forgiveness and collective guilt assignment to historical perpetrator groups depend on level of social category inclusiveness. *Journal of Personality and Social Psychology* 88, no. 2: 288–303

Zinnbauer, B. J., K. I. Pargament, B. Cole, M. S. Rye, E. M. Butter, T. G. Belavich et al. 1997. Religion and spirituality: Unfuzzying the fuzzy. *Journal for the Scientific Study of Religion* 36, no. 4: 549–64.

11

The Role of Spirituality and Religious Faith in Supporting Purpose in Adolescence

Jennifer Menon Mariano & William Damon

The contributors to positive youth development (PYD) include a range of variables that involve external community and societal supports and the individual's internal strengths and personality characteristics. Among these elements, a central indicator of youth thriving is that youngsters are engaged in pursuits that serve the common welfare and are making meaningful contributions to their communities. Thriving youngsters have concerns that go beyond their own self-centered needs and extend outward to the pursuit of goals that benefit the world beyond themselves. Illustrating this point, Lerner and his colleagues (2005) point to the 5 "C's" of adaptive youth development, which refer to the personal capacities of *competence* in domain-specific areas; a sense of personal *confidence* and self-worth; positive *connections* with individuals and institutions; *character*; and *caring and compassion* for others. Out of these five C's, they have proposed that a sixth characteristic emerges: *contribution*. Other frameworks of PYD (see, for example, the National Research Council Report 2002; Connell, Gambone, and Smith 2000; and Benson 1997, 1994) use slightly different terminologies but also point to contribution as an indicator of positive development during the period of youth.

In spite of its apparent importance in positive youth development, however, it is surprising that the intentions of young people to contribute to the world beyond the self, a phenomenon that we refer to as "youth purpose" in this chapter, have been so seldom studied in the empirical literature. Indeed, youth-development theories in general explain a number of factors influencing youth development yet neglect the study of purpose (Damon, Menon, and Bronk 2003). Because purpose appears to be such a central indicator of youth thriving (i.e., it has been connected to a number of prosocial behaviors and negatively associated with some negative ones. See, for example, Sappington and Kelly 1995; Sayles 1994; Erickson 1968; and Butler and Carr 1968), understanding more about youth purpose may help us understand more about PYD in general.

In keeping with the theme of this book, we suggest that one way of understanding the relationship between spiritual development and PYD is by examining the role of religion and spirituality in influencing young people's purposes in life. In our own program of research at the Stanford Center on Adolescence, we have examined youth purpose through a series of studies with youngsters across the United States. In this chapter, we discuss what our findings suggest about the nature of the associations between spirituality and purpose in adolescence. We also discuss what is not revealed by our study: that is, what it leaves out and fails to illumine and, consequently, what still needs to be known about this subject. Finally, we discuss what must be done before a deeper and more extensive understanding of the role of spirituality and youth purpose in particular, and spirituality and PYD in general, can be achieved.

Admittedly, we are not the first to explore the association between spirituality and purpose. Some research studies suggest a positive relationship between purpose and spirituality in adolescence. Studies by Francis and Burton (1994) and Francis and Evans (1996) found that perceived purpose in life tends to increase with church attendance and personal prayer for adolescents. A study by Francis (2000) demonstrated a modest positive relationship between Bible reading and purpose in life for the same age group. Spirituality and religiosity predict higher scores on a well-known measure of purpose, the Purpose in Life (PIL) test (Zeitchik 2000; Molcar and Steumpfig 1988; Paloutzian 1981; Crumbaugh and Maholick 1967), and students with religious beliefs tend to score higher on the Chinese version of that test (the C-PIL; see Damon, Menon, and Bronk 2003, 7).

Characteristics of these and other studies, however, would lead us to believe that there is still much more to be known about the associations between spirituality and youth purpose. The apparent limitations of the one-item measure used by Francis and colleagues ("I feel my life has a sense of purpose," measured on a five-point Likert-type scale ranging from "agree strongly" to "disagree strongly" is apparent, and the PIL test was not designed with adolescents in mind. In our own study, we, therefore, employed multiple methods and crafted the language in a way that would be appropriate for youngsters across the full span of adolescence.

We address two critical questions in this chapter: (1) What are the sources of purpose that inspire young people today, and how do religious faith and spirituality figure as such sources? and (2) What is the role of religious faith and spirituality in supporting purpose in adolescence?

DEFINING PURPOSE AND DISTINGUISHING BETWEEN PURPOSE AND SPIRITUALITY

To start with, one of the apparent challenges plaguing some of the investigations of youth spirituality and purpose is that definitions of the terms *spirituality* and *purpose* differ across studies. This makes understanding the relationship between the two challenging. The difficulty that researchers have had with this issue is evident. Peterson and Seligman's (2004) classification of character strengths illustrates this problem in their description of ways in which theorists have classified personality variables. The authors provide their own interpretation of how purpose corresponds to concepts used in these other theories.

Carol Ryff and her colleagues, for example, propose that purpose in life is one dimension of their model of well-being—a dimension which Peterson and Seligman say corresponds to the character strength of spirituality in their own classification (Peterson and Seligman 2004, 67; Ryff 1989). Michael Cawley (2000) and his colleagues' Virtue Scale includes 140 unique virtue terms, one of which is *resourcefulness*.

Resourcefulness appears along with the bracketed words *purpose* and *perseverance* in Peterson and Seligman's description, and the virtue corresponds to both spirituality and persistence in their understanding (2004, 70). In Kumpfer's (1999) taxonomy of resilience factors, purpose and meaning in life are two among a number of spiritual/motivational factors, which correspond to the strength of spirituality in Peterson and Seligman's categorization (2004, 79). In the Search Institute's developmental assets model of youth development, a sense of purpose is classified under the internal asset of positive identity, which again corresponds to spirituality for Peterson and Seligman (Peterson and Seligman 2004, 83; Leffert et al. 1998). Finally, in Peterson and Seligman's classification of strengths and virtues, purpose is considered a strength of transcendence, along with religiousness and faith.

As for defining *purpose*, most studies conflate the terms *purpose* and *mean-*

ing and do not emphasize the element of *contribution* that is central in many PYD theories. Contribution involves an element of transcendence or going beyond the self, an aspect that is also rarely considered in definitions of *purpose*. In our own study, we define *purpose* as *a stable and generalized intention to accomplish something that is meaningful to the self and engage productively with some aspect of the world beyond the self.*[1] This definition highlights the fact that purpose is a goal of sorts but is far more long term, stable, and far-reaching than short-term goals like "to get to the movie on time." It is a part of one's personal search for meaning but also has an external component, which is "the desire to make a difference in the world" (Damon, Menon, and Bronk, 2003).

DETAILS OF OUR STUDY OF PURPOSE IN ADOLESCENCE

In the following pages, we present and discuss the procedures and results of our study of youth purpose as it pertains to spirituality and religious faith. We consider what further questions need to be addressed and what problems need to be solved in order to understand better the role of spirituality and religious faith in fostering purpose in adolescence. We also discuss the implications of our findings for PYD.

Methods

To understand adolescents' potential sources of purpose, we first administered a survey measure to 444 young people in four regions across the United States, representing urban, suburban, and rural communities. Participants were recruited through schools and colleges and were representative of four age groups: Twenty-nine percent were sixth graders, 22 percent were ninth graders, 27 percent were twelfth graders, and 22 percent were college students between the ages of 20 and 22. Survey participants came from a variety of religious traditions. Fifty percent identified themselves as belonging to some denomination of Christianity. Other religious identifications were with Judaism (2 percent), Islam (1 percent), Buddhism (1 percent), Hinduism (1 percent), and an "Other" category (3 percent). The remaining participants (42 percent) either gave no response to this question or reported no religious affiliation.

In addition to asking questions about participants' religion (if any), the survey asked respondents to indicate on a five-point scale (1=Not at all dedicated; 3=Somewhat dedicated; 5=Very dedicated) their level of dedication to eighteen categories of purpose. A "category" refers to a participant or life area that individuals find important and in which they may be psychologically and actively invested. The prompt was: *We are interested in finding out what types of things you feel are most important to you, based on how much time and energy you commit to them. For the following items, please circle the number that corresponds to how dedicated you are,* and it was followed by a listing of the eighteen categories.

We drew the eighteen survey categories in part from studies of young people's sources of meaning conducted by De Vogler and Ebersole (1983, 1981, 1980) and Showalter and Wagener (2000) and made further modifications with our research team. Participants reported on the following categories: *family, country, personal growth, sports, academic achievement, good health, looking good, arts, making lots of money, lifework, general leadership, romance, political or social issues, happiness, religious faith or spirituality, community service, friends,* and *personal values.* For the current analysis, we were interested in examining participants' dedication to religious faith or spirituality and also looked at the degree of this dedication in comparison to ratings on other categories.

In the second wave of data collection, forty-eight of the participants in the survey sample were administered semistructured interviews. These participants were selected by teachers from a pool of those students who had consented to do the interviews. The interviews ranged from 30 to 50 minutes in length and asked questions about participants' sources of potential purposes; formative experiences around purpose; supports for maintenance of potential purposes; obstacles experienced in the pursuit of potential purposes; and future goals around potential purposes. In addition, when not mentioned spontaneously by participants, individuals were asked, *To what degree does religion, faith, spirituality, or God play in your life, if any?* Probing follow-up questions related to the role of these variables in influencing participants' goals were also given.

In a third part of the study, we conducted extended interviews with twelve young "purpose exemplars." These were young people who had demonstrated an extraordinary and long-term commitment to some cause. Exemplars were between the ages of 12 and 23, were recruited through Internet searches and nominations, and participated in prescreening interviews to ascertain whether they fulfilled four criteria: (1) they were dedicated to some specific cause for two years or more; (2) their rationales for involvement clearly included concerns to contribute to the world beyond themselves; (3) they had a high level of activity around their purpose; and (4) they had concrete future plans around their purpose. Two of the exemplars had purposes of religious faith. Four were dedicated to promoting social causes, including protecting the environment, providing clean drinking water for overseas communities, and curbing gun violence. Two participants were dedicated to arts-related purposes, and two more were dedicated to health promotion and advocacy. The remaining two exemplars were committed to enhancing technology and to politics.

Findings

We drew on the data of these three studies to understand better the influence of spirituality and religion on youth purpose. In asking our first question, *What are the sources of purpose that inspire young people today?* our analyses explored the degree to which adolescents identified spirituality and religious faith as a potential source of purpose in their lives. We examined mean dedication responses to each category of purpose on our survey and found that spirituality and religious faith received relatively low dedication ratings compared to most other categories. Descriptive statistics suggest that it was rated higher than only six of the seventeen other categories, but further examination showed statistically significant differences in ratings compared with four of those six categories. These four categories were *sports, general leadership, community service,* and *political and social issues.* Figure 1 shows mean dedication ratings for each category.

These survey findings suggest that few young people today are choosing purposes of religious faith and spirituality, at least not as central objects worthy of their dedication. Our interviews echo this finding to some degree. When

FIGURE 1. Mean dedication ratings by category.

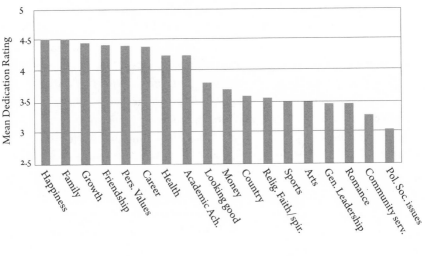

we initially asked individuals to list the things they care about in life and those things that are most important to them, we found that, of our forty-eight interview participants, only four gave replies that directly mentioned spirituality and religion. Nine of our interview participants explicitly told us that they had no interest in spirituality or religion. Five more participants said that they had attended or do attend religious services with their friends or family but did not indicate a personal interest in their religion.

For the remaining participants, however, the interview data told a more complex story not revealed by the survey data. Our findings suggest that many adolescents today are, in fact, thinking about religion and spirituality in profound ways and that, in some cases, spirituality and religion are influencing their purposes in life. One central theme emerged from the interviews that directly addressed our second question, *What is the role of religious faith and spirituality in supporting purpose in adolescence?* This theme had to do with the way that spirituality and religious faith influenced young people's cherished goals in life. Findings suggest that spirituality and religious faith may have very different functions in supporting youth purpose for some adolescents than it does for others. Spirituality has the potential to influence adolescents' purposes in life, but the way in which this occurs differs. Indeed, the relationship between spirituality, purpose, and different aspects of positive youth development may reflect different pathways, or models, for different groups of adolescents. Here, we describe these models.

Model 1: Spirituality Guides Young People toward an Intention to Contribute (Purpose), Which, in Turn, Leads to Contribution. Nineteen interview participants referred to religion's influencing their goals in life, though they spoke about this in different ways. Among the first group of youth, God was regarded as a personal guide, support, and source of encouragement toward achieving their goals in life. One sixth-grader with a longtime dream to become a teacher said, "I feel that, without God, I wouldn't feel the way I feel, like, influenced to be a teacher, because God has been there for me through my hard times and God has helped me a lot in my life so far." During hard times, she has felt that "God's there, and he's telling me, you're going to make it, you're going to achieve your goal." Another sixth-grader felt that it was God who inspired him with the desire to help people in some way—although he had not yet settled on a specific way that he would accomplish this in the future.

For young people like this, faith acts like a beacon, guiding them around obstacles that might come between them and their cherished goals. For a third participant, a 20-year-old college student who spoke about wanting to give back to her community and address minority-injustice issues through her future scholarship, the spiritual teachings imparted through the religion she was brought up in but in which she was not active was seen as a major influence. Another college student felt that his religious involvement "gives him a direction to go in."

An important point to mention here is that young people's goals may not always constitute inspiring purposes. As we noted before, purposes are long-term goals to contribute to the world beyond the self. In the case of many of the participants we describe here and throughout the paper, however, their goals were already compelling purposes or had many of the characteristics of purpose. They are, therefore, highly relevant to the theme of this chapter. For the sixth-grader first mentioned, for instance, the desire to become a teacher was motivated by her wish to help others in her community learn what she already had; and, for the second sixth-grader, the same motivation was apparent. The first college student whom we spoke with had not yet set out on her graduate career but was contemplating how she might make a social contribution through it. The second college student, who mentioned that religion gave him a direction to go in, expressed concerns to serve his family and community. He hoped to realize these goals by becoming a good history teacher, father, and community member in the small, southern, rural town where he lived.

We cannot help but observe a similarity between the function of spirituality for this group of youngsters and the function that has been attributed to purpose by other researchers. In the same way that spirituality and religion function as guides toward these youngsters' purposes, purpose has been seen to act as a guide toward other aspects of positive youth development in the literature. Researchers like Erikson (1968) and Marcia (1980) refer to a sense of drift experienced by youngsters and adults who fail to find compelling purposes worthy of their dedication in adolescence, and this drift can lead to personal and social pathologies (Damon, Menon, and Bronk 2003, 120). Frankl (1959) referred to "noogenic" neuroses that could develop in the absence of purpose and meaning. For these and other theorists, purpose acts as a compass guiding individuals toward the positive behaviors and positive personal characteristics that result in contribution, while directing one away from deviant and destructive behaviors (Damon 1995).

The role of spirituality in supporting purpose in the case of this first group

FIGURE 2. Depiction of Model 1, in which spirituality guides young people toward an intention to contribute (purpose), which, in turn, leads to contribution (positive development).

Spirituality/   Purpose   Contribution
Religious Faith   (intention to contribute)   (positive development)

is illustrated in Figure 2. The arrows in this diagram represent the active role of guiding and directing the young person, so that spirituality/religious faith guides young people toward their purposes, which, by definition, are an intention to contribute to something beyond themselves. Purpose, in turn, guides the young people toward actively making that contribution, which is one element of positive development. It may be equally possible for the direction of this guidance to occur in reverse order, but, for our purposes, the direction noted in this model most accurately represents the phenomena found in the interviews.

Model 2: Spirituality Invests Young People's Personal Goals with Value and Meaning, Which, in Turn, Contributes to These Goals Becoming Inspiring Purposes. A slightly different function was attributed to spirituality by a second group of youngsters. For this group, religious faith or spirituality provided a reason for living or for pursuing their personal goals. One 20-year-old who also had a passionate desire to teach felt that her faith was connected to this goal because "it's what I believe I'm supposed to be doing." Another twelfth-grader said that she did not follow an organized religion but noted that spirituality was something that was very important to her. She saw spirituality as playing a central role in helping people to "make peace" with themselves, so that they have a reason for living. Participants like these two girls made a direct connection between spirituality and the purpose of their lives: the first in having found her life work and calling in life and the second in a more general sense of finding a reason for being.

For this second group, therefore, the role of spirituality was to infuse the young person's chosen goals with a sense of value, worth, and meaning. In our definition, we noted that purpose, while being more specific and intentional than meaning, can certainly be a part of one's search for meaning. In the experience of these youngsters, spirituality is the ingredient that infuses their goals with meaning and, therefore, contributes to making these goals more purposeful in nature. This role of spirituality was acknowledged by other young people

FIGURE 3. Depiction of Model 2, in which spirituality invests young people's personal goals with value and meaning, which, in turn, contributes to these goals becoming inspiring purposes.

Spirituality/ ————————▶ Meaning and Value ————————▶ Personal Goals = Purpose
Religious Faith

in our sample who, though they were not religious, viewed faith as a way of establishing a sense of meaning and coherence in their lives. One college student said that she sometimes "feels the need for a framework" because she has many existential questions and felt that religion could provide this support for her. Another college student said that faith helps you know that things in life "are happening for a reason." Figure 3 shows how spirituality/religious faith infuses goals with a sense of meaning, which, in turn, makes these goals more likely to become compelling purposes for youngsters. This diagram leaves out the myriad other possible variables that may contribute to purpose in adolescence.

Model 3: Spirituality Supports Young People's Intentions to Develop Character (Moral Purpose), Which, in Turn, Supports Character Development. A third group of young people associated faith with being a good person. The goal of character development is purposeful in nature because it requires individuals to transcend their lower selves in the aim of becoming better people. Implicit in developing one's character is the desire to become someone who can make more valuable contributions to society. Though not all of the participants who talked about this theme claimed that moral goals were central in their lives and though not all of those who spoke about it were religious or spiritually oriented, those who did mention this theme believed that religious faith or spirituality were closely related to character-development goals. One male ninth-grader told us that, in relation to his personal conduct, he did not feel that he could "do anything without thinking about God," because he always felt that God was watching him. Another boy of the same age said, "God makes me a better person because everything I do I think what would he think." A male college student told us that, before coming home from the army and starting to attend church more, he felt he was "going downhill" from a moral point of view. A sixth-grader said that he admired his father and uncle because they were "good church-going people," and another ninth-grader said that she was not religious but believed strongly in the values of filial piety, which came from her religion. One ninth-grade girl talked about her annoyance with organized religion but admired the values of

FIGURE 4. Depiction of Model 3, in which spirituality supports young people's intentions to develop character (moral purpose), which, in turn, supports character development.

Spirituality/   Moral Purposes   Character Development
Religious Faith ⟶ (intention to develop character) ⟶ (positive development)

Christianity that had to do with personal conduct, and a twelfth-grader said he was not religious himself but thought religion was a good way of "keeping your morals straight." A single mother of college age said that one of her purposes in life was to help her son grow into a good man—a goal that she associated strongly with taking him to church. "We need to be spiritual," remarked another twelfth-grade girl, "so that we can love."

Figure 4 depicts a model showing the relationship between spirituality and religious faith, the moral purposes of character development, and the actual achievement of character development described by these interview participants. As we note above, "character" is considered one of the five C's of positive youth development by Lerner et al. (2005), and we, therefore, view it as an aspect of positive development.

Model 4: Involvement in a Religious Community Provides the Young Person with a Community of Shared Purpose, Which, in Turn, Reinforces His or Her Own Purposes. A fourth model suggests that the influence of spirituality on one's goals is directed by involvement in a religious community. For one twelfth-grader, for example, going to church indirectly supported one of her already cherished personal goals, which was to become a successful architect and businesswoman. Specifically, in observing the economic disparity apparent between the two church communities she attended, she became even more inspired to be financially successful so that she could make a contribution to alleviating this disparity. Furthermore, while this adolescent had previously been socializing with a group of young people involved in risky behaviors, she now had a new group of friends who shared and supported her goals. As a member of a new community who shared her goals, she was able to shore up her energies to pursue her purpose. According to her, "Church plays a big role" in her life because it helps the community, which is a shared interest. As compared to her former social group, "People at church all have the same goals."

This finding suggests a model in which spirituality is defined as a socially experienced phenomenon, taking place in a religious community. That com-

FIGURE 5. Depiction of Model 4, in which involvement in a religious community provides the young person with a community of shared purpose, which, in turn, reinforces his or her own purposes.

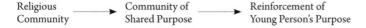

Religious Community ⟶ Community of Shared Purpose ⟶ Reinforcement of Young Person's Purpose

munity is also a community with shared purpose. Figure 5 depicts this relationship. Specifically, involvement in a religious community provides a community of purpose for the young person, and this, in turn, provides reinforcement of the young person's purposes.

Studying Young Purpose Exemplars[2]

Up to this point, we have articulated four models depicting the relationship of spirituality and religious faith to youth purpose and to other aspects of positive youth development. These relationships were apparent in a group of young people who varied in their commitment to a purpose or purposes in life. But how about young people who are highly purposeful? Do spirituality and religious faith play different roles in supporting the purposes of these exemplary youngsters? To determine this, we examined interviews with our twelve purpose exemplars.

Many of our exemplars looked very much like our interview participants when they discussed spirituality and religious faith in relation to their goals in life. For three of them, spirituality and religion were not important at all, were not that important for the time being, or were somewhat important but not directly connected to their goals. Like a number of our interview participants, this group made no spontaneous mention of spirituality or religious faith when asked about them, or they showed little interest in spiritual or religious matters.

Two more of the exemplars did not show a strong personal interest in religion, but the influence of spirituality or religion was apparent in their lives to some degree. Paul Beale, a 17-year-old creator and promoter of jazz music, was not religious but said that he took some of his musical ideas from religion and culture. According to 12-year-old Ricky Holmes, who had been raising money to build wells for communities in poor countries since he was 6 years old, God had given each person free will, and Ricky didn't see any reason not to use it to contribute something good.

The interviews of three more exemplars, Susan Susman, Gabriel Dawson, and Justin Polk, showed very different themes. In the case of 21-year-old Susan

Susman, a negative relationship between religious community and her purpose was demonstrated. She described herself as a spiritual person, but it was
apparent that certain medically related restrictions adopted by her religious
community actually did not support her passion to promote adolescent health.
Twenty-two-year-old Gabriel Dawson, who was highly dedicated to politics,
spoke about wanting to find a religion in which he could have a strong belief
because he admired others who were very sure-footed in that domain. He also
noted, however, that he was wary of individuals who are so religiously committed that they become dogmatic and single-minded to the extreme. Justin Polk,
a jazz musician, was not interested in religion but referred to having a sense of
faith that events relating to his work and to his life in general were happening
for a reason.

Our remaining four exemplars' interviews were consonant with the four
relational models proposed by the forty-eight interview participants. One
exemplar, 16-year-old Bonnie Blake, spoke about religious faith in a way similar to our second group of interview participants. For this group, depicted by
Model 2, the role of spirituality and religious faith in supporting purpose was
to infuse the young person's goals with meaning, value, and worth. An environmental activist who had spearheaded a nationally recognized oil-refining
program in her rural Texas town, Bonnie felt that her faith was instrumental in
providing a reason for her to protect the environment. She told us:

What I believe is that God created this, and God created this for us to take care of it. . . .
So I guess that's one of the big parts of why I'm so passionate about what I do. Because
this is given to me, for me to take care of, so I need to do my part for it to be there later
for somebody else.

Bonnie Blake was also similar to the third group of interview participants,
represented by Model 3, for whom spirituality and religious faith were associated with the moral purpose of character development. She spoke at length
about the virtues she struggled to gain in her life, including humility, obedience, responsibility, and being open and honest with others. For Bonnie, character was associated with the fact that "I base my life upon being a Christian."
Eighteen-year-old Nancy Valens, who was involved in fundraising for cancer
research, was not interested in religion but said that her belief in being a good
person and having a social conscience came from the religion she was brought
up in, therefore also supporting Model 3.

We observed a phenomenon in two of our religiously oriented exemplars, however, that were not present in any of the forty-eight interview participants or the other ten exemplars to the same degree. The experiences of 19-year-old Maria Lopez, a committed Christian, and Alex Pinski, a dedicated member of the Bahá'í Faith, both exemplified all four of our relational models of spirituality/religious faith and purpose.

For Alex, the intention to be of service, to his faith and, in a more general way, to humanity, was suggested and reinforced by his religious beliefs. His intention had led to his contributing to many youth-service endeavors. In Maria's case, the personal goal of becoming a physician and using this skill as a service was reinforced by a religious experience during her Quinceañera in which a church elder made a prophesy that she would one day serve as a missionary in Latin America. This example reflects Model 1, in which spirituality/religious faith supports and inspires the young person's purpose. Though it is unclear whether Maria will go on to make a contribution in this way, it certainly seems she is setting herself up to do so. When asked why she wants to become a doctor, Maria said:

Because . . . it was prophesied over me when I was younger. . . . They were praying over me, and they felt God told them something about my life, in the future. So they were saying that God has called me to Latin American nations. I desired to do it on my own. It's not like I would just do it because you're obligated to do it just because they say it. They say it to confirm what's already a desire in your heart.

Both Alex and Maria exemplify Model 2, where spirituality and religious faith are associated with character-development goals and where these goals, in turn, support character development. Maria's goal of honoring and obeying her parents, for example, even under very trying circumstances, was reinforced by her religious beliefs and related to her religious walk. She had clearly achieved some of her character-development goals, as was evidenced by the admiration she received from people in her church because of the person she is. Alex said that "to be a Bahá'í is to be the embodiment of all human virtues" and to serve his faith meant "to be a friend to all humanity . . . to be loving, to be kind, to be courteous." Furthermore, Alex viewed the assistance of God as a necessary aid to his character development, and both exemplars mentioned their love of God as a primary motivator in their obedience to religious laws of conduct.

Model 3, in which spirituality infuses personal life goals with meaning and value, was shown in Maria's life. An example of this was her choice to obey her parents' wishes in spite of the fact that they unreasonably restricted her social life. While many of her friends at church expected her to rebel under such conditions, she purposefully chose to obey her parents. Her faith gave meaning to that experience so that she grew to see the experience as a valuable part of her spiritual growth. Alex was unsure what career he might pursue but said that "the reason I exist is because of God" and believed that, whatever one does in life, it was important to incorporate God into it so as to "live spiritually enforced."

Model 4, in which a religious community becomes a community of purpose in which the young person's purposes are reinforced, was also clear for both participants. Maria demonstrated long-term activity in her church as a part of her "worship team" and in various other capacities. For Alex, who was doing a full "year of service" at a Bahá'í school at the time of the interview and had been involved in various other service and projects with the Bahá'í community over a period of some years, he was mixing with other young people his age who shared his religious commitment and worldview. Much like the experiences of the twelfth-grader whose religious community had supported her desire to become a successful professional, for Alex, involvement in his religious community provided a positive reinforcement of his faith commitment and was in opposition to the risky behaviors evinced by his former group of high school friends.

Model 5: Spirituality/Religious Faith, Moral Goals, and Personal Goals Are Inseparable and Form a Coherent Purpose in the Young Person's Life. In addition to these themes, we observed a strong coherence between Maria's and Alex's spiritual/religious goals, moral goals, and personal goals that was unlike any other interview participant. These goals were one and the same, though religious faith played the central motivating and directing role. When Maria spoke about her career goals and her goals to have a family of her own one day, it was always in the context of discussing her religious commitments. Maria hoped to marry a dedicated Christian. Alex envisioned having a Bahá'í family. Alex was unsure what career he would pursue but saw any career as being in service of his faith. Both participants felt that the realization of their own personal goals was up to the will of God; they, therefore, remained open to the fact that their goals in life might change or be altered accordingly. The cases of Maria and Alex, thus, suggest a fifth model of the relationship between spirituality and purpose (see Figure 6).

FIGURE 6. Depiction of Model 5, in which Spirituality/Religious Faith, moral goals, and personal goals are inseparable and form a coherent purpose in the young person's life.

CONCLUSIONS

Our findings suggest that, though relatively few young people are specifically choosing purposes of religion and spirituality, those who do find that spirituality and faith support their purposes in a number of ways. In this chapter, we described five models depicting the multiple support systems that are suggested by young people's experiences.

It was clear from these models that spirituality can play very different roles in supporting youngsters' purposes. Yet, a number of questions and issues remain unresolved, and these have implications for the study of spiritual development and positive youth development.

First, our survey raises questions about the importance of ascertaining not only the presence of young people's religious and spiritual commitment but also the nature and depth of that commitment. These issues are important for studying both spirituality and purpose, as both are concerned with issues of commitment. Our survey was not designed to tap the depth of young people's commitment to specific sources of purpose, merely the presence of that commitment.

An ongoing interest in the research literature regarding different ways of being religious illustrates this point. Allport and Ross (1967), for instance, made a distinction between mature and immature faith by showing how religion could be used in different ways. According to them, "intrinsic religion" is a devout, genuine, and heartfelt sense of faith, while "extrinsic religion" is more concerned with the use of religion as a means to another end, such as participation in a religious community, for social status, or for one's own personal aims. A more recent update to this formulation has been presented by Bateson, Schoenrade, and Ventis (1993), who added a third way of being religious. Their "Quest" concept implies incorporating the characteristics of complexity, certainty in the face of doubt, and a tentativeness reminiscent of humility into one's approach to religion, which is not captured by Allport and Ross's measures. Individuals with this orientation experience an "open-ended, responsive dialogue with existential questions raised by the contradictions and tragedies of life" (Bateson, Schoenrade, and Ventis 1993, 169). Bateson and his colleagues also note that a desire to gain social approval should not be viewed as being completely at odds with a religion-as-end orientation, for social influence may, in fact, be important in developing that orientation.

We find that the questions these scholars have asked about religiosity and its healthy expression are very much the same as those we must ask about purpose and PYD. One may ask, for example, *How does religion enhance or inhibit PYD?* A tentative answer may look toward the ways that young people engage with religion. In this paper, we have viewed purpose as an indicator of PYD. Though we have delineated some ways in which having a purpose may support young people's positive development, a next step in this investigation will require a deeper understanding of the ways that young people are purposeful. Purposeful young people may indeed be contributing to something beyond themselves, but the way that they make that contribution—whether it is for self-serving reasons and social approval or that contribution is an end in itself—may be important for their development. Moreover, whether that contribution is done in the spirit of tentativeness and humility could have implications for their own positive development and the development of those around them.

Furthermore, we expect that an end orientation should align more closely with purpose than a means one because it is more akin to the aspect of contribution implicit in our definition of *purpose*. However, we also cannot ignore the potential importance of the role of social influence in the development of purpose. Our survey measure neither illumined the religious orientations of our survey respondents nor revealed the ways in which they are purposeful. Future survey measures of religiosity and purpose should leave ample opportunity for exploring this and may best be used in addition to interviews and informant reports.

A second set of questions was raised by three of our exemplars whose experience with spirituality and religious faith did not fit any of our five relational models. Susan Susman had not found her religious doctrine to be supportive of her concerns for adolescent health. There may be a myriad of other ways in which religious communities and worldviews discourage young people from pursuing their purposes, depending on what those purposes are. Any study of spirituality and purpose will certainly want to investigate such cases. Such an investigation could help us understand the conditions under which youth purpose is either fostered or discouraged.

Gabriel Dawson, though seeking confirmation in a faith of his own, expressed distaste for closed-minded and dogmatic orientations to religious faith. Investigations of spirituality and purpose will do well to make distinctions between adaptive and maladaptive orientations to religious purposes. Given recent concerns about young people's commitment to religious causes that reflect destructive tendencies, such as dedication to religiously inspired terrorist groups, this kind of investigation is both important and timely.

Both Susan Susman and Justin Polk separated faith and spirituality from religion. Susan Susman said she was spiritual but not religious, and Justin Polk referred to having a sort of faith that things are meant to happen the way they do. In this chapter, we addressed the issue of defining *purpose*, but finding a workable definition of terms such as *spirituality*, *religiosity*, and *faith* would be useful for future studies of religion and purpose. We perceive a need to distinguish between different aspects of spirituality and religious faith, for instance, such as participation in a religious community, internal and external religiosity, and other personal aspects of the spiritual experience. Distinguishing these terms from other psychological constructs used to describe the adolescent's life experiences could certainly contribute to establishing the legitimacy of spiritual development as a domain of developmental science.

Clearly, the models described in this chapter need to be confirmed and investigated by other research studies and by theory. The directionality of the support suggested in each model, for instance, is under question and could be studied. Also, we have yet to explore the developmental elements of spirituality as it supports and influences youngsters' purposes. Age differences in how interview participants demonstrated each of the models was not apparent in this study, nor did survey findings reveal a coherent pattern of ratings of dedication to categories by age.

This tentativeness regarding understandings about how spirituality and purpose develop together across adolescence was apparent in the comments made by some participants. The female college student who felt it was important to take her son to church told us that she was waiting to see just how committed she was going to remain to her religion in the future. Bonnie Blake talked about her formal religious involvement as something that "just changes and goes back and forth" throughout life. A relevant problem for the study of spiritual development and the development of purpose is, therefore, understanding how the two converge in individuals' lives and throughout one's youth. We note that, while purpose and spirituality may develop side by side for some individuals (for exemplars Alex and Maria, purpose and spirituality were one and the same thing, for example), our other participants show that this is not always case. As is shown in the diversity of ways that purpose and spirituality converge in the lives of our study participants, there will be diverse paths to positive youth development via the association between purpose and spirituality during this period of life.

Finally, an issue that is not addressed here, and which is highly relevant for the study of positive youth development in general, is under what circumstances an involvement in the spiritual life actually leads to contribution. The nature of some spiritual practices when not balanced with ones of a social nature could incline young people to withdraw from the social world. How do we encourage young people to be oriented to spiritual and religious practices in a way that is beneficial for both themselves and those around them? This question, and others discussed in this chapter, needs to be addressed before we can determine with greater accuracy how spiritual development functions to promote or diminish purpose as an aspect of positive youth development.

ACKNOWLEDGMENTS

We are grateful to the John Templeton Foundation for supporting this research. We are also grateful to the many researchers who contributed to the conceptualization and data collection for this project. We would like to thank Kendall Cotton Bronk, Tanya Rose, Mollie Galloway, Matthew Andrews, Matthew Bundick, Norma Arce, and Karen Rathman for their contributions.

NOTES

1. This definition has been evolving through the course of our and our colleagues' work on the subject.

2. In this chapter, we have used pseudonyms rather than real names when we discuss the twelve youth purpose exemplars from our study.

REFERENCES

Allport, G. W., and J. M. Ross. 1967. Personal religious orientation and prejudice. *Journal of Personality and Social Psychology* 5: 432–43.

Bateson, C. D., P. Schoenrade, and W. L. Ventis. 1993. *Religion and the individual: A social-psychological perspective.* New York: Oxford University Press.

Benson, P. L. 1994. *All kids are our kids: What communities must do to raise caring and responsible children and adolescents.* San Francisco: Jossey-Bass.

Butler, A. C., and L. Carr. 1968. Purpose in life through social action. *Journal of Social Psychology* 70: 243–50.

Cawley, M. J., J. E. Martin, and J. A. Johnson. 2000. A virtues approach to personality. *Personality and Individual Differences* 28: 997–1013.

Connell, J. P., M. A. Gambone, and T. J. Smith. 2000. Youth development in community settings. In *Youth development: Issues, challenges, and directions,* 281–324. Philadelphia: Public/Private Ventures.

Crumbaugh, J. C. 1968. Cross-validation of purpose-in-life test based on Frankl's concepts. *Journal of Individual Psychology* 24: 74–81.

Crumbaugh, J. C., and R. Henrion. 1988. The PIL test: Administration, interpretation, uses, theory and critique. *International Forum for Logotherapy: Journal of Search for Meaning* 11, no. 2: 76–88.

Crumbaugh, J. C., and L. T. Maholick. 1967. An experimental study in existentialism: The psychometric approach to Frankl's concept of noogenic neurosis. In V. E. Frankl, ed., *Psychotherapy and existentialism,* 183–97. New York: Washington Square Press, Inc.

Damon, W. 1995. *Greater expectations: Overcoming the culture of indulgence in our homes and schools.* New York: Free Press.

Damon, W., J. Menon, and K. C. Bronk. 2003. The development of purpose during adolescence. *Applied Developmental Science* 7, no. 3: 119–28.

De Vogler, K. L., and P. Ebersole. 1980. Categorization of college students' meaning in life. *Psychological Reports* 46: 387–90.

———. 1981. Adults' meaning in life. *Psychological Reports* 49: 87–90.

———. 1983. Young adolescents' meaning in life. *Psychological Reports* 52: 427–31.

Dufton, B. D., and D. Perlman. 1986. The association between religiosity and the purpose-in-life test: Does it reflect purpose or satisfaction? *Journal of Psychology and Theology* 14, no. 1: 42–48.

Erikson, E. H. 1968. *Identity: Youth and crisis.* New York: Norton.

Francis, L. J. 2000. The relationship between Bible reading and purpose in life among 13–15-year-olds. *Mental Health, Religion and Culture* 3, no. 1: 27–36.

Francis, L. J., and L. Burton. 1994. The influence of personal prayer on purpose in life among Catholic adolescents. *The Journal of Beliefs and Values* 15: 2, 6–9.

Francis, L. J., and T. E. Evans. 1996. The relationship between personal prayer and purpose in life among churchgoing and non-churchgoing 12–15 year olds in the UK. *Religious Education* 91, no. 1: 9–21.

Frankl, V. E. 1959. *Man's search for meaning: An introduction to logotherapy.* Boston: Beacon.

French, S., and S. Joseph. 1999. Religiosity and its association with happiness, purpose in life, and self-actualization. *Mental Health, Religion and Culture* 2, no. 2: 117–20.

Kumpfer, K. L. 1999. Factors and processes contributing to resilience: The resilience framework. In M. D. Glantz and J. L. Johnson, eds., *Resilience and development: Positive life adaptations,* 179–224. New York: Kluwer/Plenum.

Leffert, N., P. L. Benson, P. C. Scales, A. R. Sharma, D. R. Drake, and D. A. Blythe. 1998. Developmental assets: Measurement and prediction of risk behaviors among adolescents. *Applied Developmental Science* 2: 209–30.

Lerner, R., J. V. Lerner, J. B. Almerigi, and C. Theokas. 2005. Positive youth development, participation in community youth development programs, and community contributions of 5th grade adolescents: Findings from the first wave of the 4-H Study of Positive Youth Development. *Journal of Early Adolescence,* 25, no. 1: 17–71.

Marcia, J. E. 1980. Identity in adolescence. In J. Adelson, ed., *Handbook of Adolescent Psychology*, 159–87. New York: Wiley.

Molcar, C. C., and D. W. Stuempfig. 1988. Effects of world view on purpose in life. *Journal of Psychology* 122, no. 4: 365–71.

National Research Council. 2002. *Community programs to promote youth development.* Washington, D.C: National Academy Press.

Paloutzian, R. F. 1981. Purpose in life and value changes following conversion. *Journal of Personality and Social Psychology* 41, no. 6: 1153–60.

Peterson, C., and M. E. P. Seligman. 2004. *Character strengths and virtues: A handbook and classification.* New York: Oxford University Press.

Ryff, C. D. 1989. Scales of psychological well-being. University of Wisconsin Institute on Aging. *Journal of Personality and Social Psychology* 57: 1069–81.

Sappington, A. A., and P. J. Kelly. 1995. Self perceived anger problems in college students. *International Forum for Logotherapy* 18, no. 2: 74–82.

Sayles, M. L. 1994. Adolescents' purpose in life and engagement in risky behaviors: Differences by gender and ethnicity. Doctoral dissertation. University of North Carolina at Greensboro, 1994. *Dissertation Abstracts International* 55, 09A: 2727.

Showalter, S. M., and L. M. Wagener. 2000. Adolescents' meaning in life: A replication of De Vogler and Ebersole 1983. *Psychological Reports* 87: 115–26.

Zeitchik, G. 2000. The construct validity of the purpose in life test: Quantifying Victor Frankl's "will to meaning." Doctoral dissertation, Adelphi University, 2000. *Dissertation Abstracts International* 61, 09B: 5049.

12

From "Worm Food" to "Infinite Bliss"

Emerging Adults' Views of Life after Death

Jeffrey Jensen Arnett

Religious beliefs serve many positive functions, from psychological security to community solidarity, but perhaps religion's most important function is to take the sting out of death. The evolution in humans of a substantial frontal cortex resulted in a uniquely human existential predicament. The capacity to anticipate the future is a valuable ability for enhancing survival, as it allows us to foresee perils such as food shortages and attacks by enemies, but it also allows each of us to foresee that we will die one day, as have all others before us. Faced with this disagreeable and inevitable prospect, human beings have long looked for a way to avoid its implications of extinction and nullity, and they have developed comforting answers through religious beliefs. All the major Eastern and Western religions include beliefs about some type of life after death, ranging from reincarnation to heaven and hell. In addition, innumerable tribal religions, although fabulously diverse, have in common that they include some type of belief about the continuation of life after death.

In the United States, beliefs about life after death remain strong in the early twenty-first century. According to research by the Gallup organization (Gallup and Castelli 1989), 71 percent of American adults believe in some kind of life after death, and 81 percent believe that "we will all be called before God at Judgment Day to answer for our sins." A national Harris poll (2003) of American adults found that 84 percent believed in "the survival of the soul after death," 82 percent believed in heaven, and 69 percent believed in hell.

Adolescents' afterlife beliefs are similar to adults' beliefs in many respects. The National Study on Youth and Religion (Smith and Denton 2005) found that 49 percent of 13–17-year-olds believe there is "definitely" life after death, with another 37 percent responding "maybe." Thirteen percent reported believing "definitely" in reincarnation, 36 percent "maybe." Seventy-one percent responded that they "believe in a Judgment Day when God will reward some and punish others." There seems to be a contradiction in 71 percent's believing in a Judgment Day and only 49 percent "definitely" believing in any kind of life after death, suggesting that adolescents' afterlife beliefs are still in flux and may not be internally consistent.

Although there has been substantial research on the religious beliefs of adults and adolescents, less is known about the religious beliefs of emerging adults (roughly, age 18 through the mid-20s), including their afterlife beliefs. Research on college students has shown that they are quite religious (Bartlett 2005). Seventy-nine percent believe in God, and 81 percent report attending religious services at least occasionally. However, only 42 percent of college students categorize their religious beliefs as "secure," with the rest describing themselves as "seeking" (23 percent), "conflicted" (15 percent), "doubting" (10 percent), or "not interested" (15 percent). Furthermore, most emerging adults are not college students, so college-student surveys leave many questions about religious beliefs among the majority of emerging adults.

In one of the few studies of religious beliefs among a noncollege sample of emerging adults, Arnett and Jensen (2002) found that this group's beliefs were highly diverse, falling into four roughly equal categories: conservative believers, liberal believers, deists, and agnostics/atheists. Surprisingly, little relation was found between their religious training as children and adolescents and their current beliefs. The data used in the present chapter are taken from the same study as Arnett and Jensen (2002).

To date, few studies have been published on emerging adults' afterlife beliefs, and none with noncollege samples. There are several reasons to anticipate that the afterlife beliefs of emerging adults may be considerably different from the beliefs of adolescents or adults. First, in contrast to adolescents, emerging adults are more cognitively mature and, hence, more likely to think critically about the religious beliefs they have been taught by their parents (Labouvie-Vief 2006; Perry 1970/1999). This capacity for critical thinking may influence how they think about afterlife questions. Second, in contrast to adults, emerging adults have not yet entered family roles as spouse and parent, and taking on those roles tends to lead to higher religious participation (in order to provide religious training for the children), which may have an influence on afterlife beliefs. Third, emerging adulthood is a period when identity explorations are heightened (Arnett 2004, 2006), and one aspect of these explorations may be reconsideration of religious beliefs, including afterlife beliefs. Finally, emerging adulthood is a period of learning to stand alone as a self-sufficient person (Arnett 1998), independent of parents, and part of this process may be reconsideration and perhaps rejection of parents' religious beliefs, including afterlife beliefs. In studies of criteria of adulthood, "decide on own beliefs and values" has consistently ranked near the top in importance (Arnett 1998, 2001, 2003). This suggests that emerging adulthood may be a time of reconsidering afterlife beliefs and establishing beliefs that are one's own rather than those of one's parents.

In this chapter, I present the results of a study that included questions on emerging adults' afterlife beliefs, focusing on qualitative interview responses. This will be followed by some thoughts on the relation between afterlife beliefs and positive youth development.

STUDY BACKGROUND

Data for this study were taken from a larger study of emerging adulthood (Arnett 2004). Data were collected in New Orleans, Los Angeles, San Francisco, and Columbia, Missouri. In Missouri, participants were found through addresses from high school enrollment lists from three to thirteen years prior to the study. At the other sites, potential participants believed to be in the age range for the study were approached in public places and asked to fill out a brief survey. At the bottom of the survey, they could indicate whether they would agree to participate in a more extensive study. They were offered $50 for taking part.

Participants were ages 20–29. They were diverse in ethnicity: 31 African Americans, 33 Asian Americans, 26 Latinos, and 127 whites. Their social-class backgrounds were similarly diverse, as indicated by their mother's educational attainment: 36 percent high-school degree or less, 20 percent some college, and 44 percent college degree. They were also diverse in their own educational attainment: 18 percent high-school degree or less, 44 percent some college, and 39 percent college degree.

The study included questionnaires as well as an extensive interview. The interview included several questions on religious beliefs, including, "What do you think happens to us when we die?" The material in this chapter is based on responses to this question.

FINDINGS

Responses were coded into the following categories: (1) no afterlife; (2) don't know; (3) something, but not clear what; (4) heaven only; (5) heaven and hell; (6) other (mainly reincarnation or return to an energy source). Overall, 11 percent believed in no afterlife, 21 percent were coded "don't know," 15 percent believed there was some kind of existence after death but were unclear as to the nature of it, 15 percent believed in heaven only, 25 percent believed in heaven and hell, and 13 percent were coded as "Other." The responses were consistent across ethnic groups. A chi-square test of afterlife beliefs in relation to ethnicity was not significant. Similarly, chi-square tests showed no relation between afterlife beliefs and educational attainment or socioeconomic status (as represented by mother's education).

To obtain further insights into emerging adults' afterlife beliefs, below I explore their responses for each category.

No Afterlife: "When you die, you die."

Eleven percent of participants did not believe in any kind of afterlife. Often this observation was made with mordant humor. "I think we either turn into ashes when we're cremated or we become worm food," said Andrew (age 22). "We push up daisies," said Loren (age 20). "I don't believe in reincarnation. I don't believe in the heaven or hell at all, and I don't really think that a soul lives on, either." "I think we just become fertilizer," said Tracy (age 24).

Others had a more sober perspective. "I really think there is only one life and that's why you have to make the most of it," said Lindy, age 22. "If there is an afterlife, I'll find out then, but right now I don't believe there's an afterlife. I really think that when you die, you die." Kim (age 26) wistfully imagined heaven but ended up coming down to earth:

I mean, the concept of a heaven is a beautiful concept, in that, if you accept that, at some point you can get in touch with your family and other people and so on. I think that's a beautiful idea, and having a community who's just all for one, a truthful, honest community. That's a great concept, but I don't think I believe it. I mean, I do believe that you're just going to be fertilizer.

Kim's comments are useful for drawing our attention to the fact that the promise of something desirable after death exerts an extremely strong psychological pull, even for many people who believe in no afterlife. Although the existence of life after death is impossible to prove, it is also impossible to refute, leaving considerable space for the imagination to create something alluring out of our desires. Coming down definitely on the side of no afterlife is rare, as we will see in the sections below.

Don't Know: "No one has the right answer."

The "don't-know" responses, 21 percent of the total, fell into four subtypes: confirmed agnostic, uncertain, avoidant, and oblivious. The confirmed agnostics believed that an answer to the question of what happens after death was simply unknowable. For the living, there is no way of answering that question, in their view. "I'm one of those people that, you know, there's got to be some kind of proof," said Keith, age 24, "and if there's no proof, then you can't make a judgment." They had considered the claims and the possibilities and concluded that none of them was valid. Kent, age 28, reflected:

I think that we're all going to die and what happens, nobody knows. I mean, does this electrochemical thing just quit and that's the end of it? Or does it actually go somewhere? Because that's all it is is electricity. So what happens to it? Does it just stop? Is it like a battery? What happens? Nobody knows, because once you die, you can't come back and tell anybody. I mean, sure, you might see this white light and go towards it, and there might be power in that white light. But who knows?

Others in the "don't-know" category grappled with the question of life after death and ended up uncertain, unable to answer it. "It just scares me to think about death, like, where do we really go?" wondered Helen, age 20. "Do we reincarnate or is there a heaven? And then I think, if there is a heaven, how can it hold all of us? There are just so many beliefs that I don't know which one to believe in." Arthur, age 20, found himself similarly baffled: "It's just too confusing to even dwell on it, because you'll never get an answer, right? I mean, you can speculate and speculate, and no one has the right answer. So I don't even bother."

Some in this category were avoidant, i.e., they found the question terrifying to contemplate and so tried to avoid thinking about it. "That's a terrible thing to think about now!" exclaimed Korena, age 23, when asked the question about what happens when we die. "I don't want to think of what happens when I die!" Tammy, age 21, also tried to avoid the topic. "I don't really think about it. It's too morbid for me. I'm too much of an optimist to think about it. I mean, I choose not to think about it."

There were also some who were oblivious, dismissing the question as irrelevant to them in their youthful time of life. "I don't think about dying," said Jeff. "I'm 24. I don't think about that stuff." Frances, age 24, also preferred to focus on the here-and-now: "I don't really give it much thought, because it's not really that important to me," she said. "When I'm gone, I'm gone. I don't really care what happens to me when I'm gone."

In sum, there was a variety of different "Don't Know" responses from the emerging adults, but all of them had in common that they neither believed nor disbelieved in life after death.

Something, But Not Clear What: "I think that we kind of go on."

Fifteen percent of emerging adults believed in some kind of afterlife but were unclear what it might entail. The emerging adults in this category were similar to those in the "don't know" category in that they were uncertain about what lies after death and skeptical that anything could be known for certain. However, unlike those in the "don't know" category, they had at least tentatively decided that there is some kind of life after death, although they remained vague about the nature of it. "I don't think we know," said Jonna, age 24. "I believe that there's something there, but I don't think we can know exactly what it is." "I don't know exactly," began Karen, age 21, but she added, "I definitely don't think that it's just, like, the end. I think I believe in people's spirits like remaining somehow and maybe having a sort of impact on life, like mortal life. I guess I don't want to believe that that's just it, and I don't. So something with like spirits being around. That's pretty much like the vague idea that I have about it."

Often, the emerging adults in this category conceded that their belief in some kind of life after death was motivated by fear that there may be nothing and the wistful hope that there might be something after all. They found the prospect of personal extinction unpalatable and the belief in some kind of continued existence more appealing, but their beliefs were tentative at best. "Selfishly, I think that we kind of go on, and our spirit goes on," said Jennifer, age 22, "just because it's kind of a depressing thought to think that it just ends there. But I don't know. I guess I've always just kind of believed that our spirit continues in some other place." "I mean, no one on earth has any concrete evidence," Brian, age 21, admitted, "but I hope and what I believe is that there is something after. I'm not sure what it is, but I don't think it's that you just don't exist anymore."

Hopeful agnostics, you might call the emerging adults in this category. They were uncertain about what lies beyond death and concerned that there might be nothing after all, but they persisted in the vague belief that there might be some kind of continued and hopefully pleasant existence.

Heaven Only: "Infinite Bliss"

Fifteen percent of the emerging adults believed in heaven only. Some of them sounded close to those in the "something-but-unclear-what" category, in that they sounded uncertain about what to believe, but they ultimately decided that there must be some form of heaven after death. "I don't know," answered 27-year-old Joni at first when asked about life after death, but she continued, "The thing is you get to have it, and you live life in heaven. That's my afterlife. That's what I would want. That's my perception. I would want to make it to heaven. You know, be a part of the riches and all that." "I believe your soul goes to heaven," said Jeff, age 24. "I'm not quite sure where it is or what form it is or how it works, but I believe it's out there somewhere."

Others in this category were more confident that heaven awaits after death. "I think that we go to heaven," said Miriam, age 23. "I think that we are united with God in infinite bliss." Some were confident that heaven awaits not only for themselves but for all. "I think everybody goes to heaven because every-body is God's children," said Alisha, age 22. "And even those criminals who go around killing people are forgiven because I think God forgives everybody. Sometimes it's hard to believe, but I really do think God forgives everybody for what they do, and I think everybody deserves, in some way or another, to go to heaven." Similarly, Marita, age 21, believed in an inclusive heaven. "I believe that there is a heaven, and I think that most people are there, despite what they've done or what they think they've done. I think that there are very few people so evil or so awful to others that they wouldn't be worthy of some type of heaven."

However, some in this category were reluctant to believe that everyone would make it to heaven. They did not believe in hell, but they found the concept of heaven for all problematic in some ways. "That's something that I struggle with," said Stacey, age 23. "I'm not a 'fire and brimstone' Christian, but there are evil people in this world. I'd hate to think that I'd be walking around in heaven someday and run across Adolf Hitler and give him a high-five, you know. It's a difficult question."

Stacey described herself as a Christian, and some of the emerging adults who believed in heaven for all mentioned aspects of Christianity, but more often the belief in heaven only was not phrased in any particular theological language. It was simply a hopeful belief that death would be followed by something good, "infinite bliss" or at least something more pleasant than the struggle of life on earth.

Heaven and Hell: "You're gonna bust hell wide open."

Unlike any other category of afterlife beliefs in this study, the belief in heaven and hell was drawn from a specific creed, the Christian faith. However, even here there were variations on the theme. Some stated standard Christian beliefs in heaven and hell, but others modified the standard beliefs in individualized ways.

Some of those who embraced the standard Christian beliefs in heaven and hell were blunt and direct. As Brenda (age 22) put it, "I believe if you're saved, you're goin' to heaven. If you're not, you're gonna bust hell wide open." Angie (age 23) voiced a similar view. "I think of what's in heaven. They've got mansions in heaven and don't die, don't get sick, don't be sad. [But if you're not a Christian] you're going to hell. You're going to burn forever. People don't realize that they're going to hell, and they're going to be tormented forever." For some of these emerging adults, the firmness of their beliefs in heaven and hell inspired an urgency to try to convert others to the faith. Wynne (age 27) was a Christian, but her parents were not. "That means if we don't save them before they die, then they will end up going to hell, unfortunately. And that's a problem."

Others admitted their belief in heaven and hell more reluctantly and cautiously. They recognized that this belief may be offensive to non-Christians who are deemed to be going to hell. Stuart (age 22) observed, "I believe you'll go to heaven if you're a Christian. If you're not, then you go to hell. But I don't tell that to people, you know, because that's just very unpleasant to hear. That's what Christianity is, though." Chris, age 24, recounted an interaction with a Jewish friend. "He asked me one time, 'Well, according to what you're saying, if I died, I'd go to hell. Do you believe that?' And I had to answer 'Yes.' And it's tough; it really is, because you don't want to believe that, but according to what I've read in the Bible, that's the way I believe it's going to be."

Even for Christians, the destination of heaven or hell was not necessarily tied to faith but rather to whether a person lived a good moral life. "I do believe that you're judged on what you do in your life," said Laura, age 23. "If you treated people badly, I think you're going to get paid back for it some day;

I really do. . . . I think that, if you don't make the peace with him, you're going downtown." Simon (age 25), although Catholic himself, did not believe that being Catholic was an important criterion:

I think it has a lot to do with what you do with your life. You don't have to be a Catholic; you don't have to be x religion. There are different ways to get up to the mountain, and I think every faith is seeking to get to that metaphorical top of the mountain. There's different ways to go about it, but whether you get up there depends on how you live your life. [What if you live a bad life?] I don't think you get there. [Where do you go?] You take the down button! And you go to a bad place.

It is interesting to note that some emerging adults in this category used mordant humor, as Laura and Simon did, with euphemisms such as "going downtown" and "take the down button." This is an element of similarity between them and the emerging adults who believed there is no afterlife. Perhaps in both cases the humor is used to conceal anxiety and discomfort, in the "no-afterlife" emerging adults due to the prospect of extinction and in the heaven/hell emerging adults due to admitting a belief that many of those deemed to be going to hell may find offensive.

Other Beliefs: Reincarnation and Energy Forces

Thirteen percent of emerging adults stated afterlife beliefs that did not fall into any of the previous categories. About half of the responses in this category concerned reincarnation, and most of the rest concerned some idea about an energy force to which the soul returns. When reincarnation was mentioned, it was not in the context of Buddhism or Hinduism, the two major religions that hold reincarnation as their afterlife belief. Rather, it was a vague belief that we return to earth in some form. "I don't know, for some reason I sort of believe you come back again," said Elias (age 20). "Because I have dreams where I see myself, like, being in a place, my first time there, like I've been there before. Like a *déjà vu* sort of thing." Travis (age 23) stated his belief in reincarnation only half-seriously: "I think it's more fun to believe in reincarnation. Like, you can come back as somebody at any point in time in history or in a different world. Who knows? I think it's a lot more fun to believe in that one. I hate to think you just die and that's it." Scott (age 26) at first seemed more certain. "I always thought that there was obviously reincarnation. Your soul, the older it is the more wise it is." But then he added, "Who knows?"

Beliefs in returning to some kind of energy force were diverse. "I believe you just go back to the One," said Tina, age 27. "I think we are just fragments of the light and that at some point you go back to that." Some of these beliefs were drawn from popular culture. "I feel that there is a Star Wars thing, 'the Force,' there's just this planetary aura, that everyone's thoughts and actions and feelings generate this energy," said Charles, age 24. "And when you die, the energy that you are, the nonphysical part of you, is dispersed back into that aura and kind of gets recycled. It becomes part of a million other people that are being created at that time or a little bit later." In general, however, unlike their religious beliefs more generally, emerging adults' afterlife beliefs tended to follow one of the major categories above, rather than being highly individualized.

AFTERLIFE BELIEFS AND POSITIVE YOUTH DEVELOPMENT

The framework of this book is "positive youth development," and, at first glance, afterlife beliefs may not seem to fit very well with this theme. The literature on positive youth development tends to focus on this-worldly topics such as close relationships with parents and the benefits of involvement in youth organizations (Lerner, Bretano, Dowling, and Anderson 2002). Afterlife beliefs seem to be a long way from such topics.

However, afterlife beliefs have definite benefits, at least for those who have some kind of firm faith that something exists beyond this world. Although this was not a specific topic of my interviews with emerging adults, it was evident in many responses, underlining the salience of it. Elaine (age 22) answered the question about afterlife beliefs this way:

When I was a little girl I'd be like "Oh my gosh! They'll put me in a box and put me under the ground and I'll never be able to wake up!" In high school, too, I was afraid of death. It was so scary to me. So I was fearful when I was younger. And then with Christianity, I know that there is more and I know that there is a new life afterwards, so I don't have to worry about that. It's supposed to be a fabulous place. We'll live with God for the rest of our lives. I know it will be great.

For some, the promise of some kind of afterlife was necessary in order to bear the suffering of this world. "I hope there is some kind of life after death," said Mike (age 24). "I mean, if this whole rat race is all it is, then that's going to really suck." Chris had a similar view, describing his afterlife in practical rather than ethereal terms. For him, a wonderful afterlife is something humans need in order to give meaning and purpose to their lives. "I hope most of us go on to heaven," he said. "I think that, if we were just to die and that would be the end of it, just fertilize the ground, . . . there's little hope for what we're doing here, you know, why we're here."

For Dylan, the hope of an afterlife relieved the pain of losing someone he loved. "I hope for a lot of reasons there's a higher place," he said. "I hope my grandfather is there. I mean, certainly he deserves to be there about as much as you could deserve to be there." In future research, it would be good to explore this theme further, to see if there is a relation between afterlife beliefs and experiencing the death of a loved one. It may be that losing a loved one promotes reflection about afterlife beliefs and increases the likelihood of believing in some form of afterlife.

Emerging adults who believed in both heaven and hell saw benefits from their belief on earth as well. They took seriously the peril of going to hell for those who do not believe, and they thought that, if they lived right, they might be able to save others through example. "I don't try to push it because I think you can turn people away like that," said Chris (age 24). "I try to live a life that will make other people go 'Hey, what's he got that I haven't got?'"

In contrast, for those who have no definite belief in an afterlife, the absence of belief was sometimes a source of anxiety and distress. "I've always been really afraid of death," said Russell (age 22). "It just scares me to think, to really stop and ponder the fact that someday I won't exist. . . . If I could believe in something, I would because I think I would be a lot happier person if I believed in an afterlife."

CONCLUSION: AFTERLIFE BELIEFS AND DEVELOPMENT IN EMERGING ADULTHOOD

In sum, emerging adults' beliefs about life after death are diverse, as their religious beliefs are (Arnett and Jensen 2002). Their beliefs span a wide range, from no-afterlife beliefs to certain heaven and hell, with a wide range of variations in between. Nevertheless, their afterlife beliefs are more conventional and less individualized than their overall religious beliefs. Arnett and Jensen (2002), working with a different set of interview questions on this sample, found highly original, "do-it-yourself" religious beliefs among many emerging adults; but, for afterlife beliefs, nearly all held fairly conventional views of either believing in heaven and/or hell or not believing in them.

What are the implications of the findings here for development in emerging adulthood? First, the richness of emerging adults' qualitative interview responses to the afterlife question is notable. Whatever their views about life after death, emerging adults nearly always have something interesting to say in response to the question, and their responses are often rich in insight and irony.

This is in contrast to the responses of adolescents to religious questions. Smith and Denton (2005), who included interviews with hundreds of adolescents as part of their National Study of Youth and Religion (NSYR), concluded that adolescents are "remarkably inarticulate" (27) on religious topics. Emerging adults' afterlife beliefs are often tentative, as they grapple with the enormity of the question; but, even in their uncertainty, they often exhibit a capacity for mature self-reflection.

A second, related implication of the findings here for development in emerging adulthood is that they demonstrate that many emerging adults are engaged in forms of identity exploration (Arnett 2004, 2006; Côté 2006). Erikson (1950, 1968) specified love, work, and ideology as the three pillars of identity formation. The identity explorations of emerging adults in love and work are clear, since most of them change love partners and education/work directions several times from their late-teens to their mid-20s. However, less attention has been given to the identity explorations of emerging adults with respect to ideology, including religious beliefs. From the responses of emerging adults presented here, it is evident that many of them are still in the process of forming their beliefs about life after death. Twenty-one percent of their responses were coded as "don't know," and another 15 percent were in the "something, but not clear what" category, for a total of over one-third whose afterlife beliefs seemed unsettled. In addition, many in the other categories stated their beliefs tentatively and with qualifiers: "I think . . ." "I'm not sure, but . . ." "I don't know, but I guess I believe . . ." The pervasiveness of their uncertainty appears to be much higher than for adults overall. In a 2002 survey by the Pew Research Center, only 3 percent of adult respondents identified themselves as agnostics or atheists (Kohut and Rogers 2002).

Of course, this contrast may be sharpened by the difference between asking people to choose one of several preordained responses, as in the Gallup and other surveys, and asking people open-ended questions about their beliefs, as was done here. Hopefully, this chapter demonstrates the value of a qualitative approach to investigating religious beliefs. There are complexities and subtleties to people's religious beliefs that cannot be captured from surveys but that are necessary for understanding what they believe. Furthermore, interviews allow people to express insights, observations, and even humor that illuminate their beliefs and also connect them to the interviewer and to readers as a fellow human being. For investigations of the question of life after death, a quintessentially human existential question, this connection is crucial to a complete understanding.

REFERENCES

Arnett, J. J. 1998. Learning to stand alone: The contemporary American transition to adulthood in cultural and historical context. *Human Development* 41: 295–315.

———. 2001. Conceptions of the transition to adulthood: Perspectives from adolescence to midlife. *Journal of Adult Development* 8: 133–43.

———. 2003. Conceptions of the transition to adulthood among emerging adults in American ethnic groups. *New Directions in Child and Adolescent Development* 100: 63–75.

———. 2004. *Emerging adulthood: The winding road from the late teens through the twenties*. New York: Oxford University Press.

———. 2006. Emerging adulthood: Understanding the new way of coming of age. In J. J. Arnett and J. L. Tanner, eds., *Emerging adults in America: Coming of age in the 21st century*, 3–20. Washington, D.C.: American Psychological Association.

Arnett, J. J., and L. A. Jensen. 2002. A congregation of one: Individualized religious beliefs among emerging adults. *Journal of Adolescent Research* 17: 451–67.

Bartlett, T. 2005. Most freshmen say religion guides them. *Chronicle of Higher Education* LI (33), A1: A40–41. April 22.

Côté, J. 2006. Emerging adulthood as an institutionalized moratorium: Risks and benefits to identity formation. In J. J. Arnett and J. L. Tanner, eds., *Emerging adults in America: Coming of age in the 21st century*, 85–116. Washington, D.C.: American Psychological Association Press.

Erikson, E. H. 1950. *Childhood and society*. New York: Norton.

———. 1968. *Identity: Youth and crisis*. New York: Norton.

Gallup, G., Jr., and J. Castelli. 1989. *The people's religion: American faith in the '90s*. New York: Macmillan.

Harris poll: The religious beliefs of Americans. 2003. Downloaded Nov. 15, 2006, http://www .findarticles.com/p/articles/mi_m2843/is_4_27/ai_104733222/print.

Kohut, A., and M. Rodgers. 2002. *Americans' struggle with religion's role at home and abroad*. Washington, D.C.: Pew Research Center for the People and the Press.

Labouvie-Vief, G. 2006. Emerging structures of adult thought. In J. J. Arnett and J. L. Tanner, eds., *Coming of age in the 21st century: The lives and contexts of emerging adults*, 235–56. Washington, D.C.: American Psychological Association.

Lerner, R. M., C. Bretano, E. M. Dowling, and P. M. Anderson. 2002. Positive youth development: Thriving as the basis of personhood and civil society. In R. M. Lerner, C. S. Taylor, and A. von Eye, eds., *New directions for youth development: Pathways to positive development among diverse youth*, 11–34. San Francisco: Jossey-Bass.

Perry, W. G. 1970/1999. *Forms of ethical and intellectual development in the college years: A scheme*. San Francisco: Jossey-Bass.

Smith, C., and M. L. Denton. 2005. *Soul searching: The religious and spiritual lives of American teenagers*. New York: Oxford University Press.

PART 4

Social and Cultural Contexts of the Spirituality–PYD Relation

13

Immigrant Civic Engagement and Religion
The Paradoxical Roles of Religious Motives and Organizations

Lene Arnett Jensen

Between 1960 and 2002, the proportion of the U.S. population that was foreign-born more than doubled from 5.4 percent to 11.5 percent. Currently, about 20 percent of children and adolescents in the United States are foreign-born or have a parent who is, and this number is predicted to continue to rise (Portes and Rumbaut 1996, 2001; Suarez-Orozco and Suarez-Orozco 2001). As the number of immigrants in the United States has reached unprecedented levels, public and academic debates have started to address the issue of their commitment to and engagement in the civic life of their new society. Some have argued that immigrants, with their ties to foreign cultures, are unlikely to become engaged in American civil society (Huntington 2004). Recent research, however, has demonstrated variation among immigrant groups, with a number of groups showing substantial civic engagement (Jensen 2008; Lopez and Marcelo 2008; Stepick, Stepick, and Labissiere 2008). A next research question to address, then, is what are the individual motives and institutional contexts linked to immigrant civic engagement. The present chapter addresses this question in regard to religious motives and organizations.

THE ROLE OF RELIGION IN CIVIC ENGAGEMENT

Putnam (2000) argued that, in the United States, religion is a crucial source of civic engagement. He observed that nearly half of all associational memberships in the U.S. are church related, half of all personal philanthropy goes through religious institutions, and half of all volunteering occurs in a religious context. Religion rivals education as a powerful correlate of most forms of civic engagement, and it is an especially strong predictor of volunteering and philanthropy. Putnam suggested that the tie between religion and civic engagement reflects religious values. Based on survey analyses, Putnam also argued that affiliation with religious organizations may be just as important as religious values in explaining volunteerism and philanthropy.

Stepick and Stepick (2002) noted that religious involvement is often important to immigrants to the U.S., including immigrant youth. They also argued that religious involvement may well encourage immigrant civic engagement. Stepick and Stepick, however, emphasized that, while the connection between religion and immigrant civic engagement seems likely to be important, "research on immigrant youth, church and civic engagement is virtually nonexistent" (Stepick and Stepick 2002, 250). Furthermore, there appears to be no research on what the nature might be of such a connection, that is, the relevance of affiliation with religious organizations and of religious values and motives.

Civic Engagement and Positive Outcomes

Before proceeding to the specific nature of the present study, it should be noted that most observers agree that civic engagement in the United States—whether by immigrants or nonimmigrants—is typically positive. From a societal standpoint, Putnam (2000) has demonstrated how civic engagement provides a wealth of formal and informal social contacts and networks. In turn, such social capital is linked to better functioning communities, including schools and neighborhoods. From an individual standpoint, too, Putnam (2000) noted that the breadth and depth of one's civic and social engagements predict life satisfaction and physical health.

Developmental psychologists have also recently pointed to the benefits of civic engagement for youth development (e.g., Flanagan 2004; Sherrod, Flanagan, and Youniss 2002a; Youniss and Yates 1997). Larson (2000), for example, noted that the time youth spend in structured voluntary activities, such as civic ones, is characterized by high levels of both intrinsic motivation and concentration—something that is uncommon in other contexts, including school and peers. Lerner (e.g., 2004; Lerner, Alberts, and Bobek 2007; Lerner, Dowling, and Anderson 2003) has also emphasized the connection between the establishment of civic identity in youth and positive development along a number of dimensions, including a sense of competence. For immigrant youth, in particular, research has shown that civic engagement both affirms existing cultural ties and creates new social networks (Jensen 2008).

The Present Study

The research questions of the present study were four: (1) To what extent are immigrant adolescents and adults civically engaged? (2) To what extent does their engagement occur through religious organizations? (3) To what extent are religious or spiritual motives linked to the presence or absence of engagement? (4) What is the specific nature of their religious or spiritual motives?

Since there is limited research with immigrants on this topic, qualitative interviews were conducted that would tell us about immigrants' own conceptions of their civic engagement. One of the times when qualitative research is particularly helpful is precisely when we need to understand the categories emerging in new situations and the indigenous meanings associated with those categories (such as immigrants' motives for civic engagement). Furthermore, qualitative research is particularly helpful when connections among different phenomena are not well understood (such as the relation between civic engagement and religion) (Fisher et al. 2002; Jessor, Colby, and Shweder 1996).

Since the two main sources of current immigrants to the U.S. are Asia and Latin America, data were collected for one group from each of these two parts of the world, namely, India and El Salvador. In 2001 and 2002, the largest number of immigrants to the U.S. from Asia came from India, and the second largest number of Latino immigrants to the U.S. came from El Salvador (U.S. Census Bureau 2005).

Also, immigrants from India and El Salvador arrive to the U.S. under notably different circumstances, and they arrive with access to markedly different resources. For example, over 60 percent of immigrants (aged 25 years or older) from India report having attained college degrees, whereas the comparable figure for Salvadorans is less than 5 percent (Zhou 1997). The marked difference between the two groups can serve two research purposes. First, it helps to capture a larger possible set of conceptions and connections pertaining to civic engagement and, hence, broaden the present findings. Second, to the extent that there are similarities in the findings from these two otherwise very different groups, such findings are likely to be particularly robust and common to diverse immigrants.

From each cultural group, first-generation adults and second-generation adolescents were included. In order to understand present and future civic engagement, it is important to understand the engagement of not only adults but adolescents (e.g., Flanagan 2004; Flanagan and Faison 2001; Flanagan and Sherrod 1998; Flanagan and Tucker 1999; Jennings 2002; Youniss, McLellan, and Yates 1997; Youniss and Yates 1997).

The present study addressed both political and community engagement. As Sherrod, Flanagan, and Youniss (2002a, 2002b) have argued, for today's youth and culturally diverse populations, research on citizenship needs to pertain not only to political and legal considerations but also to more general involvement with others in the community. Also, research with youth has shown disengagement from formal political activities (Galston 2001) but high engagement in community activities and volunteering (e.g., Flanagan and Faison 2001; Youniss and Yates 1997). Furthermore, emerging research with immigrants notes the importance of considering both community and political involvement (e.g., Jensen and Flanagan 2008).

Method

Participants. The participants were a total of eighty immigrants residing in the Washington, D.C., metropolitan area. The sample consisted of two immigrant groups: Asian Indians (n = 40) and Salvadorans (n = 40). Within each of these two immigrant groups, there were twenty adolescents ages 14–18 years (M = 15.25, SD = 2.86) and twenty parents (M = 43.74, SD = 4.05). (There was either a mother or father for each adolescent.) The parents were first-generation immigrants (i.e., they arrived in the U.S. in their late teens or after). The parents' average age of entry into the United States was 24.54 (SD = 4.61). The adolescents were second-generation immigrants (i.e., they were born in the U.S. or arrived prior to starting elementary school). Their mean age of entry was 2.69 (SD = 2.06). (Different researchers use somewhat varied definitions of immigrant generations. Here, we follow Zhou and Bankston 1998.)

The adolescent groups had even distributions of girls and boys (nine female Asian Indians and twelve female Salvadorans). In the two groups of parents, there was a predominance of mothers (fourteen female Asian Indians and sixteen female Salvadorans). The participation of more mothers than fathers in research on families is common.[1]

As Table 1 shows, most Asian Indian and Salvadoran parents were married. As expected, the two groups of parents differed on a number of demographic characteristics. The Asian Indians had higher levels of education and income

TABLE 1. **Parent Demographics**

	Asian Indian	Salvadoran	Difference
Marital Status (%)			
Never Married	0	5	
Married	100	75	
Separated	0	5	
Divorced	0	10	5.71 ns
Widowed	0	0	
Other	0	5	
Education (%)			
Some Elementary/Junior High	0	16	
Completed Elementary/Junior High	0	11	
Some High School	0	16	
High School Diploma or GED	0	16	40.25 ***
Some College	0	26	
College Degree	32	5	
Post-Graduate Education	68	11	
Yearly Family Income (%)			
< $15,000	0	15	
$15,000–$29,999	0	10	
$30,000–$69,999	0	50	61.93 ***
$70,000–$99,999	21	25	
$100,000–$199,999	68	0	
> $199,999	11	0	
Occupation (%)			
Unskilled Work	0	0	
Service	0	11	
Clerical, Sales	5	22	
Business Owner or Manager	10	11	
Professional or Technical	75	17	14.55 *
Homemaker	5	22	
Not Employed	0	0	
Other	5	17	

X^2 values are indicated for marital status and occupation, F values for education and income.
***$p < .001$, *$p < .05$, ns = not significant.

compared to Salvadorans. The Asian Indians were particularly likely to hold professional or technical occupations, whereas Salvadorans held a broad range of occupations.

Participants were initially recruited through local religious institutions (Catholic churches and Hindu temples) and subsequently by means of snowballing off of participants recruited through these institutions.[2] In accordance with recruitment criteria, all Asian Indian participants were of Hindu religious background (rather than, for example, Muslim or Sikh) and all Salvadorans were of Catholic background (rather than, for example, Pentecostal). Of all families contacted, 59 percent agreed to participate.

Procedure. Participants took part in a one-on-one, semistructured interview ($M = 77$ min., $SD = 21$) addressing civic engagement and other topics. In an effort to decrease socially desirable answers, the civic engagement questions were asked at the end of the interview. The expectation was that participants by then would feel comfortable and, hence, be most honest on the topic. Also, the questions required detailed answers (e.g., about the specific nature of engagement), making it harder to embellish.

The interview language for all Asian Indians and the Salvadoran adolescents was English (a language in which they were fluent). The Salvadoran parents preferred to be interviewed in Spanish. The interviewers who conducted the Spanish-language interviews were of Salvadoran background and fluent in English and Spanish. For the Spanish-language interview schedule, standard back-translation procedures were used.

Interviewers received extensive training in interview techniques. This included learning about the cultural and religious backgrounds of participants. Furthermore, two of the interviewers had resided in India. Almost all interviews (97.5 percent) took place in the homes of participants. This also increased the likelihood that participants would feel at ease during interviews.

At the outset of the interviews, written informed consent was obtained from parents on behalf of themselves and their adolescents. The adolescents provided oral assent. At the conclusion of an interview session, each participant received compensation in the form of $25. Participants were also debriefed in the sense that they were asked if they had questions or thoughts about the interview.

Subsequent to interviews, the audio-recorded interviews were transcribed verbatim by professionals. For the Spanish-language interview, professionals first transcribed into Spanish and then translated into English. These transcribers were of Salvadoran background and fluent in both languages.

Materials. With respect to civic engagement, participants answered a series of eight interview questions. These assessed participants'

1. engagement in political activities (including organizational venue, if engaged);

2. motives for being politically engaged or not being engaged;

3. views on whether political engagement is important for people in general;

4. motives for why political engagement is or is not important for people in general;

5. engagement in community activities and volunteering (including organizational venue, if engaged);

6. motives for being communally engaged or not being engaged;

7. views on whether communal engagement and volunteering is important for people in general;

8. motives for why communal engagement and volunteering is or is not important for people in general.

Participants also completed a questionnaire (again in Spanish for Salvadoran parents) that included demographic and other questions.

A Narrative Analysis and Discussion

Coding. In order to code the data, three researchers reviewed all interviews. This was done blind to participants' age, cultural background, and other demographic information. As the interviews were reviewed, coding manuals and a qualitative database were gradually constructed. The manuals were continuously refined in the process of reading interviews in order to account for all interview materials and clearly define coding categories. The qualitative database recorded both the coding categories and the verbatim response for each participant answer. By sorting all verbatim answers in the database according to coding categories, the coherency of categories was continuously assessed and the coding manuals continuously refined.[3] Once the coding manuals had been completed, inter-rater reliabilities were assessed on 20 percent of all interviews. They are reported below for each coding category. The researchers resolved discrepant coding through discussion.

In order to address the civic engagement of the immigrant participants and the religious component of engagement, the data were coded in four ways:

1. *Engagement of Self and Others.* A distinction was drawn between whether participants themselves were engaged (at the political and community levels) and whether they regarded engagement as important for people in general. Interrater reliability was 100 percent.

2. *Types of Organizations.* For every civic behavior that participants described for themselves, coders assessed the type of organization through which it

occurred. There were eight categories: cultural, medical, political, religious, school, social service, sport, and other. (A category of environmental organization was never used.) Interrater reliability was 85 percent.

3. *Motives.* Every motive that participants provided to explain engagement or nonengagement for self and others was coded into one of three "Ethics of Autonomy, Community, or Divinity" (Jensen 1991, 1997, 1998, 2004; Shweder 1990; Shweder, Much, Mahapatra, and Park 1997). Briefly, the Ethic of Autonomy defines the moral agent as an autonomous individual. Moral reasoning within this ethic includes references to an individual's rights, interests, and well-being and to equality between individuals (e.g., "Volunteering makes you feel good," "It helps you get into college"). The Ethic of Community defines the moral agent in terms of membership in social groups. Moral reasoning within this ethic includes references to a person's obligations to others, promoting the interests of groups, and interpersonal virtues (e.g., "It's my responsibility to do it," "We can make a difference when we all vote together"). The Ethic of Divinity defines the moral agent as a spiritual entity. Moral reasoning within this ethic includes reference to spiritual virtues and to divine authority, lessons, and examples (e.g., "Service to the community is service to God," "That is what has given me spiritual strength"). Interrater reliability coding with the three ethics was 91 percent (Cohen's Kappa).

4. *Religious or Spiritual Themes.* All Ethic of Divinity motives were then further analyzed in order to identify themes. Themes were identified both on the basis of the preexisting subcategories in the Three Ethics Coding Manual (Jensen 2004), as well as any new concepts introduced by participants in response to the present topic. As detailed below, four religious or spiritual themes pertaining to civic engagement were identified. Interrater reliability was 83 percent.

Engagement of Self and Others. Table 2 shows that all participants considered it important for people to be civically engaged at the community level, and almost all also held this view for political activities. When it came to the immigrants' own engagement, 81 percent were engaged at the community level and 30 percent at the political level. Community engagement was an almost-universal activity among Asian Indian adolescents (95 percent), and Salvadoran parents stood out in regard to politics, where 50 percent were engaged. Overall, it was the rare person who was not engaged. Furthermore, all participants thought that some kind of civic engagement was important.

TABLE 2. **Self's and Others' civic engagement (percent)**

Engagement	Adolescents		Parents		
	Salvadoran	Asian Indian	Salvadoran	Asian Indian	All
Self's Political	20	20	50	30	30
Self's Community	80	95	75	75	81
Others' Political	90	90	100	75	89
Others' Community	100	100	100	100	100

Types of Organizations. Our next question addressed the types of organizations through which the immigrants' own civic activities occurred. Table 3 shows that half of all participants were engaged through religious organizations. Only school also served as a venue of engagement for close to 50 percent. Engagement through religious organizations was common across both age groups and both cultural groups. For other types of organizations, there appeared to be more age and/or cultural group variation (e.g., twice as many parents were engaged through political organizations as adolescents, and almost twice as many Asian Indians were engaged through school as Salvadorans). Religious organizations, then, were a common context of engagement for immigrant participants.

TABLE 3. **Types of organizations for civic engagement (percent)**

Organization	Adolescents		Parents		
	Salvadoran	Asian Indian	Salvadoran	Asian Indian	All
Religious	45	50	65	40	50
School	40	70	25	50	46
Political	20	20	50	30	30
Social Service	20	45	10	20	24
Medical	10	35	5	0	13
Cultural	0	10	0	5	4
Sport	5	0	0	5	3
Other	15	35	5	20	19

Percentages do not add to 100 as participants could report more than one kind of civic behavior and, hence, more than one organization.

TABLE 4. **Use of ethics of divinity, autonomy, and community motives (percent)**

	Divinity	Autonomy	Community
Engagement			
Self's Political	8.7	52.2	69.6
Self's Community	10.5	71.9	78.9
Others' Political	1.4	42	72.5
Others' Community	10.1	50.6	82.3
Non-Engagement			
Self's Political	0	90.6	26.4
Self's Community	0	92.9	35.7
Others' Political	0	87.5	50
Others' Community	n/a	n/a	n/a

(a) No participants held that it was not important for other people to be engaged at the community level; hence, no one provided reasons to explain this position.
(b) Percentages do not add to 100 as participants could provide more than one motive.

Motives. Did the fact that many immigrants were engaged through religious organizations also mean that their behaviors were motivated by religious or spiritual considerations? The answer to this question seems to be "rarely." Nor did they appear to think that it was important for others to be motivated by such considerations. Table 4 shows that participants never spoke of religious or spiritual motives to explain why they or others should *not* be civically engaged. For engagement, 9–11 percent invoked divinity motives to explain their own behavior, and 1–10 percent did so to account for the engagement of other people. The use of the Ethics of Autonomy and Community motives was far more prevalent. (The details regarding Autonomy and Community motives will not be discussed here, as the present focus is on religion and spirituality.)

Looking across age and cultural groups (not shown in Table 4), Salvadoran parents in particular spoke of religious and spiritual considerations. Averaged across the four engagement behaviors (self and others, and political and community), 15.6 percent of Salvadoran parents invoked Divinity motives. The comparable figures for the other groups were 7.6 percent for Indian parents, 4.5 percent for Salvadoran adolescents, and 1.4 percent for Indian adolescents.

Religious or Spiritual Themes. While religious or spiritual motives were fairly uncommon, it still seemed useful to have an understanding of how the participants who did have such motives spoke of them. Emic research on people's civic motives has been highly limited (e.g., Friedland and Morimoto 2005; Pearce and Larson 2006), and, as far as we know, there has been no research addressing divinity considerations.

The present content analysis of the motives for engagement identified four religious or spiritual themes: (1) divine inspiration, (2) service to the Divine, (3) building religious foundations, and (4) spiritual virtues.

Divine inspiration was where participants spoke of how God (or gods) approves of rendering service to others, especially those in need. One person spoke explicitly of divine approval, saying, "Jesus is looking at you and saying 'Good job!'" Others spoke of God's teachings and examples. One Salvadoran immigrant explained, "Jesus taught us to help those in need." Another said, "The first commandment of God says to love God above all else and your neighbor as yourself. So this is a way of loving thy neighbor—helping him." Coming from a Hindu perspective but expressing a similar concept, a participant explained, "Hinduism says that we're all the same, we're all connected. You can feel that more if we all work together." Still others felt the intervention of God in their lives to awaken them to service:

I came from poverty, right. And I came from a large family, so I started having a large family, and it wasn't just like that that I was able to watch over people. That was a change that the Savior made for me: He converted me to where I started seeing the needs that fellow man has.

In the eyes of these participants, God is guiding them to civic service, and they are fulfilling a valuable purpose set forth by God (cf. Colby and Damon 1992).

A second theme, *service to the Divine*, involved a related conception of civic engagement. While the first theme emphasized civic engagement as a divine goal, this theme highlighted how such engagement is a means to pay homage to the Divine. One parent explained that, "in Hinduism, we say that service to the community is service to God." Using similar language, another person said, "By helping others, we are doing service to God." *Divine inspiration* and *service to the Divine*, then, seem to be two sides of the same coin where civic engagement is either a means or an end linking people to divinity.

Building religious foundations was a third theme. Here, the focus was on nurturing or building religious or spiritual values in others. A Salvadoran parent explained that "[it] is to help others that are in need, not just in an economic way but spiritually." A Salvadoran adolescent was similarly motivated:

I try to influence the little kids in the right direction. Especially little kids, they need you growing up. Like making them want to learn about God and stuff. Because I think that's a good thing: it gives you something to base your beliefs on—a foundation.

Also linking civic engagement to spiritual recruitment (and perhaps even to church recruitment), a parent said, "It makes me feel good that, because of us, more people are coming to church." In this view, then, civic engagement helps to build religion or spirituality in individuals and communities.

The fourth and final theme was *spiritual virtues* where participants felt motivated by virtues such as spiritual strength, faith, and humility. Combining feelings of empowerment and modesty, one immigrant parent described how

I've gotten involved and that is what has given me spiritual strength. . . . And that is what drives me. At school, I have done many things. And it's not because I have done it alone, but with everyone else.

An adolescent spoke of faith as a motive, saying, "I believe in doing good, and helping others is doing good. 'Do unto others . . .'—it's my faith to do good." Here, then, civic engagement was an expression of a moral commitment by persons who saw themselves and their behaviors in a spiritual light.

CONCLUSIONS

The present immigrant adolescents and parents from El Salvador and India regarded some form of civic engagement as important for all people. Moreover, they were engaged themselves, more so at the community than the political level. Furthermore, as previously shown, the majority of the immigrants' civic behaviors are focused on issues that do not pertain specifically to their cultural group or even to immigrants generally (Jensen 2008).

Lopez (2003) found that 75 percent of high-school seniors in a national survey reported community service or volunteering within the past twelve months. Here, 87.5 percent of the adolescents (all but one of whom was in high school) described being engaged. The present sample is, of course, not representative of all immigrants (it is a small sample from one metropolitan area, partly recruited through religious institutions). Nevertheless, these socioeconomically diverse immigrant participants were civically engaged.

The civic activities of half of all the immigrant participants occurred in the context of religious organizations, a number similar to what Putnam (2000) noted for the general American population. This religious affiliation turned out to be far more important for civic engagement than immigrant participants' religious motives. Few participants spoke of religious or spiritual motives when explaining their political or community engagement and the importance for others of such engagement. The present study, thus, highlights the importance of distinguishing individual religious motives from religious contexts in examining civic engagement.

A next step will be to examine further the meanings and functions of religious organizations in regard to immigrant civic engagement. In the words of immigrants, why is it that they get involved through religious organizations when individual religious motives for such involvement are infrequent? Also, do they stay involved through religious institutions in spite of lack of religious or spiritual motives?

In conclusion, religious or spiritual motives were rare. Religious organizations, however, played an important role in positive engagement with society for the present immigrants to the U.S. In fact, religious organization, unlike other institutions, pulled participants across age, immigrant generational status, and cultural background.

ACKNOWLEDGMENTS

I am profoundly grateful to the immigrant adolescents and parents who took time out of their busy lives to be part of this research. I also thank Renata Cerqueira, Michelle Diaz, Jeanne Felter, and Silvia Juarez for their research assistance, as well as Dean Hoge, Michael Foley, and James Youniss for encouragement of the project. Finally, I am grateful to the Center for Information and Research on Civic Learning and Engagement (CIRCLE), the Pew Charitable Trusts, and the Research Council of Denmark for their generous support.

NOTES

1. The nature of the data was dyadic, consisting of adolescent-parent pairs, but the present focus on religious organizations entailed a reduction of the dataset that rendered dyadic analyses unfeasible.

2. The fact that initial recruitment occurred through religious institutions may entail a stronger presence of religion in the civic engagement of the present immigrant participants than in other immigrants. Partly, this recruitment strategy reflects the difficulty of otherwise locating and recruiting immigrant adolescents and parents who form part of coherent cultural communities (such as Hindu Indians).

3. In grounded-theory analysis, this process is what is referred to as "open coding" and the integration of categories (based on their properties and dimensions). The process is also referred to as the "constant comparative method" (Glaser and Strauss 1967). Grounded-theory analysis can take different forms based on the research purpose, the nature of the data, and the approach of the researcher. Even Glaser and Strauss, the originators of grounded theory, eventually parted

company on whether various techniques are necessary or even desirable for an approach to be considered grounded theory (Glaser 1992; Strauss and Corbin 1990). Here, we followed the constant comparative method steps.

REFERENCES

Colby, A., and W. Damon. 1992. *Some do care: Contemporary lives of moral commitment.* New York: The Free Press.

Fisher, C. B., K. Hoagwood, C. Boyce, T. Duster, D. A. Frank, T. Grisso, R. A. LeVine, R. Macklin, M. B. Spencer, R. Takanishi, J. E. Trimble, and L. H. Zayas. 2002. Research ethics for mental health sciences involving ethnic minority children and youths. *American Psychologist* 57: 1024–40.

Flanagan, C. A. 2004. Volunteerism, leadership, political socialization, and civic engagement. In R. A. Lerner and L. Steinberg, eds., *Handbook of adolescent psychology,* 721–45. Hoboken, NJ: Wiley.

Flanagan, C., and N. Faison. 2001. Youth civic development: Implications of research for social policy and programs. *Social Policy Reports* XV, No. 1.

Flanagan, C., and L. Sherrod. 1998. Youth political development: An introduction. *Journal of Social Issues* 54: 447–50.

Flanagan, C., and C. Tucker. 1999. Adolescents' explanations for political issues: Concordance with their views of self and society. *Developmental Psychology* 35: 1198–1209.

Friedland, L. A., and S. Morimoto. 2005. The changing lifeworld of young people: Risk, resume-padding, and civic engagement. CIRCLE working paper. *www.civicyouth.org.*

Galston, W. A. 2001. Political knowledge, political engagement, and civic education. *Annual Review of Political Science* 4: 217–34.

Glaser, B. G. 1992. *Basics of grounded theory analysis.* Mill Valley, CA: Sociology Press.

Glaser, B., and A. L. Strauss. 1967. *The discovery of grounded theory.* Chicago: Aldine.

Huntington, S. P. 2004. *Who are we? The challenges to America's national identity.* New York: Simon and Schuster.

Jennings, M. K. 2002. Generation units and student protest movement in the United States: An intra- and intergenerational analysis. *Political Psychology* 23: 303–24.

Jensen, L. A. 1991. *Coding Manual: Ethics of Autonomy, Community, and Divinity.* Unpublished manuscript, University of Chicago.

———. 1997. Different worldviews, different morals: America's culture war divide. *Human Development* 40: 325–44.

———. 1998. Moral divisions within countries between orthodoxy and progressivism: India and the United States. *Journal for the Scientific Study of Religion* 37: 90–107.

———. 2004. *Coding Manual: Ethics of Autonomy, Community, and Divinity Revised.* Unpublished manuscript, Clark University.

———. 2008. Bringing culture to the civic table: Immigrants' cultural identities as sources of engagement in civil society. In L. A. Jensen and C. Flanagan, eds., *Immigrant Civic Engagement: Applied Developmental Science.* Mahwah, NJ: Lawrence Erlbaum.

Jensen, L. A., and C. A. Flanagan. 2008. Civic Engagement in Immigrant Youth: New translations. *Applied Developmental Science* 12, no. 2.

Jessor, R., A. Colby, and R. A. Shweder. 1996. *Ethnography and human development: Context and meaning in social inquiry.* Chicago: University of Chicago Press.

Larson, R. W. 2000. Towards a psychology of positive youth development. *American Psychologist* 55: 170–83.

Lerner, R. M. 2004. *Liberty: Thriving and civic engagement among America's youth.* Thousand Oaks, CA: Sage Publications, Inc.

Lerner, R. M., A. E. Alberts, and D. L. Bobek. 2007. Thriving youth, flourishing civil society: How positive youth development strengthens democracy and social justice: A Bertelsmann Foundation White Paper. Guterslöh, Germany: The Bertelsmann Foundation.

Lerner, R. M., E. M. Dowling, and P. M. Anderson. 2003. Positive youth development: Thriving as the basis of personhood and civil society. *Applied Developmental Science* 7, no. 3: 172–80.

Lopez, M. H. 2003. Volunteering among young people. CIRCLE working paper. *www.civicyouth.org.*

Lopez, M. H., and K. Marcelo. 2008. The civic engagement of immigrant youth: New evidence from the 2006 civic and political health of the nation survey. *Applied Developmental Science* 12, no. 2: 66–73.

Pearce, N. J., and R. W. Larson. 2006. How teens become engaged in youth development programs: The process of motivational change in a civic activism organization. *Applied Developmental Science* 10, no. 3: 121–31.

Portes, A., and R. Rumbaut. 1996. *Immigrant America: A portrait.* Berkeley: University of California Press.

———. 2001. *Legacies: The story of the immigrant second generation.* Berkeley and New York: University of California Press and Russell Sage Foundation.

Putnam, R. D. 2000. *Bowling alone: The collapse and revival of American community.* New York: Simon and Schuster.

Sherrod, L. R., C. Flanagan, and J. Youniss. 2002a. Dimensions of citizenship and opportunities for youth development: The *what, why, when, where,* and *who* of citizenship development. *Applied Developmental Science* 6, no. 4: 264–72.

———. 2002b. Editors' introduction. *Applied Developmental Science* 6, no. 4: 173–74.

Shweder, R. A. 1990. In defense of moral realism: Reply to Gabennesch. *Child Development* 61: 2060–67.

Shweder, R. A., N. C. Much, M. Mahapatra, and L. Park. 1997. The "big three" of morality (autonomy, community, divinity), and the "big three" explanations of suffering. In A. Brandt and P. Rozin, eds., *Morality and Health,* 119–72. New York: Routledge.

Stepick, A., and C. D. Stepick. 2002. Becoming American, constructing ethnicity: Immigrant youth and civic engagement. *Applied Developmental Science* 6, no. 4: 246–57.

Stepick, A., C. D. Stepick, and Y. Labisierre. 2008. South Florida's immigrant youth and civic engagement: Major engagement, minor differences. *Applied Developmental Science* 12, no. 2: 57–65.

Strauss, A. L., and J. Corbin. 1990. *Basics of qualitative research: Grounded theory procedures and techniques.* Newbury Park, CA: Sage.

Suarez-Orozco, C., and M. M. Suarez-Orozco. 2001. *Children of immigration.* Cambridge, MA: Harvard University Press.

U.S. Census Bureau. 2005. Statistical abstract of the United States. http://www.census.gov/prod/2004pubs/04statab/pop.pdf

Youniss, J., J. A. McLellan, and M. Yates. 1997. What we know about engendering civic identity. *American Behavioral Scientist* 40: 620–31.

Youniss, J., and M. Yates. 1997. *Community service and social responsibility in youth.* Chicago: University of Chicago Press.

Zhou, M. 1997. Growing up American: The challenge confronting immigrant children and children of immigrants. *Annual Review of Sociology* 23: 63–95.

Zhou, M., and C. Bankston. 1998. *Growing up American: The adaptations of Vietnamese children to American society.* New York: Russell Sage Foundation.

14

Ethnic Identity and Spirituality

Linda Juang & Moin Syed

My religion, my faith in being a Catholic is just as important as my identification with
being Filipino. I honor my Catholic upbringing and feel that being a Catholic is important in my life not only for spiritual guidance but also as a connection with my family's
past roots.

female, age 22

Ethnic identity has received much attention in the last decade and a half as the
U.S. has become increasingly culturally diverse (Spencer 2006). One reason
is that a strong identification with one's ethnic background has been consistently linked to a host of beneficial outcomes, such as greater self-esteem, lower
depression, higher academic achievement, and more effective coping in the face
of discrimination (Phinney 2003, 2006; Quintana 2007). At the same time, a
recent national study showed that religion/spirituality is also an important part
of the lives of many American youth (Smith and Denton 2005). Indeed, King
(2008) emphasizes that religion/spirituality can play a key role in positive youth
development by fostering a sense of purpose, competence, caring, and civic
engagement. As the opening quotation to this chapter illustrates, some young
people view their ethnic identity as intimately connected to their religious/
spiritual identity. The overall purpose of this chapter is to explore this connection. To this end, we have three main objectives: (1) to review research pertaining to the intersection between youth's religious/spiritual and ethnic identities;
(2) to illustrate how youth think about these identities, using data from a study
of ethnic identity in context; and (3) to offer future directions for the study of
religion/spirituality in relation to ethnic identity. We hope that these objectives

will contribute to a greater understanding of how intersecting identities are negotiated and how they contribute to positive youth development.

IDENTITY FRAMEWORKS: RELIGION/SPIRITUALITY AS GROUP MEMBERSHIP

While sociologists have traditionally seen identity development as a lifelong process (Coté 1996), psychological theorists have identified adolescence (Erikson 1968) and emerging adulthood (Arnett 2008) as the major life stages for identity formation. From the psychological perspective, then, young people are cognitively and socially ready to engage in the important work of actively searching for, exploring, and forming their identities. Developmental psychologists have historically studied religious identification as an aspect of personal identity, drawing on the Eriksonian processes of identity exploration and identity commitment (Erikson 1968; Meeus et al. 1999). Empirical investigations into religious identity development have generally applied Marcia's operationalization of an aspect of Erikson's theory, known as the identity status model (Marcia 1980). The identity status model has been used in a number of domains of identity development, most commonly occupation, political orientation, and religious identification (Meeus et al. 1999). One of the fundamental features of this model, and the manner in which it has been applied, is that identity development in these domains is *personal*, in that the processes of identity formation are individual cognitive and affective projects that are generally independent of other people.

Another useful framework for studying identity development that is less frequently adopted by developmental psychologists is social identity theory. Social identity theory emphasizes aspects of an individual's identity that are derived from membership in groups (Tajfel and Turner 1986). The term *groups* can take on a wide range of meanings, from social categories (e.g., race, ethnicity, gender, social class) to groups that are artificially created in a laboratory setting. Belonging to social groups provides opportunities for individuals to derive high self-esteem by providing a sense of "we-ness." However, due to the structural characteristics of most societies, which include power differentials among groups, some groups are valued less than others (e.g., in the U.S., ethnic minorities, women, and the poor and working class). According to social identity theory, identification with a subordinate group is heightened in specific contexts where that group is in the minority. This is done as a way of preserving a positive sense of self, as choosing to identify with the group in these contexts can lead to increases in self-esteem. This heightened identification that leads to a more positive sense of self can also buffer the negative impact of prejudice and discrimination experienced as members of lower-status groups (Ashmore, Deaux, and McLaughlin-Volpe 2004; Tajfel and Turner 1986).

Empirical studies of naturally occurring social groups (i.e., groups not constructed in the laboratory) have primarily focused on gender, race/ethnicity, and social class, and, to a lesser extent, sexual orientation, physical ableness, and religious identification. Although religious identity has been primarily studied using the identity status model, individuals' religious identifications designate them as members of a particular group. Therefore, understanding religious identity from a social identity perspective may provide insight into religious identity formation.

Templeton and Eccles (2005) also proposed that religious group membership may serve as a social or collective identity but that a sense of spirituality is more appropriately understood as a personal identity. Although provocative, one problematic aspect of this delimitation is that the terms *religion* and *spirituality* do not have agreed-upon definitions and remain a contentious issue: some view them as separate and distinct, others distinct yet overlapping, and yet still others argue that there is no distinction at all (Miller and Thoresen 2003). We take an agnostic stance on this definitional issue for the purposes of this chapter and use *religion* and *spirituality* interchangeably, although we recognize that greater conceptual clarity is warranted. Nevertheless, Templeton and Eccles' (2005) proposal makes an important contribution to the identity literature by attempting to integrate identity subdisciplines (i.e., personal and social identities) that are frequently considered in isolation.

INTERSECTIONALITY: A FRAMEWORK FOR INTEGRATING MULTIPLE IDENTITIES

Templeton and Eccles' (2005) general organization of identities is one of separation, despite the fact that they argue that a particular identity may be viewed as religious/collective in one context and spiritual/personal in another. Indeed, treating various domains of identity as independent of one another has been one of the major limitations of previous studies using both the identity status model and social identity theory, as an individual's status as a woman is not independent of her status as Chinese American, middle-class, heterosexual, or Buddhist. Rather, these identities are intimately interconnected and inform one another, a concept referred to in the feminist and critical race literatures as *intersectionality* (e.g., Crenshaw, Gotanda, Peller, and Thomas 1995; Hurtado 1997). Intersectionality theorists argue that identities are not additive, hierarchical, or separable, but rather that unique configurations of identities create unique social positions that shape experiences, particularly around issues of oppression (see Hurtado and Silva 2008 for an excellent brief summary).

The existing work on intersectionality has primarily focused on race, gender, and social class and has largely been theoretical. That is, scholars have theorized on how individuals' experiences are contextualized by their social positions, using intersectionality as an analytic frame (e.g., hooks 2003). However, very little of the existing work has examined how youth develop an understanding of their multiple identities as they relate to one another. In other words, although youth may be positioned within society according to their race, gender, and class, they may not see these identities as interconnected from their own perspective (see Azmitia, Syed, and Radmacher 2008). Mattis, Ahluwalia, Cowie, and Kirkland-Harris (2005) eloquently related a similar sentiment pertaining to the study of religious/spiritual development:

Any serious cultural analysis of spiritual development must shift away from a social science of surveillance in which we limit our work to those aspects of life that are observable and easily quantifiable (e.g., behavior), and toward a study of religion and spirituality that is meaning centered. (Mattis et al. 2005, 293)

In this chapter, we build from this prior work by examining how late adolescents and emerging adults view the intersection of ethnicity and religion/spirituality. Ethnicity and religion are interconnected in many parts of the world (e.g., at the country level, 94 percent of Thailand is Buddhist, 88 percent of Ireland is Roman Catholic, and 99 percent of Iran is Muslim; United Nations Statistics Division 2006). In her ethnographic study of Cambodian Americans living in Boston, Smith-Hefner (1999) asserts that "to be Khmer is to be Buddhist." Cambodian Americans have experienced tremendous oppression and suffering through the attempted eradication of their once-national religion in Cambodia. Reclaiming and integrating their Buddhist identities have been important aspects of the Cambodian-American experience. As Smith-Hefner describes it, identification with Buddhism is necessary to feel complete as a Cambodian. However, in another study with Cambodian-American youth, Su (2006) found that, while participants frequently recounted narratives of religious involvement, these experiences were associated with feelings of both closeness and distance to their ethnicity, depending on the context of the

experience. Thus, while a Buddhist identity may be a key aspect of a Cambodian identity on a macro-level, the individual meanings constructed may vary
by person and context. This individual-level, meaning-centered approach is the
focus of this chapter.

Although our goal in this chapter is to highlight the intersection of ethnic
identity and religious/spiritual identity, we first define and discuss ethnic identity in isolation to highlight how it has been conceptualized in developmental
psychology. We will then turn our discussion toward the intersection of these
identities.

ETHNIC IDENTITY

From a social identity theory perspective, ethnic identity refers to the degree
to which an individual identifies as being a member of an ethnic group (Phinney 1990, 2006). Ethnic identity is an important part of one's self development,
especially for ethnic-minority youth living in a culturally diverse and racially
stratified society as the U.S. (Waters 1999). Factors such as prejudice, discrimination, and stereotypes are particularly relevant for ethnic minorities (García
Coll et al. 1996), and as such, may propel them to pursue a deeper understanding of their ethnic identity. Ethnic identity is complex and multidimensional,
encompassing domains such as a sense of belonging to one's ethnic group,
evaluation of one's ethnic group, and centrality or salience of ethnicity to one's
life (see Phinney and Ong 2007 for an overview of different ethnic identity
domains). The formation of an ethnic identity is a dynamic developmental
process, moving generally from an unexamined stage (a lack of understanding
about the role of ethnicity in one's life), to exploration (a period of questioning
and information seeking concerning the role of ethnic identity in one's life),
to achievement (adopting a clear and committed understanding of the ways
in which ethnicity contributes to one's identity) (Phinney 1989). Recent longitudinal work has provided support for this developmental model of ethnic
identity (French, Seidman, Allen, and Aber 2006; Seaton, Scottham, and Sellers 2006). These and other studies also show, however, that, for some individuals, identity development does not proceed in such a linear fashion but may
regress at different points in time (Syed, Azmitia, and Phinney 2007).

Having a strong and positive ethnic identity has been linked to many positive psychological outcomes, such as greater self-esteem, less loneliness, and
less depression (Phinney 2003, 2006; Quintana 2007). Further, having a strong
and positive ethnic identity may prompt youth to engage in political and civic

activities. For instance, researchers have argued that ethnic minorities who (1) identify with their ethnic group and internalize their ethnic group membership as a salient and critical aspect of their identity and (2) are aware of status, resource, and power disparities of their ethnic-minority group in society are more likely to participate in community actions for social change (Gutierrez 1990; Zimmerman 1995). Consequently, ethnic identity can be considered an important aspect of positive youth development by promoting positive mental health and encouraging community involvement.

One limitation characterizing the majority of ethnic identity studies, however, is that they have not considered how other identities may be linked to, may influence, or influenced by, one's ethnic identity. Although developmental researchers have called for some time for a more integrated approach to studying the various domains of identity (e.g., Archer 1992), we still know little of how youth perceive their multiple identities in relation to one another.

ETHNIC IDENTITY AND RELIGIOUS/SPIRITUAL IDENTITIES: DO THEY INTERSECT?

We now turn to the question of whether spirituality or religious identity is important to young people's lives in relation to ethnic identity. Although research in this area is very limited, one body of literature suggests that religion may indeed be important to understanding one's ethnic self. This is evident for a group of people whose ethnic identities may be highlighted and challenged as they move across contexts, namely, immigrants. Immigrants are an ideal group in which to study the intersection of ethnic and religious identities as migration compels individuals to reformulate their personal and group identities in a new context (Williams 2000).

The process of immigration can be stressful (Berry and Kim 1988). To ease this stress, religious communities in the U.S. have had a long history of assisting immigrants in many ways. Importantly, religious communities provide refuge and act as a source of psychological support and comfort for immigrants far away from familiar surroundings (Hirschman 2004). Religious communities can, for example, operate as an extended family. In Min's (2000) study of Korean-American families, she found that those who were involved in the church also enjoyed frequent and extensive social interaction with other families: going to lunch after services, celebrating holidays and birthdays together, engaging in sports, and attending retreats. Subsequently, active involvement in a religious community in the new country is one important way for immigrants to rebuild the cultural community they left behind. The religious community can provide a familiar anchor in the midst of novel and challenging experiences associated with adaptation to a new country.

Religious communities provide not only psychological support and a sense of belonging but also important instrumental support and resources in terms of finding housing, providing host-culture-language classes, and finding employment (Hirschman 2004). This type of support system acts as an important link to the new culture's resources and enables a smoother transition to the new culture. Importantly, the religious community can facilitate the transition to the new culture for youth as well as adults. For youth of immigrant families, participating in a religious community creates a context in which they can socialize with other youth and learn from one other how to adapt to the new culture (Bankston and Zhou 1995).

Bankston and Zhou's (1995) study of Vietnamese-American adolescents demonstrates the link between religious participation and ethnic identity. In particular, religious participation facilitated a stronger ethnic identity (measured by language use, commitment to endogamy—i.e., to marry someone Vietnamese—identifying oneself as Vietnamese, and reporting most or all friends are Vietnamese). This stronger ethnic identity, in turn, predicted positive youth development. More specifically, adolescents with stronger ethnic identities did better in school, had a stronger belief in the importance of future education, aimed for college, and avoided risky behaviors such as alcohol and drug use. Thus, involvement in a religious community facilitated the transition and adaptation to American culture through strengthening the adolescent's ethnic identity, highlighting the potential overlap between an adolescent's ethnic and religious selves.

In addition to facilitating adaptation to a new culture, involvement in a religious community preserves one's traditional culture. Herberg (1960) asserted that religion is one area where immigrants do not have to change. For instance, immigrants must learn a different language, adopt different customs, and create new relationships; however, they do not have to change their religion. Participating in a religious community, then, provides continuity and an emotional connection with the old culture (Hirschman 2004). Thus, preserving one's religion and traditional culture (or ethnic identity) can be an important source of comfort for immigrants.

For youth, in particular, religious communities can play a key role in the maintenance of ethnic identity to ensure that traditional cultural values are passed onto the next generation (Bankston and Zhou 1995; Min 2000). One way to rejuvenate and maintain a strong ethnic or cultural identity is engaging in a religious community where cultural customs and traditions are practiced (Bankston 2000). Some churches hold, for instance, traditional-language classes (such as Chinese Saturday school) for their children. Some churches celebrate traditional cultural festivals and holidays. Youth who are exposed to these cultural practices have the opportunity to incorporate these cultural elements into their own identity. When ethnic and religious group memberships overlap, as is often the case in the U.S. where ethnic churches are a growing phenomena (Jeung 2005), youth may develop their religious and ethnic identities in tandem. For immigrant families, then, religious community involvement can be a vehicle through which youth's ethnic identity is fostered and sustained.

POSSIBLE PATTERNS OF ETHNIC AND RELIGIOUS/ SPIRITUAL IDENTITY INTERSECTIONS

Based on available literature, we describe three ways in which ethnic and spiritual/religious identities may connect. Ethnic and religious identities can be fused, as in Bankston's (2000) study of Laotian refugees in Louisiana, which illustrates that, for some individuals, religious identity is tightly bound to their ethnic identity (see also Smith-Hefner 1999). One Laotian in Bankston's study stated, "In Laos, my religion was Buddhism. My religion taught me to be a good person, but I never thought my religion made me Lao. In America, Buddhism is a Lao religion, and that makes me Lao. The ones who become Christian—they are still Lao, but it isn't the same" (as cited in Bankston 2000, 366). In Bankston's analysis he states, "In Laos, religious ceremonies and festivals serve to bind people together as members of villages or family groups but not as members of an ethnicity. It is precisely the resettlement in a non-Laotian setting that makes ethnicity the basis of a new social identity, and the transplanted religious practices serve to express this identity" (2000, 367). Thus, as illustrated by this group of immigrants, the significance of spirituality or religion can shift in relation to ethnic identity across contexts, becoming inseparable. In the case of Laotians in New Orleans, their religious and ethnic identities became fused together in this new context.

In another pattern, ethnic and religious identities are not fused together but remain distinct, with one domain possibly taking precedence over another. For example, Robinson (2003) found that most Pakistani adolescents in the

U.K. identified more with their religious identity than with their ethnic Asian identity. She reports that, as a result of the events of 9/11, young Muslims in Britain have been the targets of increasing antagonism. This increasing antagonism may influence their current identification strategies, such as by heightening the salience of their religious rather than ethnic Asian identity. In this case, the two identities (Asian and Muslim) are not fused but rather maintain a separate and hierarchical relationship to one another. Cavalcanti and Schleef (2005) found that, while a majority of Latinos who settled in Richmond, Virginia, maintain their religious identities, some do not. Interestingly, those who do not also report higher English-language use and greater contact with European Americans, indicating greater integration into the mainstream community, at least in their communication and social networks. These authors argue that one risk of becoming too immersed into one's religious and cultural community, then, may be to slow the transition and integration into mainstream society. Their study illustrates that, for some individuals, their religious identities are less important, or even nonexistent, in relation to their ethnic identities.

In yet another pattern, ethnic and religious identities are distinct domains yet may conflict with one another, such as the case when immigrants convert to new religions that are not indigenous to their home countries. In this case, immigrants can continue to have high affiliation with both their ethnic and religious identities, but these spheres are seen as separate and perhaps even in opposition to one another. An example is of Chinese Christians in the U.S. (Yang 1999). For many Chinese Christians, Christianity may not play a role in defining how "Chinese" the individual is. In fact, being Chinese and being a Christian may be somewhat conflicting or oppositional, as the principles of Christianity oppose traditional Chinese cultural values (e.g., ancestor worship). Nonetheless, Yang finds that Chinese Christians manage to live creatively with these two identities, as illustrated by this observation: "Overall, they consistently preserve Confucian moral values, selectively accept some Daoist notions, and categorically reject Buddhism" (Yang 1999, 160). These findings suggest that being selective in what particular cultural and religious beliefs and values to preserve (or reject) is one strategy by which individuals make sense of seemingly conflicting cultural and religious identities.

The three patterns of intersection that we have just discussed should be understood contextually. For instance, the receptivity of American society to certain religious and ethnic groups fluctuates over time. Subsequently, this shifting landscape can profoundly influence one's ethnic and religious identities. Ethnic and racial identity theorists have posited that experiences of discrimination or racism, termed *encounters* (Cross 1995) or *triggers* (Phinney 2003), may prompt individuals to explore the meaning of their ethnicity. Social identity theory further suggests that experiences of discrimination may strengthen an individual's ethnic identity to preserve a positive sense of self (Phinney 2003; Tajfel and Turner 1986). Likewise, adolescents who encounter religious discrimination may be prompted to explore their religious identities (and, at the same time, ethnic identities, if ethnicity and religion are intertwined) more than those who have not been challenged in their religious beliefs. These challenges may result in a stronger identification with their religious and/or ethnic backgrounds. Subsequently, it is imperative to consider characteristics of the broader context such as history, social change, and the particular community, that contribute to transformations in religious and ethnic identities. As contexts change, so do opportunities for integrating identities.

In sum, psychology researchers have traditionally examined different domains of identity in isolation. However, it would be productive to examine how different areas potentially influence and define one another (Archer 1992; Crenshaw et al. 1995; Hurtado 1997). Two important identity domains—religious/spiritual and ethnic—have rarely been considered in tandem. We know little of how youth synthesize ethnicity and religion/spirituality to arrive at a unique, coherent sense of self. The literature on immigrant adaptation, however, underscores the potential intersection of these two identity domains and suggests several patterns by which ethnicity and religion interconnect. Only by pursuing research that assesses domains in relation to one another can we move beyond a fragmented approach to identity formation.

In the next section, we report preliminary findings from the Ethnic Identity in Everyday Experiences Study. These findings give us clues to the diverse ways in which emerging adults construct identities that incorporate their religious and ethnic selves.

THE ETHNIC IDENTITY IN EVERYDAY EXPERIENCES STUDY

Spirituality is often discussed as an intraindividual, self-reflective process. However, people's notions of spirituality are created and sustained in social contexts, groups, and relationships and can be defined by others as well as themselves (Mattis and Jagers 2001). Social contexts are subjectively understood and experienced and, therefore, benefit from being examined at the individual level through qualitative methods. The work presented in this chapter is informed by a narrative approach to identity development. The narrative perspective focuses on the subjective construction of individuals' identities through the process of telling their life stories (McAdams 2001). The voices of the emerging adults that we present can be viewed as small excerpts that fit within their larger life stories. These narratives provide a developmentally situated account of how emerging adults view the intersection of ethnicity and religion.

Emerging adulthood is a ripe developmental period for asking about complex notions of identity. As Arnett (2008) points out in his qualitative study of conceptions of life after death, emerging adults seem to be more articulate than adolescents in discussing religion/spirituality. Indeed, many have argued that late adolescence and emerging adulthood are ideal for exploring complex aspects of identity (e.g., intersections) due to an increase in capacities for higher-order abstract reflection (e.g., Habermas and Bluck 2000; Harter and Monsour 1992). Furthermore, college-going emerging adults are exposed to new contexts and experiences that may serve as catalysts for identity work (Eccles, Templeton, Barber, and Stone 2003).

Study Background

The data presented in this chapter are part of the larger Ethnic Identity in Everyday Experiences Study, which is an ongoing multisite investigation (Santa Cruz and San Francisco) into how emerging adults think about and experience their ethnicities, being conducted by M. Syed, L. P. Juang, and M. Azmitia. We currently have 354 participants, 236 from Santa Cruz and 118 from San Francisco. Because there were no significant differences between the two sites on the responses to questions we examined in this chapter, the results will be presented in aggregate. The mean age of participants is 19.26 years ($SD = 1.11$) with 69 percent women. The ethnic group breakdown is 27 percent Asian American, 27 percent white, 19 percent mixed-ethnic, 18 percent Latino, 4 percent black, 2 percent Middle Eastern, and 2 percent Indian/Pakistani. The sample is socioeconomically diverse, and 83 percent of participants were born in the U.S.

Participants were recruited from undergraduate psychology classes and came into the laboratory to complete a computer-guided online survey. The survey included many more open- and close-ended questions than what is reported in this chapter and, therefore, took approximately 50–60 minutes to complete (range: 20–90 minutes).

For this chapter, we will present findings from only two questions: "Are there other social groups (e.g., gender, social class, religion, sexuality, sports, etc.) that you feel are more important or as important to who you are than your race/ethnicity?" and "Do you feel that you are similar to other members of your racial/ethnic group in your values?" Both of these questions included follow-up probes for the participants to elaborate on their Yes/No answers. We chose to focus on these two questions because of the likelihood that they would generate responses referencing religion/spirituality as it pertains to ethnicity. It is important to note that understanding religious/spiritual identification was not one of the goals of the study, and, as such, we did not ask any questions directly about religious or spiritual identification. Therefore, we are not using our findings to make claims about the prevalence of religion-ethnicity intersections or to suggest that these are the only ways that emerging adults conceptualize the intersection. Rather, we use our findings to begin to understand how some emerging adults view the intersection of ethnicity and religion in their own words and to provide directions for future avenues of research.

Importance of Religion/Spirituality as a Social Group

In response to the question of whether there are social groups other than ethnicity that they considered important for their identity, 62 percent of emerging adults said "yes." Of these 62 percent, 42 percent ($n = 93$) mentioned religion or spirituality, or lack thereof, as important to who they are. Participants who mentioned religion did so in varied ways. Many participants asserted that their religious background is the most important aspect of their identity, as a mixed-ethnic (white and Filipina) woman (age 18) described: "I am also a Christian and believe that this defines me more than anything else because it has the most influence on my beliefs, values, actions, and thoughts." "Religion, because it guides me," stated a Chinese-American woman (age 19). Lastly, a Korean-American woman (age 19) asserted, "I am Christian, and my faith is more important than my race/ethnicity because it is a part of who I am that I feel affects my life more profoundly than anything else." Contrary to the theoretical tenets of intersectionality, these emerging adults discussed their multiple identities in hierarchical terms, with their religion being more important to them than their ethnicities.

In contrast, some emerging adults chose to discuss their religious identification as related to, but not more important than, their ethnic background, making explicit the link between their ethnicity and religion. The quote that appeared at the beginning of the chapter nicely illustrates this point:

My religion, my faith in being a Catholic is just as important as my identification with being Filipino. I honor my Catholic upbringing and feel that being a Catholic is important in my life not only for spiritual guidance but also as a connection with my family's past roots. (female, age 22)

A similar sentiment was shared by this Mexican-American female (age 19):

My religion forms a great part of who I am. At the same time, it is a great part of my racial group as well. Ever since I was a little girl, my mom brought me up with all sorts of religious beliefs that now it forms a great part of me.

These emerging adults view their religion and ethnicity as interconnected, rather than independent, isolated domains of their identity that can be discussed in separation. For some, their ethnicity and religion is literally the same thing, such as for Jewish emerging adults. Jewish adolescents and emerging adults are infrequently considered an ethnic group by ethnic identity researchers; they are typically regarded as white. However, some Jewish emerging adults do not identify as white at all: they identity as Jewish. One 19-year-old female stated, "Religion, not so much the spirituality, but the culture of Judaism, is extremely important to me." Another 19-year-old female concurred: "Even though I am not religious, I do identify myself with being Jewish culturally. For me, this is an important part of who I am." These quotes represent an interesting situation in which an intersecting identity, Jewish, is being pulled apart into its components, namely, that it is both a religion *and* a culture.

A few emerging adults also illustrated how religious identification can be group-based, as opposed to an individual process. "I feel that my gender and religion are as important as my race. They set me apart from a number of people and make me more of my own person," said a Mexican-American female (age 18). Similarly, an 18-year-old white female stated, "Being a female is definitely a big one. Religion is important because I don't believe in that, and it separates me from people who do." These statements show evidence of an ingroup/outgroup mentality that is consistent with a social identity theory approach to religious identity (see also Templeton and Eccles 2005), and also highlight how religion can either provide a sense of belongingness or create distance from other groups of individuals.

Connections to Others through Religious/Spiritual Values

The question pertaining to whether emerging adults viewed their values as similar to other members of their ethnic groups evoked responses mentioning religion for 16 percent $(n = 57)$ of the participants. Responses to this question, more so than the responses described above, illustrated themes of interconnectedness between ethnicity and religion:

I feel that I am very similar to those in my race because a majority of Filipinos are Catholic, so our religious views are typically the same. Filipinos are known to be very religious and I believe I fall into that category. (female, age 19)

Some emerging adults, such as the one above, affirmed that they are similar to other members of their ethnic group because of their religion, whereas others viewed religion as a point of difference. These emerging adults viewed their ethnic groups as being connected to a particular religion but did not feel that they personally fit with that. For example, a 22-year-old white and Latino male explained:

Usually not, because most Latinos are Catholic and very religious. They have very strong values that coincide with the morals and values of the church. I am spiritual, but I do not go to church. I have strong values, but they are not defined by the church.

A similar statement was made by a 21-year-old female: "Not really, most of the African-American people that I have met are religious, and I do not fall into that category." Again, these quotes suggest that, for emerging adults, religion/spirituality can provide feelings of both closeness and distance with their ethnic group (Su 2006). One source of distance was specified by this Mexican-American male (age 20), for whom the distinction had less to do with morals and values and more about sources of beliefs: "I see most are really religious, while I have really taken a skew from that direction and rely more on the sciences and proof than religion and faith." This quote provides insight into some possible historical context for religious development in the U.S. With current national debates over evolution versus creationism or "intelligent design," more emerging adults may be moving away from identifying with religions that they view as incompatible with their scientific worldview.

The preceding emerging adults' responses show a general awareness of the intersection of ethnicity and religion, but they did not feel that this intersection applied to them. However, many emerging adults, particularly from Mexican and Filipino backgrounds, did feel that the intersection was relevant to their lives: "Yes, I definitely have the values of my Mexican culture, since I am Catholic. And being Catholic in a Mexican family is huge. My parents did impart these values upon me" (female, age 19). The following Mexican-American participant made it clear that the intersection is not just an individual understanding but applies at the group level as well:

I've been brought up to think that Catholicism is the right religion, don't live with a man if you are not married, no sex before marriage, school comes first, respect EVERY-ONE, and love a lot. Everyone I know shares these same values. (female, age 18)

By stating that everyone she knows shares the same values as she, the young woman is extrapolating the intersections that she personally feels to all other members of her ethnic group. A similar statement was made by this 19-year-old male: "All the Filipino people that I know are very religious. All of my Filipino friends in high school went to church every weekend, and we share the same beliefs."

The homogeneity described by the Mexican-American and Filipino participants was also related by a white male (age 18):

I don't know how to answer this because there are many white people who share conflicting beliefs. There is no universal "white" set of values. Unlike a group like Latin Americans who are overwhelmingly Christian, there are many varying religious and political perspectives, so I don't feel as though my set of values are more or less similar to the standard white person's.

This participant makes the fallacious assumption that there is more diversity within his own background than in the background of others. Incidentally, this was a common response to this question from the white participants, that there is no set of "white values." Although the above participant was the only one to make an explicit reference to another group, many of the responses implicitly indicated that there is less heterogeneity in other ethnic groups' values.

Finally, one participant spoke to the important distinction between religion and spirituality for some emerging adults: "I am not religious as some Irish families or people might be, but I do feel a very vigorous and prevailing spirit underneath myself which sometimes I associate with my heritage" (female, age 18). The distinction made here is in some ways similar to the thoughts related earlier by the Jewish participants. An ethnic background may provide a source of spirituality and identity, even if the religious aspect is not embraced or practiced.

Discussion

Following the concept of intersectionality, theoretically, ethnicity and religion/spirituality are seen as overlapping social identities that cannot be separated. However, as we believe is evident from the quotes presented, emerging adults display a wide range of conceptions about how their own ethnicity and religion are related as well as for how they are related for other people. Of course, it is important to distinguish between their subjective conscious awareness of their identities, as indicated by their responses to the questions, and how their identities appear from an outside perspective. For example, a plausible argument could be made that the Mexican-American and Filipino emerging adults who claimed that they were different from other members of their ethnic groups because they were not Catholic showed just as much evidence of intersecting identities. The fact that they have rejected what they consider an aspect of their ethnic background may have implications for the development of their ethnic identity. Because the lack of religious identification is, in a sense, a form of religious identification, one could argue that ethnic and religious identities intersect for all people. The question that we have addressed in this chapter is the degree to which they are aware of and can verbalize the connection.

Themes that arose in the qualitative data presented in this chapter are consistent with themes in the extant literature that we identified previously. For example, several of the emerging adults mentioned the importance of their family as a context for religious identification. Having a strong family connection with one's religious practices and ethnic identity can open the door for increased social interaction and belongingness with others, as found in Min's (2000) study of Korean-American families. Even without explicit mention of family involvement with religious identification, many emerging adults spoke in terms of belonging to a spiritually based group. Thus, it seems clear that religious identity can be understood from a social identity perspective as well as a personal identity approach. Doing so can highlight the way in which religious identification can help to create and foster positive social relationships through group belongingness. This may be particularly true for religious minorities, for which feelings of collective identity may help increase self-esteem and act as a buffer against the adverse consequences of discrimination (Bankston and Zhou 1995; Sellers, Caldwell, Schmeelk-Cone, and Zimmerman 2003; Tajfel and Turner 1986).

Related to the connection emerging adults see between their families, ethnicities, and religions, there is also a sense that religious identification serves to preserve a connection to their cultural background (Hirschman 2004). Those who espoused views reflecting the intersecting nature of their identities frequently invoked themes of connection, socialization, and preservation. This was especially prominent with emerging adults who referenced the role of parental socialization in their religious identity formation. Some parents may be socializing their children to view religion/spirituality as part of their culture. This may have implications for youth's capacity to reject the beliefs imparted by the parents, as doing so would signify a rejection of part of their cultural identity.

In the introduction, we described three patterns of intersection based on literature with immigrants. In our study of emerging adults, which includes both immigrant and nonimmigrants, we found evidence for two of these patterns. For instance, a Jewish participant evidenced the first pattern, a fused religious and cultural identity with complete overlap between the two. This participant viewed the Jewish faith as *both* a religion and culture. Others described their multiple identities in distinct hierarchical terms, illustrating the second pattern. For instance, some emerging adults stated that their religious identities played a more powerful and salient role in their lives than their ethnic identities. Still others reported that there was no other social group besides ethnic group that was as important to their overall identity, suggesting that their ethnic identities played a more powerful and salient role in their lives than their religious/ spiritual identities. We did not find the third pattern of intersection, with participants describing their religious and ethnic identities as being distinct and conflictual or in opposition to one another. Nonetheless, the responses of the emerging adults in our study showcase at least two ways in which religious and ethnic identities are integrated.

Although we were unable to address this issue directly with our data, Azmitia, Syed, and Radmacher (2008) have proposed that awareness of how multiple identities are interrelated is a developmental issue that warrants further research. Unlike many theories of identity that implicate adolescence as the formative period, Azmitia and colleagues suggested that understanding the interconnectedness of multiple identities is a developmental task primarily located in emerging adulthood. Drawing from a four-year longitudinal study from the beginning to the end of college, they found that emerging adults showed a gradual awareness of how their experiences were contextualized by multiple identities, although this awareness was not very sophisticated until their senior year, if at all. Thus, it may be that adolescence and emerging adults must first come to a resolution regarding their identities in isolation before they can meaningfully understand how they intersect. This exciting development for future research in identity will undoubtedly blossom in the years to come.

AVENUES FOR FUTURE RESEARCH

The voices of emerging adults presented in this chapter illustrate the diversity in the ways that young people think about their ethnic and religious/spiritual identities. This diversity should be acknowledged and embraced by future researchers interested in this topic. Accordingly, our suggestions for future research center on taking a deep consideration of the contexts of youth development. Examinations of contexts of development are difficult undertakings due to their complex and multilayered nature (Bronfenbrenner 1979; Lerner 1996). Recognizing this complexity, we provide some concrete recommendations as a means to begin to peel back those layers and move toward a greater understanding of identity development.

One of the many avenues researchers could pursue is to study the intersection of ethnic identity and spirituality in contexts that vary in cultural support. For instance, in some communities in the U.S., there are vibrant ethnic enclaves that offer rich cultural support for maintaining one's ethnic identity. Oftentimes, the church is one of the important institutions of these communities. In this context, adolescents receive not only religious/spiritual support but also strong cultural support. In other communities in the U.S., however, there are very few cultural supports to maintain an ethnic identity. However, there may be a few ethnic churches. In this context, the religious community may be an even more important source of cultural support for ethnic adolescents who do not have cultural supports outside of church. In other words, the salience of a religious community to one's ethnic identity may change in relation to the cultural richness of the surrounding context.

College-going emerging adults potentially have two contexts in which support could be provided: the college and the home community. The participants in this study from the Santa Cruz sample are in a context that does not offer a great deal of cultural support for ethnic minorities, as it is a primarily white, middle- to upper-middle-class community. These emerging adults may be able to receive enough cultural support from their home community to maintain a strong religious/spiritual identity or may band together with others in similar situations through campus organizations. Exploring how emerging adults navigate these two contexts to foster their religious/spiritual development would be a useful research endeavor.

In general, a deeper examination of the social contexts of development will enable a better understanding of how ethnic identity and religious/spiritual identity intersect. Our findings point to the importance of family, especially parental socialization, in the development of a religious identity that is part of a broader ethnocultural identity. It is important to explore how parents socialize their children in this regard and to consider other potentially influential socialization agents, such as peers. Although the research is limited, there is some evidence that the strength of adolescent ethnic identity is positively related to involvement with same-ethnicity peers (Kiang, Harter, and Whitesell 2007; Phinney, Romero, Nava, and Huang 2001). Likewise, religious/spiritual identity formation may also vary with exposure to religiously homogenous or heterogeneous peer groups.

Another interesting area for future research would be to explore whether parents and adolescents who do not share similar views in terms of spirituality and religion will experience more problems. For instance, some parents who emigrated to the U.S. from Taiwan are Buddhist or atheist, but their children may become socialized into a Christian church. What are the implications for this incongruence in terms of the parent-adolescent relationship or for the adolescents' mental health? Our prior work indicated that parent-adolescent discrepancies in beliefs about appropriate levels of parental control were related to greater family conflict, which, in turn, was related to a higher prevalence of depressive symptoms for adolescents (Juang, Syed, and Takagi 2007). However, we did not address the sources of the discrepant beliefs. For example, adolescents who do not share their parents' spiritual or religious beliefs may be more likely to reject their parents' cultural values and beliefs. Following our findings on parental-control beliefs, this discrepancy could have implications for adolescent mental health. Furthermore, it may have implications for adolescents developing ethnic identities, as they may associate particular beliefs and values with their ethnic background, which, as we highlighted earlier, could be associated with feelings of distance from their ethnicity. In sum, examining whether adolescents' religious/spiritual life is consistent with the demands of their multiple contexts, such as the family, peer, community, and majority-culture expectations, would be fruitful.

Lastly, it would be valuable for future research to consider other domains of identity in relation to ethnic and religious, such as gender or social class. In the Ethnic Identity in Everyday Experiences Study, gender and social class were the two most commonly named groups that emerging adults identified with, in addition to their ethnic group. For instance, one 27-year-old recognized that gender colored her responsibilities as a person of Filipino background:

"Being a Filipino and a woman, there are a lot of expectations and responsibilities, which we must properly reflect and uphold in our background." Another 20-year-old Mexican-American female expressed how her many identities overlap and define who she is: "I believe they all go together. The fact that I am a female, Catholic, middle-class, bisexual and Latina . . . They are all very important and inseparable." For some emerging adults, then, multiple group identities were important to their understanding of selfhood. It is a ripe time for researchers to address this complexity and approach identity development from an interconnected perspective. Only then will we discover the ways in which youth integrate their multiple identities into a coherent sense of self.

CONCLUSION

Ethnic and religious/spiritual identities are two important identity domains in the lives of emerging adults. In this chapter, we attempted to highlight how, for many youth, these two identities intersect. We see the development of strong ethnic and religious identities as a part of positive youth development. Identifying strongly with ethnic or religious/spiritual groups can be an important way for youth to integrate into a caring, supportive community with other youth and adults. Identifying strongly with ethnic or religious/spiritual groups can also provide a sense of belonging and pride that leads to positive mental health. Finally, identifying strongly with ethnic or religious/spiritual groups can prompt youth to transcend themselves, encouraging them to consider their meaning, roles, and responsibilities within the larger community. In an increasingly multicultural society such as the U.S., ethnic identification is becoming more important, even for ethnic-majority youth. At the same time, religion/spirituality continues to play an important role in youths' lives. As such, examining how the intersection of these domains offers strength, reassurance, or sometimes conflict, to youths' sense of self, is needed.

REFERENCES

Arnett, J. J. 2008. From "worm food" to "infinite bliss": Emerging adults' views of life after death. In R. Lerner, E. Phelps, and R. Roeser, eds., *Positive youth development and spirituality: From theory to research*, 231–43. West Conshohocken, PA: Templeton Foundation Press.

Archer, S. L. 1992. A feminist's approach to identity research. In G. Adams, T. Gulotta, and R. Montemayor, eds., *Adolescent identity formation*. Newbury Park: Sage.

Ashmore, R. D., K. Deaux, and T. McLaughlin-Volpe. 2004. An organizing framework for collective identity: Articulation and significance of multidimensionality. *Psychological Bulletin* 130, no. 1: 80–114.

Azmitia, M., M. Syed, and K. Radmacher, eds. 2008. On the intersection of personal and social identities: Introduction and evidence from a longitudinal study of emerging adults. In M. Azmitia, M. Syed, and K. Radmacher, eds., *New directions for child and adolescent development: Intersections between personal and social identity* 120, 1–16. San Francisco: Jossey-Bass.

Bankston, C. L. 2000. Sangha of the South: Laotian Buddhism and social adaptation in rural Louisiana. In M. Zhou and J. V. Gatewood, eds., *Contemporary Asian America: A multidisciplinary reader*, 357–71. New York: New York University Press.

Bankston, C. L., and M. Zhou. 1995. Religious participation, ethnic identification, and adaptation of Vietnamese adolescents in an immigrant community. *Sociological Quarterly* 36: 525–34.

Berry, J. W., and U. Kim. 1988. Acculturation and mental health. In P. R. Dasen, J. W. Berry, and N. Sartorius, eds., *Health and cross-cultural psychology: Towards application*, 207–36. Newbury Park, CA: Sage.

Bronfenbrenner, U. 1979. *The ecology of human development.* Cambridge, MA: Harvard University Press.

Cavalcanti, H. B., and D. Schleef. 2005. The case for secular assimilation? The Latino experience in Richmond, Virginia. *Journal for the Scientific Study of Religion* 44: 473–83.

Côté, J. E. 1996. Identity: A multidimensional analysis. In G. Adams, R. Montemayor, and T. Gulotta, eds., *Psychosocial development during adolescence*, 130–80. Thousand Oaks, CA: Sage.

Crenshaw, K., N. Gotanda, G. Peller, and K. Thomas, eds. 1995. *Critical race theory.* New York: Free Press.

Cross, W. E., Jr. 1995. In search of blackness and Afrocentricity: The psychology of black identity change. In H. W. Harris, H. C. Blue, E. E. H. Griffith, eds., *Racial and ethnic identity: Psychological development and creative* expression, 53–72. New York: Taylor and Francis/Routledge.

Eccles, J., J. Templeton, B. Barber, and M. Stone. 2003. Adolescence and emerging adulthood: The critical passage ways to adulthood. In M. H. Bornstein, L. Davidson, C. L. M. Keyes, and K. A Moore, eds., *Well-being: Positive development across the lifespan*, 383–406. Mahwah, NJ: Lawrence Erlbaum Associates.

Erikson, E. H. 1968. *Identity: Youth and crisis.* New York: Norton.

French, S. E., E. Seidman, L. Allen, and J. L. Aber. 2006. The development of ethnic identity in adolescence. *Developmental Psychology* 42: 1–10.

Garcia Coll, C., K. Crnic, G. Lamberty, B. H. Wasik, R. Jenkins, H. V. Garcia, and H. P. McAdoo. 1996. An integrative model for the study of developmental competencies in minority children. *Child Development* 67: 1891–1914.

Gutierrez, L. M. 1990. Working with women of color: An empowerment perspective. *Social Work* 35: 149–53.

Habermas, T., and S. Bluck. 2000. Getting a life: The emergence of the life story in adolescence. *Psychological Bulletin* 126: 248–69.

Harter, S., and A. Monsour. 1992. Developmental analysis of conflict caused by opposing attributed in the adolescent self-portrait. *Developmental Psychology* 28, no. 2: 251–60.

Herberg, W. 1960. *Protestant, Catholic, Jew: An essay in American religious sociology.* Rev. ed. Garden City, NY: Anchor Books.

Hirschman, C. 2004. The role of religion in the origins and adaptation of immigrant groups in the United States. *International Migration Review* 38: 1206–33.

hooks, b. 2003. Reflections on race and sex. In A. Darder, M. Baltodano, and R. D. Torres, eds., *The critical pedagogy reader*, 238–44. New York: Routledge.

Hurtado, A. 1997. Understanding multiple group identities: Inserting women into cultural transformation. *Journal of Social Issues* 53, no. 2: 299–328.

Hurtado, A., and J. M. Silva. (2008). Creating new social identities in children through critical multicultural media: The case of *Little Bill.* In M. Azmitia, M. Syed, and K. Radmacher, eds.,

New directions for child and adolescent development: Intersections between personal and social identity 120, 17–30. San Francisco: Jossey-Bass.

Jeung, R. 2005. *Faithful generations: Race and new Asian American churches.* New Brunswick, NJ: Rutgers University Press.

Juang, L. P., M. Syed, and M. Takagi. 2007. Intergenerational discrepancies of parental control among Chinese-American families: Links to family conflict and adolescent depressive symptoms. *Journal of Adolescence,* 30, no. 6, 965–75.

Kiang, L., S. Harter, and N. R. Whitesell. 2007. Relational expression of ethnic identity in Chinese Americans. *Journal of Social and Personal Relationships* 24, no. 2: 277–96.

King, P. E. 2008. Spirituality as fertile ground for positive youth development. In R. Lerner, E. Phelps, and R. Roeser, eds., *Positive youth development and spirituality: From theory to research,* 55–73. West Conshohocken, PA: John Templeton Foundation.

Lerner, R. M. 1996. Relative plasticity, integration, temporality, and diversity in human development: A developmental contextual perspective about theory, process, and method. *Developmental Psychology* 32, no. 4: 781–86.

Marcia, J. 1980. Identity in adolescence. In J. Adelson, ed., *Handbook of adolescent psychology,* 159–97. New York: Wiley.

Mattis, J. S., M. K. Ahluwalia, S. E. Cowie, and A. M. Kirkland-Harris. 2005. Ethnicity, culture, and spiritual development. In E. C. Roehlkepartain, P. E. King, L. Wagener, and P. L. Benson, eds., *The handbook of spiritual development in childhood and adolescence,* 283–96. Thousand Oaks, CA: Sage.

Mattis, J. S., and R. J. Jagers. 2001. A relational framework for the study of religiosity and spirituality in the lives of African Americans. *Journal of Community Psychology* 29, no. 5: 519–39.

McAdams, D. P. 2001. The psychology of life stories. *Review of General Psychology* 5: 100–122.

Meeus, W., J. Iedema, M. Helsen, and W. Vollebergh. 1999. Patterns of adolescent identity development: Review of literature and longitudinal analysis. *Developmental Review* 19: 419–61.

Miller, W. R., and C. E. Thoresen. 2003. Spirituality, religion, and health. *American Psychologist* 58, no. 1: 24–35.

Min, P. G. 2000. The structure and social functions of Korean immigrant churches in the United States. In M. Zhou and J. V. Gatewood, eds., *Contemporary Asian America: A multidisciplinary reader,* 372–90. New York: New York University Press.

Phinney, J. S. 1989. Stages of ethnic identity development in minority group adolescents. *Journal of Early Adolescence* 9: 34–49.

———. 1990. Ethnic identity in adolescents and adults: Review of research. *Psychological Bulletin* 108, no. 3: 499–514.

———. 2003. Ethnic identity and acculturation. In K. M. Chun, P. B. Organista, and G. Marin, eds., *Acculturation: Advances in theory, measurement, and applied research,* 63–81. Washington, D.C.: American Psychological Association.

———. 2006. Ethnic identity in emerging adulthood. In J. J. Arnett and J. L. Tanner, eds., *Emerging adults in America: Coming of age in the 21st century,* 117–34. Washington, D.C.: American Psychological Association.

Phinney, J., and A. Ong. 2007. Conceptualization and measurement of ethnic identity: Current status and future directions. *Journal of Counseling Psychology* 54: 271–81.

Phinney, J. S., I. Romero, M. Nava, and D. Huang. 2001. The role of language, parents, and peers in ethnic identity among adolescents in immigrant families. *Journal of Youth and Adolescence* 30, no. 2: 135–53.

Quintana, S. M. 2007. Racial and ethnic identity: Developmental perspectives and research. *Journal of Counseling Psychology* 54, no. 3: 259–70.

Robinson, L. 2003. South Asians in Britain: Acculturation, identity, and perceived discrimination. *Psychology and Developing Societies* 17: 181–94.

Seaton, E. K., K. M. Scottham, and R. M. Sellers. 2006. The status model of racial identity development in African American adolescents: Evidence of structure, trajectories, and well-being. *Child Development* 77, no. 5: 1416–26.

Sellers, R. M., C. H. Caldwell, K. Schmeelk-Cone, and M. A. Zimmerman. 2003. The role of racial identity and racial discrimination in the mental health of African American young adults. *Journal of Health and Social Behavior* 44, no. 3: 302–17.

Smith, C., and M. O. Denton. 2005. *Soul searching: The religious and spiritual lives of American teenagers*. New York: Oxford University Press.

Smith-Hefner, N. 1999. *Khmer American: Identity and moral education in a diasporic community*. Berkeley: University of California Press.

Spencer, M. B. 2006. Revisiting the 1990 Special Issue on Minority Children: An editorial perspective 15 years later. *Child Development* 77, no. 5: 1149–54.

Su., D. 2006. *Using self-defining memories to examine the content and context of Cambodian American ethnic identity*. Unpublished manuscript. University of California, Santa Cruz.

Syed, M., M. Azmitia, and J. S. Phinney. 2007. Stability and change in ethnic identity among Latino emerging adults in two contexts. *Identity: An International Journal of Theory and Research* 7, no. 2: 155–78.

Tajfel, H., and J. C. Turner. 1986. The social identity theory of intergroup behavior. In S. Worchel and W. Austin, eds., *Psychology of intergroup relations*, 7–24. Chicago: Nelson-Hall.

Templeton, J. L., and J. S. Eccles. 2005. The relations between spiritual development and identity processes. In E. C. Roehlkepartain, P. E. King, L. Wagener, and P. L. Benson, eds., *The handbook of spiritual development in childhood and adolescence*, 283–96. Thousand Oaks, CA: Sage.

United Nations Statistics Division. 2006. Population by religion, sex, and urban/rural residence: Each census, 1985–2004. Retrieved January 31, 2007, from http://unstats.un.org/unsd/demographic/sconcerns/popchar/popchar2.htm.

Waters, M. C. 1999. *Black identities: West Indian immigrant dreams and American realities*. Cambridge: Harvard University Press.

Williams, R. B. 2000. Asian Indian and Pakistani Religions in the United States. In M. Zhou and J. V. Gatewood, eds., *Contemporary Asian America: A multidisciplinary reader*, 392–407. New York: New York University Press.

Yang, F. 1999. *Chinese Christians in America: Conversion, assimilation, and adhesive identities*. University Park: Pennsylvania State University Press.

Zimmerman, M. A. 1995. Psychological empowerment: Issues and illustrations. *American Journal of Community Psychology* 23, no. 5: 581–99.

15

Considering Context, Culture, and Development in the Relationship between Spirituality and Positive Youth Development

Na'ilah Suad Nasir

When psychology was in its infancy, it was largely concerned with the softer side of understanding the functioning of mind, personality, and development (Cole 1996). Scholars (who now read much more like philosophers than psychologists) pondered the nature of development and of the soul (e.g., James 1890). One goal was to focus on the things that the biological sciences couldn't help us understand—the internal world of the mind, the unique collective human psychology (e.g., Jung), and unconscious drives and needs (e.g., Freud). As the field of psychology developed, it began to feel inferior to the hard sciences like physics, and so it decided to focus on only what was measurable (putting the lives of millions of mice in jeopardy) and to develop formulas to quantify psychological processes. By the 1950s, this approach dominated psychology. But time passes, as it is inclined to do, and the pendulum has swung such that we are now back in a time when a focus on humans as whole, thinking, feeling beings is once again at the forefront; hence, this volume on the intersection of spirituality and positive youth development.

I relay this history here for two reasons. First, it conveys the understanding that a concern for human processes that were unobservable yet central to the human experience has long been present in the field of psychology. Thus, the focus on spirituality and positive youth development in this volume represents a continuation of the psychological tradition, not a departure from it. Second,

I relay this history as a caution, lest we approach these topics with an inferiority complex about the "soft" nature of the topics before us. These aspects of the human psyche and social life are critical to understand and greatly motivate human behavior and self-definitions for people around the world.

Focusing a psychological lens on the spiritual is certainly a novel endeavor for contemporary mainstream psychology. The early scholars (e.g., Rodgers, Maslow, James, and Jung) whose work focused on issues of spirituality did not conduct studies in the contemporary sense, though their theories and writings were clearly informed by their experiences in the field, the research literature of the time, and their observations of clinical patients. Thus, early research in this area was not forced to wrestle with some of the many thorny conceptual and methodological issues facing psychologists today engaged in the empirical task of understanding the role of spirituality in development. The charge of this volume is to identify some of these challenges and offer suggestions for researchers in their endeavor to understand better the intersection between spirituality and positive youth development.

There is no doubt that spirituality is a critical area for psychological research. It is both little studied and paramount to the subjective experience and lives of people. This point was underscored for me recently in a conversation with one of my senior colleagues—a well-known developmental psychologist—who was spending the year on sabbatical. I asked her about her sabbatical plans, and she told me about the book she was working on but also noted that she planned to take better care of herself on this sabbatical. She noted several areas of self-care, including spirituality, eating healthy, and exercise. It struck me in the course of that conversation how rare it is for academics to talk about spiritual matters, though spirituality is such a profound part of people's experiences of their lives.

Spirituality may also be important from a scientific perspective because it is associated with positive outcomes despite adversity—that is, it seems to support resilience. For example, in a study of families with children with developmental delays, researchers (Gallimore, Weisner, Guthrie, Bernheimer, and Nihira 1993) found that the families who adapted the most smoothly were those with a strong sense of their spirituality—those who saw their lives as being important above and beyond their current circumstances and who saw themselves in the context of transcendence.

Positive youth development is also an area in its infancy, though perhaps more frequently empirically studied than spirituality. Understanding the positive aspects of development—what makes us thrive, what supports deep satisfaction, what supports our growth as loving, giving, altruistic, and whole human beings—is an approach that has the potential to shift psychology from a field that looks at the pathology of the human to one that understands and celebrates our potential (Larson 2000; Seligman and Csikszentmihayli 2000). Thus, in my view, the focus on the intersection between positive youth development and spirituality is not only an important scientific endeavor but also important with respect to reenvisioning our human potential for happiness, wellness, wholeness, and greatness.

CONCEPTUAL AND THEORETICAL ISSUES

In this chapter, I highlight four key issues researchers will need to address to support research in this area that is rigorous and true to the phenomenon under study and that pushes the field in theoretically important directions. Specifically, I argue that (1) researchers will need a robust operational definition of spirituality, (2) researchers must understand both spirituality and positive youth development as highly contextual and taking shape in different ways in different contexts, (3) researchers should attend to the role of spiritual identity and how that identity has implications for behavior and psychological processes, and (4) researchers must strive to understand the intersection between positive youth development and spirituality as involving an interaction between individual characteristics and the environment or context.

Underlying these four issues is the assumption that, since the intersection of spirituality and positive youth development is a new field, initial research on this topic must be grounded in the particulars, so that we may better understand what the important issues are (rather than making *a priori* assumptions about them) and how they play out in the lives of individuals. I also detail the methodological challenges inherent in this work, as well as my suggestions for addressing them, and offer thoughts about potential starting places. I focus largely in this chapter on the study of spirituality and less so on the study of positive youth development, with some attention to the possible relation between the two. This imbalance occurs because, to some degree, the field of positive youth development has made significant headway defining and operationalizing variables that researchers can draw upon, while, in order to understand the implications that spirituality has on positive youth development, theoretical and methodological work will need to be done to develop conceptual tools to understand the development of spirituality.

Defining Spirituality

A robust operational definition is critical to launching a program of research that is rigorous and valid. Defining the term *spirituality* is complex because it involves both internal attributes (e.g., traits, personality, proclivities, beliefs) and external factors (membership in a spiritual community, values, norms) (Zinnbauer, Pargament, Cole et al. 1997). A robust definition would need to include both action in the world (what one does) and beliefs about one's role in the world (why one does it). Beliefs that indicate spirituality might include believing in a transcendent power, viewing oneself as a spiritual being, and believing that spiritual matters are central in one's life (Hill and Pargament 2003; Zinnbauer et al. 1997). However, the enactment of *spirituality* (that is, a functional definition) is less well represented in the literature and might include acting in ways that help and support individuals and social causes and actively supporting one's own spiritual growth; it may also involve belonging to a spiritual community. Thus, there are at least three critical components to an operational definition of *spirituality*—considering oneself to be spiritual, acting in altruistic ways toward others, and being committed to one's own spiritual development.

Many might agree that these elements are central to being a spiritual person; however, this definition alone does not address how to obtain a sample of "spiritual" people for the purposes of research. One can hardly go out into the general public to assess spirituality (based on the above criteria) and include only those who qualify by virtue of their answers on a screening measure. Such a procedure would be difficult to implement and impractical. One common solution has been to sample from religious communities, thus conflating religious believers with spiritual people. I would caution against this approach. Though tidy, it confounds religion, spirituality, and membership in a community of practice and does not allow one to distinguish between the effects of these three often-intertwined but conceptually quite different variables.

While it is clear that religions tend to support dedication to spiritual principles and that spiritually inclined people often seek out religious communities as places to find others who share their values (Marler and Hadaway 2002; Zinnbauer et al. 1997), spirituality and the practice of religion are not the same thing, though they have at times been confounded in the research literature (Hill and Pargament 2003; Zinnbauer et al. 1997; MacDonald 2000; Miller and Martin 1988; Saucier and Skrzypinska 2006). People choose to join religious communities for a wide range of reasons, including the need for community, access to resources, and a desire to improve their lives. Further, the definition above does not confine spirituality to the practice of religion; many spiritual people may not practice a specific religion at all, or they may do so without belonging to a religious community. Research has shown that not only is there a group of people who consider themselves to be "spiritual, but not religious" but that these people may differ in important ways (including personality traits) than religious people (Saucier and Skrzypinska 2006).

Studying the intersection between spirituality and positive youth development also requires an operational definition of *positive youth development*, though such a definition is perhaps more readily available in the research literature (Damon 2004; Larson 2000; Lerner, Brentano, Dowling, and Anderson 2002). In general, "The field of positive youth development focuses on every child's unique talents, strengths, interests, and future potential" (Damon 2004, 13). The focus for those concerned with measuring positive youth development has been conceptualizing indices of "thriving" (Lerner et al. 2002).

A methodological challenge, both for measuring spirituality and positive youth development, is individuals' tendency to over-report positive aspects of their behavior and under-report negative behaviors—a social desirability effect. Self-report measures by nature rely on individuals accurately reporting actions, values, and beliefs. Alternatives to self-report measures include peer nominations—where community members are asked to nominate "spiritual exemplars" or observational behavioral measures—which don't seem very realistic given both the practical constraints and critical internal components of spirituality.

Understanding Both Spirituality and Positive Youth Development as Highly Contextual and Taking Shape in Different Ways in Different Contexts

As noted, not only are the meanings of *spirituality* and *positive youth development* nuanced and difficult to measure (and even more difficult to standardize), but they also take shape differently with respect to context. There is much literature to support the idea that development and context are deeply intertwined (Cole 1996; Rogoff 2003; Saxe 1991; Lerner 1991; Brofenbrenner 1979, 1993). Developmental and learning processes are intricately tied to the local contexts within which they occur—through the nature of the activity itself, the goal of the activity, how it is organized, the tools that one has access to, and who else is present. For instance, what one learns about math and how one's mathematical cognition develops are greatly influenced by whether one is learning at school or at home, with peers or with a teacher, with sticks and rocks or with rulers and fraction manipulatives (Saxe 1991; Saxe, Dawson, Fall, and Howard 1995). Similarly, the development of social and emotional competencies is shaped in key ways by cultural activities and social norms.

This interaction between context and development may be particularly relevant in the moral and spiritual realms, where the content is highly subjective. Indeed, what is considered "right" or "good" is culturally defined and enacted (Schweder, Mahapatra, and Miller 1987). Thus, what is considered spiritual or positive developmentally may differ across cultural spaces and practices.[1] However, context includes more than just culture. In this section, I outline the relevance of considering three types of context: the developmental context, the local context (place, space, and group), and the historical context.

Developmental Context. In understanding the relation between spirituality and positive youth development, the developmental context is an important factor to consider. Changes over developmental time in spirituality may not be linear. That is, unlike cognition or social skills, spirituality may not increase in stepwise fashion with age. For instance, some people may experience an increased or decreased sense of spirituality during adolescence, or during this time a previously unquestioned spiritual commitment may become questioned. Others may think little about spirituality until later in development—perhaps becoming salient in young adulthood or with the birth of a child. These varying trajectories through spiritual terrain may be related to the extent to which and the way in which the child's social world organizes such experiences for the child and/or the way the young person makes sense of them.

Similarly, the development of competencies and characteristics that we would consider positive may take a variety of developmental paths. Both spiritual development and positive youth development may be affected in important ways by experiences of challenge or hardship (Larson 2000). For instance, spirituality can be sparked by difficult times, when one turns to a higher source for guidance and consolation. Such experiences may, in turn, support the development of positive competencies, characteristics, and ways of coping in the world.

Cultural and Geographic Context. The prior section made the assumption that positive developmental traits and commitment to spirituality are internal to the person being studied. However, researchers must consider the possibility that both spirituality and positive youth development may not remain stable in individuals across a variety of social situations and social and cultural practices. A person who seems spiritual, generous, and compassionate in one setting may be quite the opposite in another. Profound context effects have been reported in research across multiple domains. Not only do we behave, think, and feel differently in different social contexts, but we all move through multiple spaces constantly as we go about our days. Spirituality and positive youth development must be understood in relation to these ever-changing contextual features of people's lives—the manifestation of them may look different depending on where one looks and at what time. This is made even more complicated by the fact that there are communities where it is acceptable to be spiritually oriented and communities where it is not. I mentioned earlier that people often don't discuss being spiritual in academic contexts. This feature of the setting may make spirituality difficult to identify in university classrooms (for instance) but easy to see in religious settings or celebrations.

The nature and demonstration of spirituality and positive youth development may also shift in fundamental ways depending on the racial (and gender and class) group membership of the people. Ethnicity and gender groups and people in particular geographic locations may hold differing deep beliefs or expectations for the scope of their lives. They may make sense of actions differently; thus, instruments and protocols designed to measure spirituality and, to a lesser degree, positive youth development must get at underlying beliefs rather than surface behaviors. For instance, a construct like helping others may take different forms and may have different consequences for males and females, for individuals from various racial/ethnic groups, and for people who negotiate vastly different levels of adversity (economic and otherwise).

Historical Context. Spirituality may also take different forms depending on the historical context of the family, society, and individual. That is, the historical context makes particular spiritual paths available and others less available. Buddhism is widely endorsed and practiced in North American today, but, in the 1920s in the midwestern United States, my great-grandmother was ostracized for practicing Buddhism. Wicca was punishable by death in the 1700s but is widely practiced today. African slaves in the United States were forbidden from practicing Islam—their native religion—so they took up Christianity (which was more freely available to them) to express their spirituality (Austin 1997; Diouf 1998; Alford 2007). The point is that religious practices are differentially available over historical time, and, since people often look to religious practices to express spirituality, the forms that spirituality takes may also shift over time. Even what is considered spirituality may be historically located. Similarly, what are viewed as positive developmental traits also shift over time as well. For instance, while suppressing one's emotions may have been viewed positively fifty years ago for men, today it is viewed as positive development to appropriately express emotion.

These three types of contextual variations on spirituality, positive youth development, and the relation between the two have deep and wide consequences for our understandings of these phenomena. Our studies must be sensitive to these multiple layers of context.

Identity in Positive Youth Development and Spirituality

Key to the definition of *spirituality* is the person's subjective sense of themselves and their life as of spiritual significance. Spirituality is, thus, a form of identity—the extent to which and the way in which one identifies oneself as spiritual. The research literature on identity has highlighted the critical role of identity in development (Erikson 1950, 1980; Harter 1999; Roeser, Peck, and Nasir 2006; Spencer 1999). The coalescing of a coherent sense of identity is a core task of both adolescence and adulthood and is central to people's sense of satisfaction and well-being. In the research literature, identity is both a role that we perform and an internal sense of who we are (Nasir and Hand 2006; Wenger 1998; Gee 2001).

In sociocultural theories of identity, identity is viewed as being related to becoming an increasingly central participant in a community of practice (Wenger 1998), and this body of research points out the intertwining of identity and participation in cultural practices. Recent work has argued that identities come in "constellations" (Nasir 2006). In other words, certain identities afford and preclude other identities, by virtue of society's definitions of identities that "go together." For example, being a gang member and being a good student are identities that don't often go together and, thus, are not in the same identity constellation (e.g., people who have one don't often have the other).

This concept may be particularly useful in considering the relationship between spirituality and positive youth development. Identifying oneself as spiritual becomes part of an identity repertoire that both comes from and has implications for the other identities that one takes on. Researchers might ask, then, what other identities a spiritual identity implicates (in particular contexts, at a particular time). That is, when one takes on a spiritual identity, what other identities and practices become available? On the other hand, researchers might also ask what other identities and practices give rise to the development of a spiritual identity. Importantly, the answers to these questions may depend on what tradition the spirituality is couched within. This reinforces the importance of attending to why and how spirituality is expressed.

I noted earlier that we can conceptualize identities as being linked in important ways to the communities of practice that individuals participate in. This notion implies that local practices, such as games, home practices, and church activities may influence the identities that individuals have access to and take up. This perspective implies that the multiple life-spaces where people learn and express spiritual identities are critical to understand. The way that these spiritual identities and the expression of them change as people move across the activities of their lives may also be an important focus for research.

Interaction of Individual and Context

Many of the sections thus far have touched on the intricate relation between the individual and his or her social context. Perhaps the most oft-cited controversy in psychology and in science is the nature/nurture controversy (Garcia-Coll, Bearer, and Lerner 2004), which is concerned with where the locus of influence on development is—something hard-wired in individuals, potentially carried on the DNA, or something that our environment provides. Recent iterations of this debate are no longer polarized about whether nature or nurture shapes development more—most theorists acknowledge the significant role of both, as well as the ways in which they are deeply linked (Garcia-Coll, Bearer, and Lerner 2004; Grigorenko, 2008).

With respect to spiritual development and positive youth development, an important question is whether we assume that some people are simply more spiritual or possess more positive developmental traits or whether we assume that spirituality and positive developmental trajectories are a product of the social context and can be socialized.

By way of illustration, consider two cases. Consider the experiences of two students, each real cases that illustrate one of these assumptions. Barbara grew up in a nonreligious household, in extreme poverty, with a mother and older brothers who were heavily drug- and alcohol-dependent and who were abusive to her. At the age of 12, Barbara decided that she wanted to be Catholic—there was a Catholic church down the street from her house, which she started attending on her own, and she reports that she was drawn to the idea that, through the practice of Catholicism, she could be good and the conception of God as omnipresent and omnipotent. In this case, Barbara's own desire for "goodness" and spiritual development (as well as her independence and the proximal location of the church) supported her involvement in Catholicism. In this case, not only do we see an interaction between individual personality characteristics and the local environment, we also see an interaction between spirituality and the development of positive attributes (an interaction that is potentially mediated over time through involvement in the church community and its practices).

William's case illustrates another kind of profile. William grew up in a religious community and attended a religious school. Despite his contrary behaviors and disruptive actions, William was treated in this community with kindness, and his teachers and community members had high hopes for his development and for his future. As he grew older, William's disruptive behaviors decreased, and he began to take up some of the characteristics that his teachers and senior community members expected of him, such as kindness to his fellow classmates and generosity. These changes in the positive developmental traits that he displays co-occurred with increasing responsibility in the religious community, such as leading prayer for younger students.

In the first case, Barbara possesses traits and developmental assets that cause her to seek out a spiritual community, and in the second, the spiritual community seems to support the development of both a spiritual awareness and positive developmental competencies. These two might look similar on measures of spirituality and positive youth development at certain time points, but their trajectories are very different. And not only do trajectories differ, but the movement through social contexts over time interacts with development and individual personality and proclivities in complex ways.

Since spirituality and positive youth development can be understood both as having personality and trait components and as being taught and socialized, it will be an important focus for the field to understand better this interaction of individual characteristics and environment/context. Some people may have a natural affinity for the spiritual and may find it even in the most spiritually deficient environments. Others may be surrounded by spiritual concepts, ideas, and practices and may accept or reject them, or they may go through iterations of both over the life-cycle. Better understanding both of these trajectories and the many other versions of pathways toward spirituality and positive youth development is an important potential contribution of the proposed line of research and for increasing our understanding of how to best support the positive developmental trajectories of youth.

METHODOLOGICAL ISSUES

Much has been made in this chapter thus far of the importance of maintaining a dual lens that simultaneously keeps in perspective the individual and the context. However, how to do this methodologically is not obvious. Here, I propose the use of a cultural practice framework that takes socially organized activities as the unit of analysis.

As noted in the introduction, the methodological problem of maintaining a simultaneous view of the individual and his or her social context has been an important topic in psychology—both historically and contemporarily (Cole 1996; Rogoff 1990, 1995, 2003; Bronfenbrenner 1979, 1993; Lewin 1935, 1951) articulated the importance of understanding the multiple layers of context within which individual developmental processes were embedded with ecological theory. They argued that several cocentric layers of context were important to focus on, ranging from the society to the individual and including schools and families. However, while this theory resonated well with scholars concerned with ecological influences on development, it offered less in the way of methodologies with which to study these interactions, other than to argue that research should attempt to understand these multiple interacting layers of context. It is less clear in this theory *how* one conducts an analysis that takes these multiple levels into account.

Sociocultural theory is similarly concerned with multiple layers of context and understanding the way in which these layers intertwine (Nasir and Hand 2006). Scholars in this tradition argue that cultural activities that people participate in as a part of their everyday lives are an important focus for the study of human development (Cole 1996; Lave 1988; Rogoff 2003; Saxe 1999; Wenger 1998; Wertsch 1998). From this perspective, understanding development requires a focus on how individuals participate in particular activities and how they draw on artifacts, tools, and social others to solve local problems.

Cultural practices as a unit of analysis allow for analyses of multiple levels of context simultaneously. Leontiev (1979), a student of Vygotsky, described activities as a focus for psychological analysis in great detail. Activity theorists elaborated and built on his ideas and articulated the process by which goal-directed activity unfolds and gives rise to thinking and development. Engeström and colleagues (1999) added a second layer to the classic triangle of mediation to account for the importance of community norms, roles, and the distribution of tasks in social activity.

A focus on cultural practices supports a close look at both psychological processes and the broader contexts that surround their expression and development. Practices provide a context within which relationships are developed and maintained and where one can view both socialization and active sense-making on the part of the individual. From this perspective, cultural practices do not exert unilateral influence on the child but rather interact with individuals in a bidirectional process, whereby people play an active role in their own development.

I draw from Nasir and Kirshner (2003) to describe four critical levels of cultural practices that support the analysis of the influence of multiple levels of context on the developing individual. In the following sections, I briefly describe these four levels of analysis—institutional context, cultural practices, social interaction, and individual characteristics. It is important to note that, though I treat them as distinct levels for the purposes of analysis, in reality, these levels are deeply interwoven. At each level, I also consider how this level of practice might have implication for understanding the relation between spirituality and positive youth development.

Institutional Context

Institutional contexts provide a critical backdrop for understanding how development plays out within cultural practices (Bronfenbrenner 1979; Forman, Minnick, and Stone 1996; Moen, Elder, Luscher, and Bronfenbrenner 1995; Nicolopoulou and Cole 1993). Institutions might be churches, schools, social-service agencies, community centers, or families. Analyses of the institutional context include the guiding philosophies and histories of the host institutions as well as institutionally held cultural belief systems (Nasir 2004). Analysis at this level might also focus on what membership in the institution or organization entails and how the nature of membership might have implications for individuals' participation in particular activities within the institution. It is important to note that the institutions themselves are set in neighborhood, city, state, and even national contexts that are critical for understanding the events at hand; such settings influence who has access to what institutions as well as fundamentally shape the nature of the practices themselves (Bronfenbrenner 1979).

With respect to the study of the relation between spirituality and positive youth development, this level might focus on churches, mosques, synagogues, ashrams, and temples and on how these as institutional settings are informed by values, beliefs, and norms that have implications for the types of daily activities that members get involved in, which, in turn, will influence the development of positive competencies and characteristics. An exploration of the level of the institution may also reveal particular kinds of norms, values, and organizational structures that best support or that fail to support the positive developmental trajectories for members at different developmental stages (both with respect to age and positive development).

Cultural Activities

At the level of the cultural practice, we focus on the activity structures in the practice and the tools and artifacts that participants use to carry out the activity (Saxe 1999; Vygotsky 1978). Activity structures include the cycles of activity in a practice—that is, the components that make the activity coherent. These activity structures fundamentally shape the nature of the cultural practice, how activity is accomplished, and the roles available for participants (Hand 2008; Nasir and Hand, in press). Tools and artifacts refer to the material and symbolic goods with which we accomplish activity (Cole 1996). Material tools include physical cultural props (e.g., desks, computers), while symbolic tools are more abstract—the ideas (often collectively held) that come to structure our activity as we participate in the environment. Symbolic tools (or what have also been called ideational artifacts) include cultural belief systems that people use to understand their activity and that newcomers (including children) are socialized into through activity (Cole 1996; Nasir 2004, 2006).

In studying the relation between spirituality and positive youth development, this level might focus on the particular practices (both within spiritual communities and among spiritual people). Research questions at this level might include the following:

What implications do the practices of spiritual people have for the development of PYD characteristics?

How are activities structured and enacted by participants of different ages?

What are the variations in ways individuals embrace and enact these practices and with what consequences for PYD?

To what degree are spiritual practices (and/or PYD practices) offered by religious institutions?

Social Interaction in Practices

The third level focuses more specifically on social interaction within cultural practices. Social interaction is particularly rich for understanding how participation in cultural practices comes to mediate spirituality and how spirituality gets expressed across settings. This level may also explore how spiritual people interact and perceive interactions. An important focus here is the microprocesses that occur within interactions that shape development (Cote and Levine 2002; Nasir and Saxe 2003). It is as individuals interact and talk with one another that identities and perspectives on the world are shaped (Antaki and Widdicombe 1998; Ochs and Capps 1996; Nasir and Cooks 2007). In particular, at this level of analysis, our concern is to understand the microgenetic (moment-to-moment) development of spirituality, spiritual identities, and positive developmental competencies. One could also focus on how positive developmental characteristics are fostered or even jointly accomplished in social interaction.

With respect to a focus on understanding the interactions between spirituality and positive youth development, this level of analysis might focus on the relationships and interactions within spiritual communities, among spiritual people, and for the same individual across spiritual and other life contexts. Research questions might include the following:

What are the ways in which interactions occur in spiritual communities that may support the development of PYD?

What are the ways spiritual people interact and make sense of interactions that strengthen PYD?

What are the ways the social interactions of spiritual people are facilitated by well-developed positive competencies?

Does social interaction in spiritual settings support growth or changes in the nature of social interactions in other life settings?

Individual Characteristics and Meaning

While some social theories reduce individuals to nonagentic pawns, sociocultural theories assume an active, interactive, highly agentic individual (Cole 1996; Holland, Lachicotte, Skinner, and Cain 1998; Nasir 2005). In this level of analysis, concerns include the sense-making processes, proclivities, and characteristics of individuals that interact with and get expressed in social interactions, cultural practices, and institutions. In essence, this level captures both individual variation and how the individual makes sense of and internalizes or rejects beliefs, values, and ways of being and doing offered in cultural practices (Holland, Lachicotte, Skinner, and Cain 1998).

The way that the individual makes sense of the social context is critical to consider. Spencer and colleagues (Spencer 1999, 2006; Spencer and Harpalani 2004) use the term *phenomenology* to highlight the importance of this individual sense-making and its developmental and life consequences.

At the level of the individual, research that examines the relation between spirituality and positive youth development might explore the following:

Do people who participate in spiritual communities differ from the general population in the extent to which they display PYD characteristics?

If so, to what extent are those differences due to engagement in the practices of those communities, or are such individuals drawn to religious communities due to preexisting PYD differences?

What role does spiritual or religious identity play in mediating the relationship between participating in a spiritual community and PYD?

Are there ways in which religious/spiritual identities might constrain PYD development? Under what conditions?

These four levels, with a concern for cultural practices at the center, offer a comprehensive way to attend to both developing individuals and the way in which their development is intertwined with the multiple settings of their lives.

RESEARCH APPROACHES

Given the four levels outlined above, what might the components be of a feasible program of research that incorporates these four levels, and what might it add to our understandings of the relation between positive youth development and spirituality? In this section, I briefly outline two potential studies: one a qualitative study that focuses on exploring the lives of spiritual children and adolescents and the ways in which spirituality might support the positive youth development; and, second, a retrospective interview study of spiritual and nonspiritual adults that combines interview and survey methods.

Qualitative Study of Spiritual Adolescents

One potential study as a first step in a program of research with the goal of understanding the relation between spirituality and positive youth development would be to explore the lives of spiritual adolescents. One could include a sample of both spiritual exemplars (nominated by members of their communities) and more common spiritual adolescents (as measured by spiritual identity and spiritual practices). Methodologies might include both observations of their social interactions and practices in the context of religious communities and institutions, as well as interviews with the adolescents, their parents, peers, and other community members and leaders. Interviews could focus on both students' current level and type of participation, displays of PYD characteristics, retrospective accounts of their developmental trajectories (both with respect to spirituality and PYD), and the adolescents' own sense of the role of spirituality in their lives. Individual survey measures and other instruments could examine the personal meaning of spirituality and types of positive competencies developed. Such a study would allow for an exploration of all four levels of analysis and might result in initial hypotheses about the relation between spirituality and PYD, as well as about how these aspects of development play out in the lives of youth and how they are supported or challenged in community contexts.

Interview and Survey Study of Spiritual and Nonspiritual Children, Adolescents, and Adults

This study would sample spiritual and nonspiritual (based on the definition discussed earlier) children (age 10), adolescents (age 15), and adults (over 25), in order to explore differences in positive youth development and variations in spirituality and its role in their lives. This study has the potential to offer insight into developmental trends (at least cross-sectionally) in the intersection between spirituality and PYD, as well as to explore differences between spiritual and nonspiritual people, perhaps offering insight into the value-added effects of spirituality in people's lives. A more complex version of this study might also vary membership in a religious community, such that there would be four groups in each age cohort (spiritual, no religious community; spiritual with a religious community; not spiritual, no religious community; not spiritual, religious community). Questions might include the following:

Are there particular positive developmental competencies that are more related to spiritual development?

How does spirituality get expressed over time in different ways?

What are the pathways toward and away from spirituality?

Do spiritual people express more or different positive developmental competencies?

Survey and interview measures might also explore interpretations of social relationships, spiritual identity, and reactions to stress and adversity.

CONCLUSIONS

In this chapter, I have put forth one perspective on developing a program of research on spirituality and positive youth development. This perspective has largely focused on incorporating issues of context into such work—not context in the sense that it surrounds the important phenomena, but context in the sense that it cocreates such phenomena. Research in this area not only has the potential to contribute to what we know about spirituality and youth development but also to advance what we know about development as a process that is intertwined with culture, context, and the social world. It is both remarkable and inevitable that the field of psychology has come to the point in its own development that it is able to broach such a taboo (for academics) topic as spirituality. It is remarkable in that the study of spirituality has been both rare and marginal in mainstream psychology. It is inevitable in that spirituality as an important aspect of development and as an experience in people's lives has been a fundamental part of the human experience for centuries—it was only a matter of time before psychological science began to develop the tools and inclination to understand better this critical aspect of human life and development. The task now is to take up this challenging area of research in a way that is honest, careful, and wholistic.

NOTES

1. However, it is also important to note that, while these may be defined with respect to culture, they may share underlying universal commonalities. See Turiel et al. 1987 for a discussion of universal and relative dimensions of moral development.

REFERENCES

Alford, T. 2007. *A prince among slaves*. New York: Oxford University Press.

Antaki, C., and D. Widdicombe. 1998. *Identities in talk*. Thousand Oaks, CA: Sage.

Austin, A. 1997. *African Muslims in antebellum America: Transatlantic stories and spiritual struggles*. New York: Routledge.

Bronfenbrenner, U. 1979. *The ecology of human development: Experiments by nature and by design*. Cambridge, MA: Harvard University.

———. 1993. The ecology of cognitive development. In R. H. Wozniak and K. W. Fischer, eds., *Development in Context: Acting and thinking in specific environments*, 3–44. Hillsdale, NJ: Erlbaum.

Cole, M. 1996. *Cultural psychology: A once and future discipline*. Cambridge, MA: Belknap Press.

Cote, J., and C. Levine. 2002. *Identity formation, agency, and culture*. Mahwah, NJ: Erlbaum.

Damon, W. 2004. What is positive youth development? *The Annals of the American Academy of Political and Social Science* 591, no. 1: 13–24.

Diouf, S. 1998. *Servants of Allah: African Muslims enslaved in the Americas.* New York: NYU Press.

Engeström, Y., R. Miettinen, and R.-L. Punamaki. 1999. *Perspectives on activity theory.* Cambridge: Cambridge University Press.

Erikson, E. 1950. *Childhood and society.* New York: W.W. Norton.

———. 1980. *Identity and the life cycle.* New York: Norton.

Forman, E., N. Minnick, and C. A. Stone, eds., 1996. *Contexts for learning: Sociocultural dynamics in children's development.* London: Oxford.

Gallimore, R., T. Weisner, D. Guthrie, L. Bernheimer, and K. Nihira. 1993. Family response to young children with developmental delays: Accommodation activity in ecocultural context. *American Journal of Mental Retardation* 98, no. 2: 185–206.

Garcia-Coll, C., E. Bearer, and R. Lerner, eds. 2004. *Nature and nuture: The complex interplay of genetic and environmental influences on human behavior and development.* Mahwah, NJ: Lawrence Erlbaum.

Gee, J. 2001. Identity as an analytic lens in educational research. *Review of Research in Education* 25: 99–125.

Gutiérrez, K. D., and B. Rogoff. 2003. Cultural ways of learning: Individual traits or repertoires of practice. *Educational Researcher* 32, no. 5: 19–25.

Hand, V. 2008. *Competent opposition: The construction of competing participant frameworks within a low-track mathematics classroom.* Unpublished manuscript, University of Colorado, Boulder.

Harter, S. 1999. *The construction of the self: A developmental perspective.* New York: Guilford Press.

Hill, P., and K. Pargament. 2003. Advances in the conceptualization and measurement of religion and spirituality. *American Psychologist* 58, no. 1: 64–74.

Holland, D., W. Lachicotte, D. Skinner, and C. Cain. 1998. *Identity and agency in cultural worlds.* Cambridge, MA: Harvard University Press

James, W. 1890. *Principles of psychology,* Vol. 1. New York: Henry Holt & Co.

Larson, R. W. 2000. Toward a psychology of positive youth development. *American Psychologist* 55, no. 1: 170–83.

Lave, J. 1988. *Cognitive in practice: Mind, mathematics, and culture in everyday life.* Cambridge: Cambridge University Press.

Lave, J., and E. Wenger. 1991. *Situated learning and legitimate peripheral participation.* Cambridge: Cambridge University Press.

Leontiev, A. N., and D. B. Elkonine. 1979. The child's right to education and the development of knowledge of child psychology. *Quarterly Review of Education* 9, no. 2: 125–32.

Lerner, R., 1991. Changing organism-context relations as the basic process of development: A developmental contextual perspective. *Developmental Psychology* 27, no. 1: 27–32.

Lerner, R., C. Brentano, E. Dowling, and P. Anderson. 2002. Positive youth development: Thriving as the basis of personhood and civil society. *New Directions for Youth Development* 95: 11–34.

Lewin, K. 1935. *A dynamic theory of personality.* London: McGraw-Hill.

———. 1951. *Field theory in social science.* New York: Harper Brothers.

MacDonald, D. 2000. Spirituality: Description, measurement, and relation to the five factor model of personality. *Journal of Personality* 68, no. 1: 153–97.

Marler, P., and C. Hadaway. 2002. Being "religious" or being "spiritual" in America: A zero-sum proposition. *Journal for the Scientific Study of Religion.* 41, no. 2: 289–300.

Miller, W. R., and J. E. Martin. 1988. Spirituality and behavioral psychology: Toward integration. In W. R. Miller and J. E. Martin, eds., *Behavior therapy and religion: Integrating spiritual and behavioral approaches to change,* 13–23. Newbury Park, CA: Sage.

Moen, P., G. Elder, K. Luscher, and U. Bronfenbrenner, eds. 1995. *Examining lives in context:*

Perspectives on the ecology of human development. Washington, D.C.: American Psychological Association.

Nasir, N. 2004. "Halal-ing" the child: Reframing identities of opposition in an urban Muslim school. *Harvard Educational Review* 74, no. 2: 153–74.

———. 2006. Identity constellations: The interweaving of ethnic, religious, and academic identities in a Muslim school. Unpublished manuscript. Stanford University.

Nasir, N., and J. Cooks. 2007. Becoming a hurdler: How learning settings afford identities. Unpublished manuscript, Stanford University.

Nasir, N., and V. Hand. 2006. Exploring socio-cultural perspectives on race, culture, and learning. *Review of Educational Research* 76, no. 4: 449–75.

———. 2008. From the court to the classroom: Managing identities as learners in basketball and classroom mathematics. *Journal of the Learning Sciences* 17, no. 2: 143–79.

Nasir, N., and B. Kirshner. 2003. The cultural construction of moral and civic identities. *Applied Developmental Science* 7, no. 3: 138–47.

Nasir, N., and G. Saxe. 2003. Ethnic and academic identities: A cultural practice perspective on emerging tensions and their management in the lives of minority students. *Educational Researcher* 32, no. 5: 14–18.

Nicolopoulou, A., and M. Cole. 1993. Generation and transmission of shared knowledge in the culture of collaborative learning. In E. Forman, N. Minick, and C.A. Stone, eds., *Contexts for learning: Sociocultural dynamics in children's development*, 283–314. London: Oxford.

Ochs, E., and L. Capps. 1996. Narrating the self. *Annual Review of Anthropology* 25: 19–43.

Powell, L., L. Shehadi, and C. Thorsen. 2003. Religion and spirituality: Linkages to physical health. *American Psychologist* 58, no. 1: 36–52.

Roeser, R., S. Peck, and N. S. Nasir. 2006. Identity, well-being, and achievement in school contexts. In P. Alexander and P. H. Winne, *Handbook of educational psychology*, 2nd ed. Mahwah, NJ: Lawrence Erlbaum.

Rogoff, B. 1990. *Apprenticeship in thinking: Cognitive development in social context.* Oxford: Oxford University Press.

———. 1995. Observing sociocultural activity on three planes: Participatory appropriation, guided participation, and apprenticeship. In J. V. Wertsch, P. del Rio, and A. Alvarez, eds., *Sociocultural studies of the mind*, 139–64. Cambridge: Cambridge University Press.

———. 2003. *The cultural nature of human development.* Oxford: Oxford University Press.

Saucier, G., and R. Skrzypinske. 2006. Spiritual but not religious? Evidence for two independent dispositions. *Journal of Personality* 74, no. 5: 1257–92.

Saxe, G. B. 1991. *Culture and cognitive development.* Mahwah, NJ: Erlbaum.

———. 1999. Cognition, development, and cultural practices. *New Directions for Child and Adolescent Development* 83: 19–35.

———. 2002. Children's developing mathematics in collective practices: A framework for analysis. *Journal of the Learning Sciences* 11, nos. 2–3: 275–300.

Saxe, G., V. Dawson, R. Fall, and S. Howard. 1995. Culture and children's mathematical thinking. In R. Sternberg and T. Ben-Zeev, eds., *The nature of mathematical thinking*, 119–44. Hillsdale, NJ: Erlbaum.

Schweder, R., M. Mahapatra, and J. Miller. 1987. Culture and moral development. In J. Kagan and S. Lamb, eds., *The emergence of morality in young children*, 1–83. Chicago: University of Chicago.

Seligman, M., and M. Csikszentmihalyi. 2000. Positive psychology: An introduction. *American Psychologist*, 55 no. 1: 5–14.

Spencer, M. B. 1999. Social and cultural influences on school adjustment: The application of an identity-focused cultural ecological perspective. *Educational Psychologist* 34, no. 1: 43–57.

———. 2006. Phenomenology and ecological systems theory: Development of diverse groups.

In R. Lerner, ed., *Theoretical models of human development*, 6th ed., 829–93. Vol. 1: *Handbook of child psychology*. New York: Wiley.

Spencer, M. B., and V. Harpalani. 2004. Nature, nurture and the question of "how?": Phenomenological variant of ecological systems theory. In C. Garcia-Coll, E. Bearer, and R. Lerner, eds. *Nature and nurture: The complex interplay of genetic and environmental influences on human behavior and development*, 53–77. Mahwah, NJ: Lawrence Erlbaum Associates.

Turiel, E., M. Killen, and C. Helwig. 1990. Morality: Its structure, functions, and vagaries. In J. Kagan and S. Lamb, eds., *The emergence of morality in young children*, 155–244. Chicago: University of Chicago.

Turner, R. P., D. Lukoff, R. T. Barnhouse, and F. G. Lu. 1996. Religious or spiritual problem: A culturally sensitive diagnostic category in the DSM-IV. *Journal of Nervous and Mental Disease* 183: 435–44.

Vygotsky, L. S. 1978. *Mind in society: The development of higher psychological processes*. Cambridge, MA: Harvard University Press.

Wenger, E. 1998. *Communities of practice: Leaning, meaning, and identity*. Cambridge: Cambridge University Press.

Wertsch, J. V. 1998. *Mind as action*. New York: Oxford University Press.

Zinnbauer, B., K. Pargament, B. Cole, M. Rye, E. Butter, T. Belavich, K. Hipp, A. Scott, and J. Kadar. 1997. Religion and spirituality: Unfuzzying the fuzzy. *Journal for the Scientific Study of Religion* 36, no. 4: 549–64.

16

Application of the Ecological Model

Spirituality Research with Ethnically Diverse Youths

Guerda Nicolas & Angela M. DeSilva

Understanding the role of spirituality in positive youth development is an important focus of adolescent research today. Unfortunately, however, this research does not always allow for an examination of the sociocultural contexts of the adolescents being examined. As the U.S. population of children and adolescents becomes more diverse with respect to race and ethnicity, researchers must employ research methods that allow for a better understanding of the sociocultural contexts of youths. Without a comprehensive understanding of the sociocultural factors in adolescents' lives, it is difficult to understand truly the impact that spirituality has on their development. Therefore, this chapter provides an overview of important ways that researchers can conduct culturally sensitive research on spirituality and positive youth development.

Today, America is more ethnically diverse than ever before, and the number of ethnic minority youths is increasing rapidly. In 1990, 31 percent of children in the United States were from racial- and ethnic-minority groups (Hollmann 1993); by the year 2015, it is projected that 48 percent of children will be from racial- and ethnic-minority groups (Lewit and Baker 1994; Perry and MacKun 2001). Currently it is estimated that racial- and ethnic-minority adolescents represent 39 percent of all adolescents in the United States and more than half of the adolescent population in the largest cities in the United States (i.e., New York, Los Angeles, Chicago, Houston, and Philadelphia; U.S. Census Bureau 2005). Hispanic (17 percent) and Black (15 percent) adolescents are the largest minority groups, composing over one-third of the adolescent population, with other minorities accounting for 7 percent of the adolescent population (U.S. Census Bureau 2006). It is projected that, by the year 2040, racial- and ethnic-minority adolescents will constitute the majority of the nation's adolescent population, with the population of Hispanic adolescents increasing by nearly 60 percent (to constitute 27 percent of the total adolescent population), the percentage of Black adolescents remaining essentially unchanged, and the percentage of White adolescents steadily declining (U.S. Census Bureau 2004b).

The population data on ethnic-minority adolescents do not account for the heterogeneity found among specific racial groups. For instance, the Black population in this country consists of not only African Americans but also immigrants from the Caribbean, South America, and Africa. Asian Americans and Pacific Islanders comprise at least forty-three separate subgroups who speak over one hundred languages, and American Indian/Alaskan natives consist of more than five hundred tribes with different cultural traditions, languages, and ancestry (Pollard and O'Hare 1999). The Hispanic community in the United States is also quite diverse, with individuals emigrating from a variety of countries such as Mexico, Puerto Rico, Cuba, and Central and South America. Given the increasing rates of ethnic-minority adolescents and their different experiences in society, it is imperative that we broaden our research to take into account adolescents' sociocontext. What are some of the areas that researchers need to consider in creating a research program with a diverse sample? What are the implications for not addressing the unique personal and contextual factors of ethnic-minority youths in research programs? Through the lens of the ecological model, we will address these questions throughout this chapter.

THE ECOLOGICAL MODEL AND SPIRITUALITY

According to Bronfenbrenner, creator of the ecological model (1979), individuals are significantly affected by interactions among a number of overlapping ecosystems: microsystems (i.e., family, peer groups, classroom, neighborhood, and religious setting [e.g., church, temple, mosque]); exosystems (i.e., broader community and the educational system); and macrosystems (i.e., cultural values, political philosophies, economic patterns, and social conditions). According to the ecological model, these systems collectively compose the social context of human development, including adolescents (Lerner 2004; Lerner, Brown, and Kier 2005; Miller, Lerner, Schiamberg, and Anderson 2003; Theokas et al. 2005; Theokas and Lerner 2006). Therefore, it is essential that each of these systems be taken into account when trying to understand various facets of youths' lives.

Historically, research on child and adolescent development has explored the influence of the family but failed to assess the influence of other factors, such as culture, family, community, and school (Steinberg and Scheffield-Morris 2001). Although the family is central to most aspects of child and adolescent development, Bronfenbrenner's ecological model emphasizes the importance of looking at the ways in which the individual and the family interact with the cultural groups, institutions, and systems in which they exist. Just before the turn of the century, researchers started to recognize the importance of (1) examining a more diverse group of adolescents when studying development and (2) understanding adolescent development in context (Lerner 2004; Lerner, Brown, and Kier 2005; Miller, Lerner, Schiamberg, and Anderson 2003; Steinberg and Scheffield-Morris 2001; Theokas et al. 2005; Theokas and Lerner 2006). In fact, Steinberg and Scheffield-Morris (2001) noted that "compared with studies conducted prior to the mid 1980s, recent research is more contextual, inclusive, and cognizant of the interplay between genetic and environmental influences on development" (101).

Similar to other areas of development, social context has traditionally been neglected in research on spiritual and religious development (Regnerus, Smith, and Smith 2004). When social context has been addressed, it has focused primarily on the influence of the family on the adolescent (Regnerus, Smith, and Smith 2004; Stark and Fink 2000). Recently, however, researchers have recognized the importance of examining spirituality from an ecological perspective, whereby the role of the larger social context is assessed in relation to spiritual development (e.g., Gunnoe and Moore 2002; King, Furrow, and Roth 2002; Martin, White, and Perlman 2003; Regnerus, Smith, and Smith 2004). In a recent study titled *Social Context in the Development of Adolescent Religiosity*, Regnerus and his colleagues (2004) found that, although parents were the primary influence in adolescents' church attendance, friends and school environment also played an important role. For example, when church attendance was low for adolescents' friends, the "anticipated probability" that the adolescent would attend church on a regular basis drastically decreased. Regnerus and his colleagues believe that their results "affirm the importance of an ecological approach to the study of religious development in youth, one that considers understanding the multiple social contexts in which youth live as essential to research on adolescent religion and spirituality" (Regnerus, Smith, and Smith

2004, 36). Although scholars have recently emphasized the importance of utilizing an ecological perspective to understand religious and spiritual development, it is equally important to do so in order to understand the ways in which spirituality facilitates positive youth development. The following section of this chapter illustrates the application of the ecological framework in conducting spiritual research with ethnic-minority adolescents.

OUR EXPERIENCES CONDUCTING SPIRITUAL RESEARCH WITH ETHNICALLY DIVERSE COMMUNITIES

The importance and utility of the ecological perspective has been well documented in research on adolescent development in general (e.g., Lerner 2004; Lerner, Brown, and Kier 2005; Miller, Lerner, Schiamberg, and Anderson 2003; Steinberg and Scheffield-Morris 2001; Theokas et al. 2005; Theokas and Lerner 2006) and spirituality more specifically (e.g., Regnerus, Smith, and Smith 2004). We believe that, in order to truly employ the ecological perspective with ethnic-minority adolescents, the researcher must: (1) utilize a community-based approach; (2) be attentive to the recruitment, selection, and reporting of the sample; and (3) carefully address issues of selection and administration of measurements. Using examples from our research on spirituality with ethnic-minority adolescents, the following section provides a summary of ways that we have addressed these areas in our research.

Community-Based Approach

Many scholars agree that culturally competent research with ethnic- and linguistic-minority adolescents can occur only within the context of their community (Triandis 1992; Trimble and Fisher 2006). Given that many adolescents develop and practice their spirituality within the context of their community (Regnerus, Smith, and Smith 2004), culturally competent spirituality research with ethnic-minority adolescents must be developed and implemented within the context of the adolescents' community. This is exemplified in the work of Triandis (1992), who argues that "to do research that is ignorant of or insensitive to the major features of the local culture often means to do poor research and thus wastes the time of local subjects, as well as the funds, and that is unethical" (232). Without community involvement in the development and implementation of a research project, "any results from the study are questionable at best, and harmful to the communities at worst" (Mohatt and Thomas 2006, 107). In their research in a Haitian community, Nicolas and her colleagues (2006) found that engaging community members in the development of the research methodologies, the meaning ascribed to research questions and participant responses, and the analysis and interpretation of the data leads to the creation of long-lasting partnerships with the community members and obtaining research results that are meaningful to the researchers and the participants in the project.

A Community-Based Approach to Understanding Spirituality with Ethnic-Minority Adolescents. Approximately four years ago, we began a research project to understand the role of spirituality in the lives of a diverse group of adolescents. Our goal was to build on the current literature in this area by sampling a diverse (e.g., ethnic background, socioeconomic status, religious background) group of adolescents. Given our plan to interview adolescents from different ethnic communities, it was essential that we first develop connections with individuals from each of these communities. This was especially true as researchers in various disciplines have started to document the centrality of developing partnerships with communities when conducting research with ethnic-minority adolescents (e.g., Brody et al. 1994; Rumberger and Larson 1994; Tharp and Yamauchi 1994). In one of the ethnic-minority communities with whom we have partnered, we first established a relationship with leaders in the community where we provided support and consultation around a project they were conducting, and they provided feedback about a number of areas relating to our project (i.e., the items of the measures, the recruitment and interviewing process, the compensation of the participants, etc.). Feedback from such community connections resulted in alterations to our research design. More specifically, although we planned to post fliers in various schools as a means of recruitment, one of the community leaders recommended establishing connections with classroom teachers as a means for recruiting students. The community members suggested that we have several compensation prizes for the adolescents, suggesting gift cards to music stores in addition to gift cards to video stores, pointing out that the parents of the adolescents in their particular community were predominantly immigrants and did not have memberships at movie rental stores, which could limit adolescents' use of the gift card. Finally, we collaborated with community members around several of the questions we planned to ask regarding spirituality and incorporated their feedback into the open-ended questions that we asked adolescents. For example, it was recommended that we include an open-ended question about whether there was anything the study did not ask the adolescent regarding his/her spiritual experience but that he/she thinks is important. In response to their suggestion, we decided to end the series of open-ended questions with the following question—*Is there a question about religion or spirituality that I have not asked you but you think is important for us to know?*

Prior researchers working in ethnic-minority communities have found advisory boards useful when developing research measures and methodologies (e.g., Brody et al. 1994). Therefore, in addition to collaborating with community leaders, we developed an advisory board comprising adolescents from diverse ethnic backgrounds from the various communities where participants were recruited. We consulted with the adolescent advisory board around issues similar to those for which we consulted community leaders—research questions, recruitment procedures, interview administration, and participant compensation. However, the adolescents often offered a slightly different perspective from the community leaders, and this allowed us to integrate feedback from different community sources into the project.

In addition to utilizing the experiences and expertise of community members in the design phases of our research, our study was conducted within the communities where the participants live. This approach is recommended by scholars who conduct research with ethnic-minority adolescents (e.g., Cauce, Ryan, and Grove 1998). In order to accommodate participants better and help them feel more comfortable with the research process, we conducted interviews in schools, churches, community centers, and homes. This enabled us to join the adolescents in their context and gain a better understanding of the intersections of their context and spirituality.

Finally, a vital part of the community-based approach was our continued collaboration with the communities subsequent to the completion of our project. In an effort to accomplish this plan, we scheduled "days of dialogue" where members of our research team went into the communities where we collected data to share our results and obtain feedback from community members. This provided community members the opportunity to offer feedback around the research process and interpretations of the adolescents' responses to our research questions. Overall, employing a community-based approach to our research ensured that we conducted research that was culturally sensitive, meaningful, and beneficial to the communities in which we worked.

Composition of Sample

In addition to developing partnerships with communities, researchers need to be mindful of the sample they recruit and also the information obtained from and about the sample. A common error found in cross-cultural research is a small sample size. Although small samples make the research more manageable (e.g., easier recruitment and less funding required), they reduce the quality and generalizability of the research. Glover and Pumariega (1998) state that research "across different regions or subgroups often requires a much larger sample [than research within groups]" (291). When conducting research comparing different groups, it is important to have a large enough sample such that there are enough individuals in each of the different groups to make comparisons among the groups (Coakley and Orme 2006).

When conducting research with different ethnic groups, it is also important to match the sample on individual demographic characteristics and also sociocontextual factors. As many characteristics as possible must be matched in order to rule out variations between ethnic groups that are not related to spirituality (Van de Vijver and Leung 1997). For example, it is important to examine microlevel variables such as level of education, number of years living in the United States, and religious background, as they may have an influence on the experiences or influence of spirituality. Additionally exolevel and macrolevel variables such as the religious community, the neighborhood, the school setting, and the larger cultural context must also be evaluated as possible confounding variables. An individual's experience in his or her neighborhood, for instance, can significantly contribute to the spiritual values and experiences reported, thereby contributing to variation between ethnic groups (Beldona, Inkpen, and Phatak 1998; Benson 2004; Furrow, Ebstyne King, and White 2004; Regnerus, Smith, and Smith 2004). If variables such as these are not accounted for, differences in spirituality may be reported that are actually a function of neighborhood experiences. Consideration of microsystem, exosystem, and macrosystem factors is essential for conducting competent research with diverse groups of adolescents as it ensures that the results obtained can accurately be attributed to cultural differences and not other individual and contextual factors (Schaffer and Riordan 2003).

Participants in Our Spirituality Project. From the start of the project, we were committed to ensuring that our sample was diverse with respect to age, socioeconomic status, religious background, and race. This is common practice when conducting research with ethnic-minority adolescents (Brody et al. 1994; Miller, Forehand, and Kotchick 1999; Rumberger and Larson 1994; Tharp and Yamauchi 1994). Throughout the recruitment process, it was imperative that we tracked the number of participants in each ethnic group in order to make sure that we collected data from each of the prominent ethnic groups in the United States (e.g., Asian, Black, Hispanic, Native American, and White) and that the numbers of participants from each group remained balanced. This allowed for valid cross-cultural comparisons about spirituality.

In order to assess the individual and sociocontextual factors of the adolescents with whom we worked, we designed a demographic form that addressed many of the variables important to understanding ethnic-minority populations. Our demographic form tapped into the following areas: (1) age; (2) sex; (3) ethnic background; (4) level of education; (5) birth country; (6) age came to the United States; (7) birth state; (8) nationality; (9) parents' and grandparents' birth country; (10) languages spoken; (11) language spoken at home; (12) household composition; (13) number of siblings; and (14) religious background. Additionally, as part of the interview process, we created a series of open-ended questions addressing the adolescents' views about and experiences with religion and spirituality. Researchers often create and adapt questions to address specific constructs pertinent to the research at hand (e.g., Miller, Forehand, and Kotchick 1999). Several of the questions we developed examined the extent to which adolescents believe various social relationships impact their spirituality and religiosity. The following is a sampling of questions we asked the adolescents:

What is the reason you attend religious services?

Do your ideas about religion/spirituality have an impact on how you live your life?

Do your parents/guardians speak with you about religion or spirituality?

Are your ideas about religion/spirituality the same as those of your relatives, friends, classmates, teachers, community members, etc.?

Do you feel comfortable expressing your religious/spiritual views with your relatives, friends, classmates, teachers, community members, etc.?

By including a demographic form and open-ended questions in the interview process, we were able to evaluate the extent to which microlevel, exolevel, and macrolevel factors impact the adolescents' experiences of spirituality.

Instrumentation

The use of culturally inappropriate assessment measures represents another methodological challenge to quality research with ethnically diverse groups of adolescents (Glover and Pumariega 1998). When selecting and administering measures for a research project, it is important that researchers take into consideration the different ways in which microsystem, exosystem, and macrosystem factors may impact the responses of adolescents.

Culturally Appropriate Administration. When conducting research across different cultures, researchers must consider the different ways culture can impact the administration of research measures and questions. First, data collection procedures must be equivalent and consistent across the different ethnic groups participating in the study. For example, if research questions are read to adolescents in one community or cultural group due to literacy issues, then the same procedure must be followed with all of the participants involved in the project (Ortega and Richey 1998). Similarly, if the research questions are translated into a different language for one group of adolescents, they must be translated for all of the adolescents. Having a consistent procedure across different cultural groups ensures the reliability and validity of the measures and the study in general (Ortega and Richey 1998).

The time span of the data collection is another factor related to the administration of measures that could potentially impact research findings. Researchers should make sure that the data are collected from all ethnic groups at approximately the same time (Sekaran and Martin 1982; Yu, Keown, and Jacobs 1993) because the time span during which instruments are administered can affect the responses of participants in the project. For instance, collecting data at different points in time (e.g., Black adolescents interviewed in 2000 and White participants interviewed in 2007) may reflect cohort effects rather than actual cultural differences (Roberts and Boyacigiller 1984; Roberts, Hulin, and Rousseau 1978). If researchers cannot avoid timing differences in the collection of data, it is imperative that such differences be accounted for in the procedure and discussion sections of manuscripts from these projects.

Finally, it is crucial that researchers assess adolescents' comfort and familiarity with the format of the assessment (Arthur and Bennett 1995; Geletkanycz 1997). Individuals' ease and/or previous experience with the evaluation process can greatly impact the responses of the adolescents, which, in turn, will likely impact the results of the study. For example, completing surveys and standardized instruments is a process most individuals in Western countries are exposed to by the third grade. Thus, adolescents from Western countries may be more familiar and comfortable with paper-and-pencil assessment instruments than their peers from non-Western countries (Lonner 1990). In addition to being comfortable with the assessment measures, it is crucial that ethnic-minority adolescents have a certain level of comfort with the researchers (i.e., good rapport). Adolescents from some cultural groups may be hesitant to disclose personal information to researchers who are not from their same cultural background. For example, research suggests that, if participants are not comfortable with researchers, and particularly if they are not from the same cultural background, they may rush through projects in an effort to avoid any further contact with the researchers (Van de Vijver and Leung 1997). In their work with Native Americans, one participant made the following comment to Trimble and Mohatt (2006): "We tell [other researchers] what we want them to know, but we don't go into much detail. And sometimes a few of us will make up stories, so that they'll go away and we don't insult them by saying [no] to them" (332). Researchers need to pay close attention to these issues and develop strategies for how to reduce any discomfort experienced by ethnic-minority adolescent participants. Establishing and upholding rapport must be an ongoing process during all phases of the project.

Accounting for Linguistic Differences. The process of translating instruments across languages can pose specific problems for researchers surveying ethnically diverse groups of adolescents (Holtzman 1968). Through the process of translation, it is often difficult to ensure that words maintain their original and intended meaning. Therefore, before administering measures to adolescents for the planned research project, measures should be translated and administered to representatives of each ethnic group. At the end of the translational process, the meaning of each item in the instrument should be the same for adolescents from different ethnic backgrounds. Unfortunately, it is often quite difficult to maintain the intended meaning of research questions. For example, in translating a scale from English to Vietnamese, one of the original items on the instrument ("I have been looking forward to things with enjoyment") became partially distorted after the back-translation process (translated from English into Vietnamese and then back to English) and actually resulted in two phrases that were not consistent with the original English version (i.e., "I have been hoping/expecting to be happy" and "I have been feeling optimistic"; Small, Rice, Yelland, and Lumley 1999). In addition, researchers must be cautious of whether questions are cognitive or emotional in nature (Ortega and Richey 1998; Ponterotto and Casas 1991). Take, for example, the following questions: "What do you think about your supervisor?" and "How do you feel about your supervisor?" Although the meaning of these two questions is similar, the former has a cognitive orientation, while the latter has an affective one. It is not clear how these factors affect the translation of items into different languages and whether an affective item in one language elicits the same response in another translation (Schaffer and Riordan 2003). Therefore, if researchers intend to translate their measures, it is highly recommended that instruments are first back-translated and tested with a pilot group, prior to administering them to the intended group of adolescents. This will help to ensure that items and instruments have equal meaning across cultures.

Consider Scale Inequality. Many instruments in developmental and religious research utilize Likert scales. However, adolescents from non-Western cultures have limited exposure to this response format, and this often leads to erroneous results for them (Riordan and Vandenberg 1994). For example, adolescents from collectivist cultures are more likely to respond with the middle range/neutral or undecided category of a Likert scale than adolescents from individualist societies (Triandis 1994). In addition, researchers have found that some participants will complete an instrument out of a sense of responsibility, irrespective of whether they agree with the items or feel that they reflect their true feelings (Javeline 1999). This is an important issue for researchers to keep in mind when selecting, administering, and interpreting measures, as reports of varying response patterns in ethnic groups should reflect genuine cultural differences, not a misapplication of scales across cultures.

Addressing Instrumentation Issues in Our Spirituality Project. Because the central goal of our research project was to ascertain the relationship between spirituality and positive youth development for a diverse sample of adolescents, instrumentation was a primary focus when planning the project. The majority of questionnaires that we wanted to use for the study were developed and normed in English. Therefore, to avoid jeopardizing the validity of the measures, we decided that all of the questionnaires would be administered in English. Consequently, all of the adolescents recruited for participation in the project were fluent in English. In addition to potential linguistic problems, we needed to consider how accurately the Likert scales would capture the true experience of the adolescents. In order to address potential confusion with the Likert scales, our interviewers were trained to provide very clear descriptions of how the scales work. We conducted a semistructured interview with each participant, allowing for a documentation of any ambiguity relating to the scale or items on the scale. We also prepared "response cards" to be given to each participant to help guide the adolescent in responding to the questions. Response cards allowed the adolescents to look at each of the possible responses to a particular question while the interviewer read the question.

Finally, we developed a series of open-ended questions to examine the spiritual and religious experiences of the adolescents. This is an approach that has been used in the past with research conducted with ethnic-minority adolescents as a way of accessing their true experiences (e.g., Cooper, Jackson, Azmitia, and Lopez 1998; Miller, Forehand, and Kotchick 1999). Consistent with previous research, our primary intention in developing these questions was to allow adolescents the opportunity to express their experiences, without having to fit them to the responses offered through the standardized questionnaires. One of the first questions we asked pertained to whether the adolescents consider religion and spirituality to be the same or different. We noted cultural trends in the responses, where Asian adolescents tended to think they are same (e.g., "I think religion and spirituality are the same, because well . . . to me it's sort of connected.") and Hispanic adolescents tended to think they are different (e.g., "I think they are different, 'cause religion is more of something . . . well, spirituality is something you feel, and religion is more something like you follow."). We believe that the combination of standardized questionnaires and open-ended questions accurately captured the spiritual and religious experiences of the adolescents in our study.

Prior to administering our questionnaires to adolescents in the community, we developed an extensive interview protocol and rigorously trained each of the interviewers. The interview protocol was designed to lead the interviewers through each step of the interview process. It provided clear guidelines (including sample scripts about what to say) about the introduction of the study (including reviewing assent/consent forms), the administration of the standard instruments (including the order of the instruments, directions to read for each instrument, and directions for the interviewers [e.g., no translating words or items on the instruments]), the administration of the open-ended interview, and the closing of the study (including debriefing the adolescents and compensating them for their efforts). In order to be able to conduct interviews in the community, each interviewer attended a series of three training sessions. Extensively training interviewers for research in the community with ethnic-minority adolescents is strongly encouraged by researchers working in communities (e.g., Evans, Mejía-Maya, Zayas, Boothroyd, and Rodriguez 2001; Miller, Forehand, and Kotchick 1999). The first session oriented interviewers to the project in general, the interview protocol, and the measures being used. The second session educated interviewers about conducting ethical and culturally sensitive research with ethnic-minority adolescents. As part of this training session, interviewers were required to read several articles published by leading scholars in the field of multicultural research (e.g., Fisher and Ragsdale 2006; Helms, Henze, Satiani, and Mascher 2005; McIntyre 1997). In the third session, interviewers were supervised in conducting mock-interview sessions (from introduction to closing). The combination of the interview protocol and three training sessions helped to ensure that each interview would be conducted in a similar and culturally sensitive manner. It is also important to note that, as part of an ongoing training process, our research team engaged in discussions about the experiences of the interviewers conducting the research in diverse communities, focusing specifically on issues of race and culture.

Infused in each training session with the researchers were skill-building strategies to facilitate comfort on the part of the adolescents with both the interviewer and the interview process. First, after introductions, the interviewers were trained to explain the interview process to the adolescents, making sure to provide the adolescents with the time to ask any questions they may have prior to answering specific research questions. Because many of the questions in the interview pertained to personal issues, interviewers informed adolescents that they could skip any questions with which they were not comfortable, and they also conducted routine check-ins with the adolescents throughout the interview process (e.g., "I know some of the questions are a little personal, so I just want to make sure that you are still feeling comfortable"). Being personable and upfront facilitated participant comfort with the interview process and helped establish rapport.

SUMMARY AND CONCLUSIONS

Although historically researchers have not utilized Bronfenbrenner's ecological model (1979) to study the development and manifestation of spirituality, it is imperative that we listen to our colleagues (e.g., Gunnoe and Moore 2002; King, Furrow, and Roth 2002; Martin, White, and Perlman 2003; Regnerus, Smith, and Smith 2004) who urged a shift in the way we explore the spiritual experiences of adolescents. The process of developing a research project that examines spirituality within the context in which it is practiced requires a strong commitment on the part of the researchers—a commitment that cannot lessen when the process takes longer than expected. Regnerus and colleagues (2004) poignantly articulated the need for this type of research when they stated that "individuals live linked lives, developing and functioning in a dynamic and reciprocal fashion with their environments, including relationships with other individuals, groups, and subcultures" (29). Failure to integrate an ecological framework into our research focusing on spirituality with ethnically diverse adolescents may lead to data that are not representative of the experiences of the adolescents with whom we are working.

This chapter provided an overview of some of the key components (community partnership, sample selection, and instrumentation) that must be addressed in conducting spirituality research with ethnically diverse adolescent populations. Although incorporating these components is essential in conducting culturally sensitive research, it is in no way an easy process. In addition to our experience, researchers with extensive experience conducting culturally relevant research advise developmental researchers with interest in conducting culturally sensitive research to have patience, commitment, and passion about the work. At the onset of developing and implementing a project with ethnically diverse communities, researchers must first establish a true partnership with the communities, a process that may be time-consuming as connections are often made in different ways depending on the particular community. Through community connections, researchers are able to receive consultation about the research project and process, as well as develop a better understanding about the type of training that the research team must have in order to conduct the project effectively. Given the time and energy often involved in developing relationships with communities, a strong commitment and passion about the work is needed to engage in the undertaking. Despite the added efforts involved in conducting culturally responsive and sensitive developmental research, doing so leads to results that are reflective of the sample from which the data are collected, as well as ensures that projects are reflective of the demographic makeup of children in the United States.

REFERENCES

Arthur, W., and W. Bennett. 1995. The international assignee: The relative importance of factors perceived to contribute to success. *Personnel Psychology* 48: 99–114.

Beldona, S., A. C. Inkpen, and A. Phatak. 1998. Are Japanese managers more long-term oriented than United States managers? *Management International Review* 38: 239–56.

Benson, P. L. 2004. Emerging themes in research on adolescent spiritual and religious development. *Applied Developmental Science* 8, no. 1: 47–50.

Brody, G. H., Z. Stoneman, D. Flor, C. McCrary, L. Hastings, and O. Conyers. 1994. Financial resources, parent psychological functioning, parent co-caregiving, and early adolescent competence in rural two-parent African American families. *Child Development* 65: 590–605.

Bronfenbenner, U. 1979. *The ecology of human development: Experiments by nature and design.* Cambridge, MA: Harvard University Press.

Cauce, A., K. Ryan, and K. Grove. 1998. Children and adolescents of color, Where are you? Participation, selection, recruitment, and retention in developmental research. In V. McLoyd & L. Steinberg, eds., *Studying minority adolescents: Conceptual, methodological, and theoretical issues,* 147–166. Mahwah, NJ: Lawrence Erlbaum Associates.

Coakley, T. M., and J. G. Orme. 2006. A psychometric evaluation of the cultural receptivity in fostering scale. *Research on Social Work Practice* 16: 520–35.

Cooper, C. R., J. F. Jackson, M. Azmitia, and E. M. Lopez. 1998. Multiple selves, multiple worlds: Three useful strategies for research with ethnic minority youth on identity, relationships, and opportunity structures. In V. C. McLoyd and L. Steinberg, eds., *Studying minority adolescents: Conceptual, methodological, and theoretical issues,* 111–25. Mahwah, NJ: Lawrence Erlbaum Associates.

Evans, M. E., L. J. Mejía-Maya, L. H. Zayas, R. A. Boothroyd, and O. Rodriguez. 2001. Conducting research in culturally diverse inner-city neighborhoods: Some lessons learned. *Journal of Transcultural Nursing* 12: 6–10.

Fisher, C. B., and K. Ragsdale. 2006. Goodness-of-fit ethics for multicultural research. In J. E. Trimble and C. B. Fisher, eds., *The handbook of ethical research with ethnocultural populations and communities,* 3–25. Thousand Oaks, CA: Sage.

Furrow, J. L., P. Ebstyne King, and K. White. 2004. Religion and positive youth development: Identity, meaning, and prosocial concerns. *Applied Developmental Science* 8, no. 1: 17–26.

Geletkanycz, M. A. 1997. The salience of "culture's consequences": The effects of cultural values on top executive commitment to the status quo. *Strategic Management Journal* 18, no. 8: 615–34.

Glover, S. H., and A. J. Pumariega. 1998. The importance of children's mental health epidemiological research with culturally diverse populations. In M. Hernandez and M. Isaacs, eds., *Promoting cultural competence in children's mental health services,* 271–303. Baltimore, MD: Paul H. Brookes Publishing Co., Inc.

Gunnoe, M. L., and K. A. Moore 2002. Predictors of religiosity among youth aged 17–22: A longitudinal study of the National Survey of Children. *Journal for the Scientific Study of Religion* 41, no. 4: 613–22.

Helms, J. E., K. T. Henze, A. Satiani, and J. Maschcer. 2006. Ethical issues when white researchers study ALANA and Immigrant people and communities. In J. E. Trimble and C. B. Fisher, eds., *The handbook of ethical research with ethnocultural populations and communities,* 299–324. Thousand Oaks, CA: Sage.

Holtzman, W. H. 1968. Cross-cultural studies in psychology. *International Journal of Psychology* 3, no. 2: 83–91.

Hollman, F. W. 1993. *US. population estimates by age, sex, race, and Hispanic origin: 1980 to 1991.* Current Population Reports, Series P-25, No. 1095. Washington, DC: U.S. Government Printing Office, February 1993.

Javeline, D. 1999. Response effects in polite cultures: A test of acquiescence in Kazakhstan. *Public Opinion Quarterly* 63: 1–28.

King, P. E., J. L. Furrow, and N. Roth. 2002. The influence of families and peers on adolescent religiousness. *Journal of Psychology and Christianity* 21, no. 2: 109–20.

Lerner, R. M. 2004. Innovative methods for studying lives in context: A view of the issues. *Research in Human Development* 1, nos. 1 and 2: 5–7.

Lerner, R. M., J. D. Brown, and C. Kier. 2005. *Adolescence: Development, diversity, context, and application.* Canadian ed. Toronto: Pearson.

Lewit, E. M., and L. S. Baker. 1994. Child indicators: Race and ethnicity—changes for children. *Future of Children* 4, no. 3: 134–44.

Lonner, W. J. 1990. An overview of cross-cultural testing and assessment. In R. W. Brislin, ed., *Applied cross-cultural psychology,* 56–76. Newbury Park, CA: Sage.

Martin, T. F., J. M. White, and D. Perlman. 2003. Religious socialization: A test of the channeling hypothesis of parental influence on adolescent faith maturity. *Journal of Adolescent Research* 18, no. 2: 169–87.

McIntyre, A. 1997. *Making meaning of my Whiteness: Exploring racial identity with White teachers.* Ithaca: State University of New York Press.

Miller, K. S., R. Forehand, and B. A. Kotchick. 1999. Adolescent sexual behavior in two ethnic minority samples: The role of family variables. *Journal of Marriage and the Family* 61, no. 1: 85–98.

Miller, J. R., R. M. Lerner, L. B. Schiamberg, and P. M. Anderson, eds. 2003. *Human ecology: An encyclopedia of children, families, communities, and environments.* Santa Barbara, CA: ABC-Clio.

Mohatt, G. V., and L. R. Thomas. 2006. "I wonder, why would you do it that way?": Ethical dilemmas in doing participatory research with Alaska native communities. In J. E. Trimble and C. B. Fisher, eds., *The handbook of ethical research with ethnocultural populations and communities,* 93–116. Thousand Oaks, CA: Sage.

Nicolas, G., S. Houlihan, R. Singer, M. Coutinho, and A. DeSilva. 2006. Igniting the fuel: Conducting cross-culturally competent research with ethnic and linguistic adolescents. Unpublished manuscript, Boston College, Chestnut Hill, MA.

Ortega, D. M., and C. A. Richey. 1998. Methodological issues in social work research with depressed women of color. *Journal of Social Service Research* 23, no. 3–4: 47–70.

Perry, M. J., and P. J. MacKun. 2001. Population change and distribution: Census 2000 brief. Census 2000.

Pollard, K. M., and W. P. O'Hare. 1999. America's racial and ethnic minorities. *Population Bulletin* 54, no. 3: 1–52.

Ponterotto, J. G., and J. M. Casas. 1991. *Handbook of racial/ethnic minority counseling research.* Springfield, IL: Charles C Thomas.

Regnerus, M. D., C. Smith, and B. Smith. 2004. Social context in the development of adolescent religiosity. *Applied Developmental Science* 8, no. 1: 27–38.

Riordan, C. M., and R. J. Vandenberg. 1994. A central question in cross-cultural research: Do employees of different cultures interpret work-related measures in an equivalent manner? *Journal of Management* 20: 643–71.

Roberts, K. H., and N. A. Boyacigiller. 1984. Cross-national organizational research: The grasp of the blind men. *Research in Organizational Behavior* 6: 423–75.

Roberts, K. H., C. L. Hulin, and D. M. Rousseau. 1978. *Developing an interdisciplinary science of organizations.* San Francisco: Jossey-Bass.

Rumberger, R. W., and K. A. Larson. 1994. Keeping high-risk Chicano students in school: Lessons from a Los Angeles middle school dropout prevention program. In R. J. Rossi, ed., *Educational reforms for at-risk students,* 141–62. New York: Teachers College Press.

Schaffer, B. S., and C. M. Riordan. 2003. A review of cross-cultural methodologies for organizational research: A best-practices approach. *Organizational Research Methods* 6, no. 2: 169–215.

Sekaran, U., and H. J. Martin. 1982. An examination of the psychometric properties of some commonly researched individual differences, job, and organizational variables in two cultures. *Journal of International Business Studies* 13, no. 1: 51–65.

Small, R., P. L. Rice, J. Yelland, and J. Lumley. 1999. Mothers in a new country: The role of culture and communication in Vietnamese, Turkish and Filipino women's experiences of giving birth in Australia. *Women and Health* 28, no. 3: 77–101.

Stark, R., and R. Fink. 2000. *Acts of faith: Explaining the human side of religion*. Berkeley: University of California Press.

Steinberg, L., and A. Scheffield-Morris. 2001. Adolescent development. *Annual Review of Psychology* 52: 83–110.

Tharp, R. G., and L. A. Yamauchi. 1994. *Polyvocal research on the ideal Zuni Indian classroom.* Educational Practice Report No. 10. Santa Cruz: University of California at Santa Cruz, National Center for Research on Cultural Diversity and Second Language Learning.

Theokas, C., J. Almerigi, R. M. Lerner, E. Dowling, P. Benson, P. C. Scales, and A. von Eye. 2005. Conceptualizing and modeling individual and ecological asset components of thriving in early adolescence. *Journal of Early Adolescence* 251: 113–43.

Theokas, C., and R. M. Lerner. 2006. Observed ecological assets in families, schools, and neighborhoods: Conceptualization, measurement and relations with positive and negative developmental outcomes. *Applied Developmental Science* 10, no. 2: 61–74.

Triandis, H. C. 1992. Cross-cultural research in social psychology. In D. Granberg and G. Sarup, eds., *Social judgment and intergroup relationships: Essays in honor of Muzafer Sherif*, 229–44. New York: Springer-Verlag.

———. 1994. Cross-cultural industrial and organizational psychology. In H. C. Triandis, M. D. Dunnette, and L. M. Hough, eds., *Handbook of industrial and organizational psychology*, vol. 4, 103–72. Palo Alto, CA: Consulting Psychologists Press.

Trimble, J. E., and C. B. Fisher, eds. 2006. *The handbook of ethical research with ethnocultural populations and communities*. Thousand Oaks, CA: Sage.

Trimble, J. E., and G. V. Mohatt. 2006. Coda: Virtuous and responsible researcher in another culture. In J. E. Trimble & C. B. Fisher, eds., *Handbook of ethical considerations in conducting research with ethnocultural populations and communities*, 325–34. Thousand Oaks, CA: Sage.

———. 2004a. 2005 Federal Poverty Guidelines. Retrieved on January 2007 from http://aspe.hhs.gov/poverty/05poverty.shtml.

———. 2004b. Population Division, Population Projection Branch. Question and Answer Center. Retrieved October 2006 from www.census.gov/population/www/projections/natdet-D1A.html.

———. 2005. American Community Survey: American FactFinder Retrieved October 2006 from http://factfinder.census.gov.

———. 2006. Housing and Household Economic Statistics Division: Current Population Survey (CPS) Table Creator. Retrieved October 2006 from www.census.gov/hhes/www/cpstc/cps_table_creator.html.

Van de Vijver, F., and K. Leung. 1997. *Methods and data analysis for cross-cultural research*. Thousand Oaks, CA: Sage.

Yu, J. H., C. F. Keown, and L. W. Jacobs. 1993. Attitude scale methodology: Cross-cultural implications. *Journal of International Consumer Marketing* 6, no. 2: 47–64.

17

Possible Interrelationships between Civic Engagement, Positive Youth Development, and Spirituality/Religiosity

Lonnie R. Sherrod & Gabriel S. Spiewak

In this paper, we examine the potential relationships between spirituality/ religiosity, civic engagement, and positive youth development. Since all three concepts can have multiple meanings, research must begin by clarifying the definitions of each construct that underlie the questions being addressed.

As experimental scientists we are trained to conceptualize independent variables that are influences on development and dependent variables that are outcomes of development. Each of these constructs—positive development, civic engagement, and spirituality/religiosity—can be either an independent or a dependent variable. One goal of the present paper is to address the possible relationships between these constructs, and, to do that, one must begin by addressing some definitional issues. To some extent, each area relates to values, identity, and their relationship to behavior.

POSITIVE YOUTH DEVELOPMENT (PYD)

The clearest construct of the three is the idea of positive youth development. Generally, PYD is an approach, not an actual construct, similar to a lifespan approach to developmental research (Sherrod, Busch, and Fisher 2004). As an approach, it offers a model of applied research and has important implications for policy that affects young people.

PYD argues that development is promoted by assets, both internal and external. Additionally, there is variability in the assets individuals bring to their contexts of growth and development, and these contexts of families, schools, communities, and societies or nations vary in the qualities that promote the development of these assets (Benson 2004; Benson, Leffert, Scales, and Blyth 1998; King and Benson 2006). This approach is applied in that it offers guidance for policy and programs as well as research. Following decades of unsuccessful research and policy to eliminate risk and prevent negative outcomes, PYD advocates for examining the strengths youth have—rather than their risks—and for designing policies and programs oriented to promoting positive outcomes—rather than preventing negative ones. This PYD approach is attractive for several reasons (Sherrod 2006a). However, here, we will address areas where further refinement in regard to clarifying certain ideas is merited.

The comprehensive multilevel nature of the approach presents a few issues about the definition of assets that merit attention (Sherrod 2006a). The approach is clear; definitional issues arise, however, in regard to the articulation and measurement of assets that promote positive development and that represent positive developmental outcomes. Recent research by Theokas and colleagues (2005) highlights the importance of both conceptual and empirical attention to the idea of developmental assets. Using the same sample that generated the original list of forty internal and external assets (Benson et al. 1998; Scale et al. 2000), these investigators find considerable colinearity among the assets. As a result, the forty assets reduce to fourteen categories, and these can be further conceptualized in terms of individual and ecological assets; ecological followed by individual assets predicted indicators of individual thriving (Theokas et al. 2005).

Past research had indicated that the more internal and external assets youth possess, the healthier and more successful is their development into adulthood (Benson 2004; Scales et al. 2000). This was a disturbing finding because research also indicated that young people have only 16.5–21.6 assets on average of forty identified, and youth in New York have only 5 percent of the optimal number of assets (Benson 2006; Benson, Leffert, Scales, and Blyth 1998). Recent research has called into question this "more is better" hypothesis, at least for certain assets such as after-school programs and sports (Zarrett et al., 2007). Nonetheless, the PYD approach clearly highlights the need for development of youth policy to promote development based on the resources available to them in their families, schools, and communities.

These contradictory findings point to the need for both theoretical and measurement attention to the idea of assets inherent to the PYD approach.

Assets can be both independent and dependent variables. To some extent

across development, assets move from being promoters of positive development to being positive outcomes of development. Of course, the processes
that link developmental influences to developmental outcomes will vary across
development. But they may also vary across type of assets. Hence, all assets are
not equal. The measurement of assets may also vary by their conceptualization
as independent or dependent variables (Sherrod 2006a). As currently conceptualized and measured, assets represent youth's perceptions of their individual
strengths as well as perceptions of the resources available to them. Multiple
measures are always desirable in research, but objective as well as subjective
assessment of youth assets is sorely needed. We may consider different assessments of assets as they are used as independent-versus-dependent variables.

A second issue in addition to the conceptualization and measurement of
assets is a clearer articulation of how the assets combine to produce positive
development (Sherrod 2006a). This issue is, of course, relevant to definition in
that one way in which assets may not be equal is how they interact or combine
to influence outcomes. Much research views assets to be cumulative; the Theokas et al. (2005) and Zarrett et al. (2007) findings would imply that this is not
the case. Developmentally one might expect assets to interact or their accumulation to plateau after a certain level is reached. The interaction of assets is an
important issue regardless of whether one considers assets as independent or
dependent variables. For independent variables, this issue asks about the processes that link influences to outcomes, while, for dependent variables, it asks
about the definition of positive development. How many assets does a young
person need to be considered thriving? There is research showing that even
affluent youth have problems (Luthar 2003). Is it possible to have too many
assets? Is there an optimal number that characterizes positive development? In
regard to after-school programs, there has been concern that younger children
may be pushed to do too much; this overscheduling hypothesis has not, however, been empirically supported (Mahoney, Harris, and Eccles 2006).

Hence, while the PYD approach has multiple attractive features, further
theoretical and empirical research is needed to define assets, attend to their variability and measurement, and the processes that link influences to outcomes.

CIVIC ENGAGEMENT IN YOUTH

Civic engagement is clearly one positive outcome of youth development. The
PYD approach proposes that five C's denote or reflect positive development:
competence, confidence, character, connection, and caring. A sixth C has been

proposed as an additional important outcome of development: contribution (Lerner 2004). Civic engagement represents contribution to the society or nation in which one lives. However, there can be multiple forms of contribution, and it is not clear if all should be considered to be forms of civic engagement in terms of participation as a citizen.

For example, young people are not afforded all the opportunities to participate as a citizen that they will enjoy as an adult; eighteen years is the legal age for voting, one hallmark indicator of civic engagement. Does this mean that early forms of political involvement, such as participation in school government or attention to Nickelodeon's "Kids Vote," can only be considered precursors to rather than actual forms of civic engagement?

One can question whether civic engagement even in adults has to involve the political, that is, relate to the polity, or can also consist of civic contributions that are not political. Flanagan and Faison (2001) distinguish civic, from the Latin *civis*, from political, from the Greek *polites*. Although the two terms have different historical roots, they have similar meaning denoting a member of the polity. But Flanagan and Faison argue that the term *political* has come to mean more specifically affairs of the state or the business of government. *Civic* maintains the broader meaning associated with being a member of the polity. Hence, they choose to use *civic* as the broader version. They then differentiate civic literacy as knowledge of community affairs and political issues, civic skills as competencies in achieving group goals, and civic attachment as a feeling or belief that you matter. They argue and present evidence that social relations, opportunities for practice, and the values and behaviors communicated by adults and social institutions determine youth's civic development in these three areas. High-school students also see the responsibilities of citizenship as consisting of polity-oriented ones (such as voting and expressing patriotism) and civic-oriented ones (such as giving back to the community and defending the rights of minorities) (Bogard and Sherrod 2008), which gives some support to the importance of this distinction in defining civic engagement.

I have argued that there are three components to this area that we need to address differentially in regard both to youth precursors and adult outcomes (Sherrod 2003; Sherrod 2006b; Sherrod, Flanagan and Youniss 2002). One is "active citizenship" or the person's involvement in what Flanagan and Faison define as political. This represents the person's knowledge about, involvement in, and commitment to politics, government, or country. Another component is concern for others, an interest in contributing to the general social good as well as helping one's neighbors. This is a type of altruism; however, it must be further differentiated as to whether it requires self-sacrifice. It is far easier to be concerned about one's neighbor and interested in helping others if it doesn't cost you anything. Finally, there is a sense of group connectedness. To whom or what does one's loyalty lie; to whom or what is the person most attached. This can be to one's self, to the family, to the community or neighborhood or some institution therein such as the church, to one's nation or country, and/ or to one's fellow man. These three components—active or typical citizenship, concern for others, and group connectedness—follow somewhat different developmental trajectories (Sherrod 2003).

The definition of civic engagement is important because it determines the indicators we measure to assess the nation's health in this regard. Being informed about politics is one indicator. Is getting information while browsing the Web (CNN pops up when you sign onto the Internet and something catches your attention) an indicator of "being informed" of the same importance as reading a newspaper? For a democracy to survive, it must have actively participating citizens. One criterion for the definition of civic engagement, of course, is whether and how the behavior contributes to the well-being of the democracy (Sherrod, Flanagan, Kassimir, and Syvertsten 2005). However, some would consider volunteering to contribute as much to national well-being as voting; others would not, so that the definition becomes controversial (Sherrod, Flanagan, and Youniss 2002). What is clear is that civic engagement represents contribution as an outcome to positive development, so that this can become an important criterion for defining it (Sherrod and Lauckhardt 2007). The definitional issue, then, becomes whether the other C's such as caring can also be considered to be forms of civic engagement.

Contribution can, however, promote development in addition to representing an outcome. That is, contribution as represented by civic engagement can be either an independent or a dependent variable. Values and individual identity determine in large part the nature of one's civic engagement. However, civic participation also contributes to the formation of values and to individual identity. Hence, the different forms of civic engagement can be both influences on and outcomes of positive youth development.

SPIRITUALITY VERSUS RELIGIOSITY

The construct that raises the most difficult definitional issues is spirituality. Religion or religious participation is clearer because it relates to objectively observable behavior tied to a specific social institution. However, in the same way that volunteering may not indicate that the person is civically engaged, participating in religion may not indicate that the person is religious. Religiosity often showcases extrinsic qualities such as religious practices (Marler and Hadaway 2002), but that does not mean that internal loyalty, commitment, stability (Fuller 2001), and virtue (Oman and Thoresen 2003) are not also important to religiosity. In fact, these latter qualities serve as foundational pillars for the common definitional attributes of religion; possessing a particularistic agenda, advancing a sociohistorical frame of reference (Roeser et al., 2006), and insisting on uniform moral conduct are all predilections associated with religiosity, meaning that religion can be more than a body of institutionalized or traditional beliefs and practices.

Spirituality is often contrasted with religiosity in being represented by internally focused pursuits, individually crafted beliefs and practices (Marler and Hadaway 2002), and a globalistic, humanistic orientation befitting one endowed with a universal spirit (Hill, Pargament, and Hood 2000). Additional concepts associated with spirituality include transcendence, selflessness (Dowling, Gestsdottir, and Anderson 2003), awareness (Roeser et al. 2006), meaning (Piedmont 1999; Hill, Pargament, and Hood 2000), sanctity and holiness (Oman and Thoresen 2003). However, many of these concepts can apply both to spirituality and religiosity (see Piedmont 1999; Zinnbauer, Pargament, and Scott 1999; Marler and Hadaway 2002). And there is some disagreement on the definition of each element; for example, Roeser et al. (2006) does consider transcendence of the ego-state to be one component of spirituality, but he does not see it as mystical. While mysticism and transcendence might be considered to be more typical qualities of spirituality than of religiosity, they certainly can apply to either. And, unfortunately, this cluster of concepts applicable to both spirituality and religiosity contains the crucial elements to positive youth development (e.g., King 2003).

Current research is also not entirely helpful in differentiating spirituality and religiosity. One national longitudinal study of adolescents reports that teenagers do not generally differentiate spirituality from religiosity (Smith and Denton 2005). Other recent studies continue to confirm the existence of a strong "spiritual-but-not-religious" sector of the population (Fuller 2001; Marler and Hadaway 2002). And the United States also boasts a long history in which demographic sectors' clear differentiation between religion and spirituality were commonplace (e.g., Fuller 2001; Wuthnow 1998).

For purposes of this paper, *spirituality* will be clumsily defined as a primarily internal pursuit of meaning or awareness, transcendence, and/or sanctity through means of selflessness and a sense of universal human connectedness, thereby offering some differentiation from *religiosity*. However, research on religious development will be recognized as a convergent construct with much to contribute in understanding the pathway from spirituality to PYD. The vast commonalities are just as crucial as the somewhat-minimal differences in this regard.

As we argued for PYD assets and for indicators of civic engagement, spirituality can be both an independent and a dependent variable. Because religiosity and spirituality protect youth from various risk or negative behaviors (Smetana and Metzger 2005), both can be considered to be important predictors of positive outcomes to development. However, they are also positive outcomes in that they represent possession of values and a positive sense of self (Templeton and Eccles 2005).

As difficult as definitional issues are to the theoretical discussion, they loom as an even-larger threat to measurement efforts. To some extent, the bidirectional nature of development makes it acceptable—even desirable—that constructs can be both independent and dependent variables. If individuals are producers of their own development, then any variable can be either depending on the research question one asks. One clever avenue of navigating through this mess would be to adopt a "big tent" for dependent variables. Outcome measures are generally more concerned with the overall presence of spirituality than with its particular components. In contrast, independent variables should constitute rigorously selected, specific components of spirituality, thereby affording the research community with precise knowledge as to the effects of the construct (see Piedmont 1999). A similar approach might prove beneficial in parsing through a familiar definitional quandary with regards to civic engagement and PYD.

In the remainder of this paper, we explore theoretical and research strategies for studying the interrelationships between civic engagement, positive youth development, and spirituality/religiosity. It is important to point out that spirituality and religiosity may contribute to the development of civic engagement as an outcome of positive youth development; however, the corollary question of how civic engagement may contribute to the development of spirituality as an outcome of PYD is equally important to consider. For a variety of reasons that we will describe, we think the latter question is the more interesting of the two.

HOW MIGHT SPIRITUALITY/RELIGIOSITY FUNCTION TO PROMOTE CIVIC ENGAGEMENT AS AN EXAMPLE OF PYD (AND VICE VERSA)?

This is a developmental question. Development may, in fact, offer one means of differentiating spirituality and religiosity, as defined above. Religion certainly may involve spirituality and often does; however, if one can be spiritual without being religious, how might spirituality arise independent of religion? Religion teaches and socializes values, beliefs, and behaviors and also can promote the development of spirituality, although it may not always do so. It is not clear if there is a present and powerful socializing agent for spirituality of the same order as that which occurs through religion. Religion and spirituality may differentiate across development as young people explore their identity, evaluate past orientations, and achieve a stable sense of self (Erikson 1968). However, there are no empirical data that address how one becomes spiritual without first being religious.

A common goal of religiosity, as one popular spiritual teacher puts it, is "the total transformation of character, conduct, and consciousness" (Easwaran 1991). Its broad aim is to arouse the heart, activate the hands, and stimulate the mind. While each domain might not be manifest in every spiritual agenda and persona, one who fully integrates each branch would probably garner wide recognition for his or her spiritual mastery. The three domains of heart, hands, and mind present exciting options for addressing development. Framing spirituality and its development into the three interconnected spheres of heart, hands, and head insightfully addresses the definitional issues while also offering guidance for articulating research questions. A model using the domains of heart, hands, and mind to interrelate spirituality, religiosity, civic engagement, and positive youth development is depicted graphically in Figure 1. This model is similar to that offered by the 4-H (National 4-H Council 2006), except that the 4-H model includes health. This similarity we think underscores the close bond that civic engagement and spirituality share under the rubric of positive youth development.

Spirituality, while dominating the affective (heart/soul) sphere, still permeates the other two spheres in composing its comprehensive definition. Similarly, religiosity is most deeply rooted in the cognitive or mind aspect, yet clearly maintains an open relationship with humanity's affective states and physical conduct. Civic engagement, in turn, is a dominant positive-development construct in the sphere of human conduct, or "hands," although, of course, ide-

FIGURE 1. Relationships and Components of Religiosity, Spirituality, and Civic Engagement

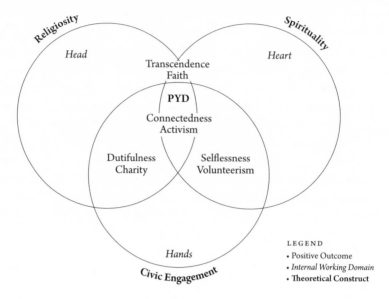

ology (head) and positive affect (heart) also contribute to the refinement of civic engagement.

The affiliation that we suggest between each of the three dimensions and heart, mind, and hands should not obscure the extensive overlap that we insisted upon earlier. In fact, their overlap allows their expression as positive development (see Figure 1). For example, an adolescent who is developmentally engaged on both cognitive (head) and conduct (hands) dimensions will likely enwrap the selfless act of giving charity in a profound sense of duty; in contrast, one who is motivated out of a humanistic selflessness—characteristic of the heart dimension—will feel far more liberated in his/her volunteerism. Indeed, charity is a fitting combination of religiosity and civic engagement, and volunteerism nicely depicts the region shared by spirituality and civic engagement. We propose that the overlap of all three dimensions captures the essence of PYD.

One can consider development by exploring what contribution might be made by each of three major developmental theories: identity and life cycle, attachment, and social learning. Through this sampling of developmental theories, we hope not only to clarify this definitional model based on heart, mind, and hands but also to illustrate the conceptual strengths of the respective PYD

FIGURE 2. Developmental Theories as Pathways Linking Religiosity and Spirituality to Civic Engagement

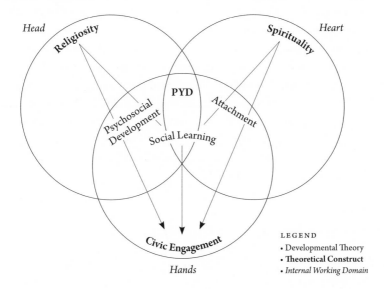

dimensions. Figure 2 adds these theoretical foundations to the initial model depicted in Figure 1.

Identity Theory and Life-cycle Approaches

Identity theorists, based in Erikson's theory of psychosocial development, are mainly concerned with the ideological component of religion. Adolescents' struggle for identity cohesion and for achievement is heavily based in cognition. In the pursuit of identity, even the spiritual notion of transcendence and relationship with a higher being is couched not in terms of emotional benefits as much as in cognitive ones. Meaning, perspective, and understanding of the "larger issues in life" predominate (King 2003). Indeed, it is the pursuit of meaning on the path to identity formation that inspires psychosocial theorists to address the contributions of religious institutions (King 2003; Markstrom 1999).

The link between identity formation and religious development is further illustrated by the closely affiliated life-cycle approach. Several studies have been conducted supporting the impact of life-cycle experiences and expectations upon religious affiliation. These studies generally focus on church membership of young and middle-aged adults and how experiences of marriage and parenting affect the intensity of affiliation (Stolzenberg, Blair-Loy, and Waite

1995; O'Connor, Hoge, and Alexander 2002). Recent qualitative data have supported a parallel phenomenon among adolescents who moderate their levels of religiosity based on their conceptions of adolescence as a unitary stage of life (Smith and Denton 2005).

Two important points emerge from this literature. First, the link between cognitive processes and religious development is transparent. Second, the psychosocial moment is ripe for identity formation through cognitive avenues—whether one's aim is to cultivate religious behaviors and attitudes or to develop civically engaged youth. In either case, the clearest path to PYD in instances where religious or civic identity predominates passes through psychosocial conceptions. Relevant research and initiatives should be made aware of these relationships and proceed accordingly.

Attachment Theory

In contrast, an emerging literature has applied attachment theory to a believer's relationship with God. Under this framework, a secure attachment relationship with God will promote emotional stability and psychological well-being (Birgegard and Granqvist 2004). This approach is further bolstered by studies demonstrating the spiritually enhancing effects that distressing situations induce (Jonas and Fischer 2006). The attachment system is predominantly controlled by affective stimuli and produces behaviors that seek proximity to the attachment figure—a transcendent being in this case. These behaviors are reminiscent of essentially spiritual characteristics; they are also linked with behaviors and even personality characteristics that relate to the promotion of civic engagement.

Once again, the intersecting network of human resources, social frameworks, and psychological theories offers a couple of insights. First, the emotional side of human development can be harnessed to promote spiritual development. Second, both spiritual and civic aims can achieve further positive youth development by appreciating the perspective of attachment theory as it relates to emotional development.

Social Learning Theory

A final theoretical model to consider is that emanating from social learning theorists. Social learning theory, in fact, may be considered to blend the previous two theories based in identity and attachment, respectively. Both identity and attachment theories concede the utility of modeling in developing religious and spiritual sentiments and their associated behaviors. This theory focuses less on the content of what is being developed and more on the process of how behavior is learned. In this sense, social learning theory may be the most useful of the three in applying religious development to civic engagement.

The case for religious and spiritual development as social cognition revolves around the "how" rather than the "what." If the four criteria for observational learning—attention, retention, reproduction, and motivation—can be adequately offered by spiritual or religious figures, then proper spiritual/religious development is likely to occur. In fact, the four criteria for observational learning are prevalent in both religiously and spiritually oriented communities. Societies tend to publicize exemplars of a religious or spiritual nature and build legends around their legacies, thereby fulfilling both attention and retention. In addition, high standards mandate the reproduction of spiritual behaviors, with great rewards motivating the effort to abide by those standards (Oman and Thoresen 2003). This sequence leading to observational learning can be replicated on either affective (Reimer 2003) or cognitive (Martin, White, and Perlman 2003; Reimer 2003) dimensions.

Taken together, these three theories offer a variety of angles with which to promote behaviors consistent with PYD. Their affiliation with the three dimensions of PYD—heart, mind, and hands—as well as the internally integrated triumvirate ought to excite future research and policy and help clarify definitional and conceptual issues. It might even define a foundation for building a field of research. A variety of permutations of the three linking different facets of PYD are open to exploration, particularly within traditional areas such as religiosity, spirituality, and civic engagement.

A SPECIAL CONCERN FOR DISENFRANCHISED YOUTH

However, considerable methodological, as well as theoretical, challenges confront further field development. One important challenge is the incorporation of diverse and disenfranchised youth in both theory development and empirical research.

The expression of heart, mind, and hands across areas such as religiosity, spirituality, and civic engagement as outcomes of positive youth development is likely to vary considerably across variables of individual address, such as gender, social class, ethnicity, and immigrant status. At the same time, this taxonomy may offer a means of addressing the development of diverse youth in a culturally appropriate manner. For example, one can ask how heart, mind, and hands may be expressed through outlets other than religiosity or civic engagement.

The family, schools, or communities, one's race or culture, and even the pursuit of success or wealth represent institutions to which one can direct one's heart, minds, and hands. Different social institutions are important in the lives of diverse youth. Immigrant youth, for example, tend to show substantial family involvement and connectedness (Fuligni, Tseng, and Lam 1999). Racial identity and racial socialization can play an important role in the lives of minority youth who may face discrimination (Hughes and Chen 1997; Spencer, Fegley, and Harpalani 2003). Both can interact or not with spirituality and religiosity. Minority youth, for example, are more likely to show devote religion (Wallace, Forman, Caldwell, and Willis, 2003). These other social institutions, as they may represent expressions of heart, mind, and hands, hence add to the permutations one must consider in asking how religion and spirituality contribute to PYD.

One interesting issue for research is whether one form of expression leads to others, as we have illustrated in our model (Figures 1 and 2) or whether feeling a sense of commitment to one institution replaces others or makes it less likely that the person will develop other attachments. For example, in our Figure 1, one might replace spirituality, religiosity, and civic engagement with family, schools/community, and ethnicity. For some youth, their hearts may reside with their family, their minds with their ethnicity in terms of identity, and their hands with their school and community. Again, the intersection might represent PYD.

Research on infant attachment to caregivers indicates that they can have multiple attachment figures; however, they often have a hierarchy of attachments (Ainsworth, Blehar, Waters, and Wall 1979). Does the same apply to youth's expression of heart, mind, and hands, and does it have implications for the outcome of their interaction? The issue is whether youth can have attachments to multiple social institutions varying by their expression of heart, mind, and hands. And if so, is there a hierarchy of such attachments?

SUMMARY AND CONCLUSIONS

The theoretical model we propose is intended to help frame future research and interventions with clear aims and purposes. The divisions between the three domains of head, heart, and hands are pervasive in human experience and behavior. Spirituality, comprising a central component of the human experience across history, is no different. Research on spirituality and religiosity reflects this broad range, with studies on resiliency in the face of trauma (Pargament 1997) appealing to spirituality much the same as studies on self-efficacy (e.g., Ai, Peterson, Rodgers, and Tice 2005) or risky health behaviors (Sinha, Cnaan, and Gelles 2007). In each instance, a different constitution of human resources is at stake: emotional resources predominate in coping with trauma, whereas cognitive attitudes often direct youth toward or away from risk behaviors.

Our model suggests that the religion and spirituality constructs we measure, the religious interventions we construct, and the spiritual treatments we offer should reflect the essence of the human behavior or experience at hand, rather than broad and murky conceptual constructs. In this vein, initiatives associating trauma with spirituality in a developmental context would do well to consult attachment-theory literature; research focusing on the impact of religion on self-efficacy among youth should be informed by social learning theory; and religious or spiritual interventions aimed at curbing adolescent risk behaviors should be informed by identity theory among other cognitive or attitudinally based theories.

Considerable theoretical, including definitional, work is necessary to advance the field of study of religious and spiritual development. We have attempted to begin this work in this chapter in our own small way. Perhaps needless to say, measurement development represents a next step for empirical research to proceed. By embedding our ideas in existing developmental theories, we hope to at least reach this goal, providing some potential for approaching multiple forms of measurement, rather than relying exclusively on subjective measures. It is often said in our field that our ideas outstrip our measurement, and nowhere is that more true than for this area. Connecting spiritual and religious development to positive youth development, the goal of this volume, provides a goal to direct both theory and measurement development.

REFERENCES

Ai, A.L., C. Peterson, W. Rodgers, and T. N. Tice. 2005. Effects of faith and secular factors on locus of control in middle-aged and older cardiac patients. *Aging and Mental Health* 9, no. 5: 470–81.

Ainsworth, M., M. Blehar, E. Waters, and S. Wall. 1979. *Patterns of attachment: A psychological study of the strange situation.* Hillsdale, NJ: Erlbaum.

Benson, P. 2004. Developmental assets and human development. Paper presented at the International Conference on Applied Developmental Science, University of Jena, Germany. October 9–12.

Benson, P., N. Leffert, P. Scales, and D. Blyth. 1998. Beyond the village rhetoric: Creating healthy communities for children and adolescents. *Applied Developmental Science* 2: 138–59.

Birgegard, A., and P. Granqvist. 2004. The correspondence between attachment to parents and God: Three experiments using subliminal separation cues. *Personality and Social Psychology Bulletin* 30, no. 9: 1122–35.

Bogard, K., and L. Sherrod. 2008 (under review). The influence of discrimination distress and parent socialization on civic attitudes among youth of color.

Dowling, E. M., S. Gestsdottir, and P. M. Anderson. 2003. Spirituality, religiosity and thriving among adolescents: Identification and confirmation of factor structures. *Applied Developmental Science*, 7, no. 4: 253–60.

Easwaran, E. 1991. *Meditation: A simple eight-point program for translating spiritual into daily life.* Tomales, CA: Nilgiri Press. Retrieved Dec. 19, 2006, from http://www.easwaran.org/nilgiri .cfm/pageid:2021

Erikson, E. H. 1968. *Identity: Youth and crisis.* New York: Norton.

Flanagan, C., and N. Faison. 2001. *Youth civic development: Implications of research for social policy and programs.* Social Policy Reports, no. 1. Ann Arbor, MI: SRCD.

Fuligni, A., V. Tseng, and M. Lam. 1999. Attitudes toward family obligations among American adolescents with Asian, Latin American and European backgrounds. *Child Development* 70: 1030–44.

Fuller, R. C. 2001. *Spiritual but not religious: Understanding unchurched America.* New York: Oxford University Press.

Hill, P. C., K. I. Pargament, and R. W. Hood. 2000. Conceptualizing religion and spirituality: Points of commonality, points of departure. *Journal for the Theory of Social Behaviour* 30, no. 1: 51–77.

Hughes, D., and L. Chen. 1997. When and what parents tell children about race: An examination of race related socialization among African American families. *Applied Developmental Science* 1: 200–214.

Jonas, E., and P. Fischer. 2006. Terror management and religion: Evidence that intrinsic religiousness mitigates worldview defense following mortality salience. *Journal of Personality and Social Psychology* 91, no. 3: 553–67.

King, P. E. 2003. Religion and identity: The role of ideological, social, and spiritual contexts. *Applied Developmental Science* 73: 197–204.

King, P., and P. Benson. 2006. Spiritual development and adolescent well-being and thriving. In E. C. Roehlkepartain, P. E. King, and P. L. Benson, eds., *The handbook of spiritual development in childhood and adolescence*, 384–98. Thousand Oaks, CA: Sage.

Lerner, R. M. 2004. *Liberty: Thriving and civic engagement among America's youth.* Newbury Park, CA: Sage Publications.

Lerner, R., A. Alberts, and D. Bobek. 2008 (In press). *Thriving youth, flourishing civil society: How positively developing young people may contribute to democracy and social justice.* A Bertelsmann Foundation white paper. Gutersloh, Germany: The Bertelsmann Foundation.

Luthar, S. 2003. The culture of affluence: Psychological costs of material wealth. *Child Development* 74, no. 6: 1581–93.

Mahoney, J., A. Harris, and J. Eccles. 2006. Organized activity participation, positive youth development, and the over-scheduling hypothesis. *Social Policy Report* 20, no. 4.

Markstrom, C.A. 1999. Religious involvement and adolescent psychosocial development. *Journal of Adolescence* 22: 205–21.

Marler, P. L., and C. K. Hadaway. 2002. "Being religious" or "being spiritual" in America: A zero-sum proposition? *Journal for the Scientific Study of Religion* 412: 289–300.

Martin, T. F., J. M. White, and D. Perlman. 2003. Religious socialization: A test of the channeling hypothesis of parental influence on adolescent faith maturity. *Journal of Adolescent Research* 18, no. 2: 169–87.

National 4-H Council. 2006. Retrieved October 9, 2007 from http://www.fourhcouncil.edu/.

O'Connor, T. P., D. R. Hoge, and E. Alexander. 2002. The relative influence of youth and adult experiences on personal spirituality and church involvement. *Journal for the Scientific Study of Religion* 41, no. 4: 723–32.

Oman, D., and C. E. Thoresen. 2003. Spiritual modeling: A key to spiritual and religious growth? *The International Journal for the Psychology of Religion* 13, no. 3: 149–65.

Pargament, K. I. 1997. *The psychology of religion and coping: Theory, research, practice.* New York: Guilford Press.

Piedmont, R. 1999. Does spirituality represent the sixth factor of personality? Spiritual transcendence and the five-factor model. *Journal of Personality* 67, no. 6: 985–1012.

Reimer, K. 2003. Committed to caring: Transformation in adolescent moral identity. *Applied Developmental Science* 7, no. 3: 129–37.

Roeser, R., M. Rao, M. Shah, Y. Hastak, A. Gonsalves, and R. Berry, R. 2006. A return to the varieties of religious experience: Research notes from India. Paper presented as part of a symposium "Theoretical Issues in the Study of Adolescent Spiritual Development" at the biennial meeting of the Society for Research on Adolescence, San Francisco.

Scales, P., D. Blyth, T. Berkas, and J. Kielsmeier. 2000. The effects of service learning on middle school students' social responsibility and academic success. *Journal of Early Adolescence* 20, no. 3: 332–58.

Sherrod, L. R. 2003. Promoting the development of citizenship in diverse youth. *PS: Political Science and Politics*: 287–92.

———. 2006a. Civic engagement as an expression of positive youth development. In R. Silbereisen and R. Lerner, eds., *Approaches to Positive Youth Development*, 59–74. Thousand Oaks, CA: Sage Publications.

———. 2006b. Promoting citizenship and activism in today's youth. In S. Ginwright and R. Watts, eds., *Beyond resistance! Youth activism and community change: New democratic possibilities for practice and policy for America's children*, 287–300. New York: Routledge.

Sherrod, L. R., N. Busch, and C. Fisher. 2004. Applying developmental science: Methods, visions, and values. In R. Lerner and L. Steinberg, eds., *Handbook of Adolescent Psychology*, 747–80. New York: Wiley and Sons.

Sherrod, L. R., C. Flanagan, R. Kassimir, and A. Syvertsen, eds. 2005. *Youth activism: An international encyclopedia.* Westport, CT: Greenwood Publishing Group.

Sherrod, L. R., C. Flanagan, and J. Youniss. 2002. Dimensions of citizenship and opportunities for youth development. *Applied Developmental Science* 6, no. 4, 264–72.

Sherrod, L., and J. Lauckhardt. 2008 (In press). Cultivating civic engagement. In J. Rettew, ed., *Positive psychology: The science of human flourishing*, Vol. 4, Westport, CT: Greenwood Publishing Group.

Sinha, J. W., R. A. Cnaan, and R. J. Gelles. 2007. Adolescent risk behaviors and religion: Findings from a national study. *Journal of Adolescence* 30, no. 2: 231–49.

Smetana, J., and A. Metzger. 2005. Family and religious antecedents of civic involvement in middle class African American late adolescents. *Journal of Research on Adolescence* 15, no. 3: 325–52.

Smith, C., and M. L. Denton. 2005. *Soul searching: The religious and spiritual lives of American teenagers.* New York: Oxford University Press.

Spencer, M. B., S. G. Fegley, and V. Harpalani. 2003. A theoretical and empirical examination of identity as coping: Linking coping resources to the self processes of African American youth. *Applied Developmental Science* 7: 181–88.

Stolzenberg, R. M., M. Blair-Loy, and L. J. Waite. 1995. Religious participation in early adulthood: Age and family life cycle effects on church membership. *American Sociological Review* 60, no. 1: 84–103.

Templeton, J. L., and J. S. Eccles. 2005. The relation between spiritual development and identity processes. In E. C. Roehlkepartain, P. E. King, L. M. Wagener, and P. L. Benson, eds., *The handbook of spiritual development in childhood and adolescence*, 252–65. Thousand Oaks, CA: Sage Publications.

Theokas, C., J. Almerigi, R. Lerner, E. Dowling, P. Benson, P. Scales, and A. Von Eye. 2005. Conceptualizing and modeling individual and ecological asset components of thriving in early adolescence. *Journal of Early Adolescence* 251: 113–43.

Wallace, J. M., T. A. Forman, C. H. Caldwell, and D. S. Willis. 2003. Religion and U.S. secondary school students: Current patterns, recent trends, and sociodemographic correlates. *Youth and Society* 35: 98–125.

Wuthnow, R. 1998. *After Heaven: Spirituality in America since the 1950s.* Berkeley: University of California Press.

Zarrett, N., R. Lerner, J. Carrano, K. Fay, J. Peltz, and Y. Li. 2007. Variations in adolescent engagement in sports and its influence on positive development. In N. L. Holt, ed., *Positive youth development through sport*, 9–23. Oxford, U.K.: Routledge.

Zinnbauer, B. J., K. I. Pargament, and A. B. Scott. 1999. The emerging meanings of religiousness and spirituality: Problems and prospects. *Journal of Personality* 67, no. 6: 889–919.

18

A Palace in Time

Supporting Children's Spiritual Development
through New Technologies

Marina Umaschi Bers

Sixteen-year-old Janet connects to Zora, a multiuser virtual city. There, during a summer workshop for youth, she has created an avatar, a virtual representation of herself, a virtual home, and a Jewish temple. A visit to Janet's virtual home on Zora reveals much about her: her favorite friends, her most-loved games, her family's history. After working on her own virtual home, she creates the Jewish temple. She makes a virtual rabbi to welcome visitors with a blessing. She invites other children to make the decorations of the virtual synagogue. There are Hebrew letters, a map of Israel, a picture of a man praying. Janet clicks on a silver mezuzah.[1] It tells her a story about the meaning of the prayers it holds. She decides to add a television to the temple. Inside it, she puts a snapshot from the movie Schindler's List that she found on the Web. The system enables her to associate objects with "values."[2] She associates the television with the value "documentation" and defines it as "very important to remember history. That way, bad things won't happen again. Holocaust survivors are getting very old now, and if someone doesn't record their stories of what happened, we are doomed to forget and repeat the horrors." While exploring the Jewish temple, Janet encounters Marie. Both girls chat via their avatars, and then Marie invites Janet to visit the virtual Baptist Church she created.

This vignette presents one of the many experiences that young people had while participating in a summer workshop in which they created a virtual city to explore cultural and religious differences using the Zora three-dimensional virtual world. Zora provides tools for users to design their world, with both private and public spaces, and to create and program the objects and characters that will populate the virtual world. In the process of creation, children can also communicate in real-time with each other (Bers 2001). Zora is one of the many contemporary examples of virtual worlds that enable children to develop a virtual community and inhabit it. However, what is different from commercial applications such as Second Self and The Sims is that Zora was explicitly developed with a theoretical and pedagogical framework that looks at the positive role that technology can play in young people's lives. Thus, it has design features that enable children to not only have fun while creating their virtual city but also to learn about their own identities and to explore the different aspects of the self, spirituality among them.

Zora is an example of a computer-based identity-construction environment (ICE). In this chapter, I will present both Zora and other identity-construction environments I have developed over the last decade of work. While the technologies and the context of their use are different from each other, a virtual world in Zora, a robotics workshop in Project InterActions, and a storytelling authoring toolkit in SAGE (Storytelling Agent Generation Environment), in the many different pilot studies, children used these identity-construction environments to explore their own sense of spirituality and, in some cases, to learn more about religious traditions.

When first thinking about how to design these computer-based identity-construction environments, one image came to my mind: the Jewish Sabbath. I am not an observant person nor do I strictly follow the laws and rituals prescribed by Judaism with respect to the Sabbath. However, I find the idea of the Sabbath fascinating. The Jewish philosopher and theologian Abraham Joshua Heschel said that the Sabbath is a "palace in time" in our modern lives: "[Its] goal is not to have but to be, not to own but to give, not to control but to share, not to subdue but to be in accord" (Heschel 1951). According to Heschel, the seventh day belongs to the realm of time, as opposed to the realm of space. In a beautiful and simple manner, he explored the many reasons that make the Sabbath a holy day. One of them has been particularly influential in the design of identity-construction environments such as Zora: the Sabbath is a time for introspection and reflection, a time for stopping the everyday work and looking back at who we are, how we are feeling, and how we are building a caring, just, and responsible community.

When designing a technological environment for learning about identity, all of these images associated with the seventh day came to mind. The Sabbath served me as a powerful "object to think with" (Papert 1980). It illuminated the kind of experience that I hoped young people would have while engaging with the technology. I wanted children to enter a very special place, "a palace in time," that would afford them similar experiences to the ones I had when entering the synagogue: engage in self-reflection, creation, creativity, communication, and participation in a community. I hoped children would collaborate with others in ongoing community projects and, at the same time, engage in personally meaningful projects. I imagined them entering into "a palace in time" where they would find tools for self-reflection and community building. These tools would go beyond the traditional prayers, words, and conversations that I found at the synagogue. I wanted results of quiet introspection and self-reflection to become tangible and manipulable. Thus, in Zora, children can create virtual temples with interactive religious objects and characters. And they can embed stories in virtual objects that can be programmed to react to users' inputs. Zora, as well as the other computer-base identity-construction environments I will later present, is my attempt at designing a virtual "palace in time," a technologically rich environment that would afford similar kinds of experiences as the Sabbath.

In this chapter, I will first provide an overview of the design history of identity-construction environments by describing an iterative process of designing, implementing, and evaluating three projects through psychoeducational interventions: SAGE, Zora, and Project InterActions. I will conclude the chapter with reflections on the positive role that technology can have in the spiritual development and the religious education of young people.

IDENTITY-CONSTRUCTION ENVIRONMENTS: A DESIGN HISTORY

Over the last decade, I have developed and conducted studies with several kinds of computer-based identity-construction environments that enable children to learn by doing. In this chapter I will present three of those that children have used to explore spirituality and religion.

1. SAGE, Storytelling Agent Generation Environment, provides tools to create wise storytellers to interact with by telling and listening to stories (Bers and Cassell 1999).

2. The Zora virtual environment, as mentioned before, provides tools to create and inhabit a three-dimensional virtual city (Bers, Gonzalez-Heydrich, and DeMaso 2001).

3. Project InterActions engages families in the design, building, and programming of a robotic artifact to represent a value or an aspect of their reli-

gious and cultural backgrounds (Bers, New, and Boudreau 2004). Within this project, a pilot study was done in a Jewish day school in Argentina during the high holidays. Families came together to make robotic creative prayers (Bers and Urrea 2000).

Identity-construction environments are specifically designed with the goal of promoting positive youth development (PYD). PYD involves cognitive, personal, social, emotional, and civic aspects of young people, which research-ers refer to as the six C's: competence (cognitive abilities and behavioral skills), connection (positive bonds with people and institutions), character (integrity and moral centeredness), confidence (positive self-regard, a sense of self-efficacy), caring (human values empathy and a sense of social justice), and contribution (orientation to contribute to civil society) (Lerner et al. 2005; Lerner 2002; Lerner, Fisher, and Weinberg 2000). Together, these characteris-tics reflect a growing consensus about what is involved in healthy and positive development among people in the first two decades of their lives (Scales, Lef-fert, and Blyth 2000). Thus, interventions such as the ones that can be devel-oped through the use of computer-based identity-construction environments to promote learning about spirituality and religion can also have an impact on positive youth development.

In my own work on PYD, I pay particular attention to technologically rich contexts and aim to understand what kinds of technology-based interventions are more likely to engage children in a learning trajectory that I call positive technological development (PTD). Today's youth use computers for learning, working, playing, communicating, dating, buying, and protesting. According to a 2005 study conducted by the Pew Internet and American Life project, more than one half of all American teens use the computer and the Internet to create media (Lenhart, Madden, and Hitlin 2005). Psychoeducational pro-grams must take advantage of the natural tendency of youth to use technology in creative ways.

Computer-based identity-construction environments such as the ones I will later describe are purposefully designed to be used in targeted interven-tions that will promote positive technological development, which, in turn, might promote positive youth development. Mirroring the six C's of PYD, the PTD (positive technological development) learning trajectory aims at helping children develop: (a) *competence* in the development of computer literacy and technological fluency, (b) *confidence* in their own learning potential and their own ability to solve technical problems, (c) *caring* about others, to be expressed by using technology to engage in collaboration and to help each other when needed, (d) *connection* with peers or adults to use technologies to form virtual communities and social support networks, (e) *character* to become aware of their own personal values, be respectful of other people's values, and assume a responsible use of technology, and (f) *contribution* by conceiving positive ways of using technology to make a better learning environment, community, and society (Bers 2006; Bers 2007).

Computer-base identity-construction environments make use of stories as essential building blocks for helping children use technology in positive ways and for carrying on the "know thyself" mandate. According to the Jewish tradition, God presented himself to Moses in the Sinai by saying, "I am that I am" (Exodus 3:14). This recursive definition of identity serves as a metaphor of our own need to discover who we are. New technologies are changing the way in which we know our own selves (Turkle 1984, 1995). I will present some examples of how three different technologies and psychoeducational programs were developed with this purpose.

SAGE: STORYTELLING AGENT GENERATION ENVIRONMENT

SAGE is an authoring environment for children to create their own wise storytellers to interact with by telling and listening to stories. Children can engage with SAGE in two modes: (a) by choosing a wise storyteller from a library of already-existing characters and sharing with him or her what is going on in their lives, and (b) by designing their own sages and programming the conversational interaction between storyteller and potential users, as well as creating the database of inspirational stories offered by the storyteller in response to user's problems (Bers and Cassell 1999). The sage storyteller "listens" and then offers a relevant tale in response

The SAGE architecture has three components:

1. *Computation module* is in charge of parsing the user's story to extract nouns and verbs, expanding these keywords through WordNet, a hierarchical semantic lexical reference system, and performing a match between the user's personal story and an inspirational story in the database.

2. *Authoring language* serves to design the personality of the interactive characters, the conversational flow between user and storyteller, and the database of stories offered by the system in response to the user's story. In the

database of stories, users can write or record the inspirational tales that will be offered by the sage. They can index them with personal and moral values that they consider good descriptors of the main story points.

3. *Interface* is the layer through which a user communicates with the system, composed by the computer screen and a programmable interactive soft toy that behaves as the storyteller's assistant. It takes different personalities according to the hat it wears (see Figure 1). For example, by putting on a yin-yang hat, the toy behaves as a Taoist sage, and by replacing it with a yarmulke, the toy becomes the disciple of a rabbi who tells Hasidic stories. The toy moves its ears, shakes its body, and blinks its eyes to indicate attentiveness. These body movements can also be programmed by the children.

Over the years, I conducted several studies to examine children's interactions with the sage storytellers and their ability to build their own meaningful characters. My earliest research showed that children engaged deeply with sages that I had designed, such as a Hasidic rabbi and a Buddhist scholar.

FIGURE 1. The Hasidic SAGE on the computer screen and the interactive, stuffed storyteller assistant.

When interacting with them, children revealed aspects of their inner lives and the problems they face every day. For example, below is an extract from a conversation log between a ten-year-old and the Hasidic rabbi:

SAGE: You know Abraham, a story is more than a story. A story has power. Nu, what is your story? What is in your mind?

USER: Once I was learning math, and I could never understand it.

SAGE: *Oy vey!* Go on, Abraham.

USER: I kept on trying, but no matter how much I studied, I could not get it.

SAGE: I see, Abraham. What else do you have to share with this old rabbi?

USER: I have the same problem with English, Hebrew, and all other subjects.

SAGE: Well, your situation reminds me of a story. Remember that there is power on the stories. The Gerer rabbi said: "Exile contains redemption within itself, as seed contains the fruit. Right work and real diligence will bring out the hidden reward." Abraham, do you see the point of this story related to your own story?

USER: Yes. So now I see that I just have to keep on trying no matter what.

SAGE: I hope it was helpful. *Zie gezunt*, be well. May the Lord, source of life, bless you, Abraham, with peace, and may you find your inner peace. Amen.

In this example, the child, Abraham, shares his learning problems with the Hasidic SAGE who tells him a comforting story from the database by matching nouns, verbs, and values.

In the next example, the same format is used by a Taoist SAGE who tells a comforting story to Conor, who is also experiencing problems with school:

SAGE: *Nin hau*. What is your name?

USER: Conor.

SAGE: My name is Chuang-Tzu. I am a Taoist sage who lives by the Way, the Tao, wandering free and easy. How old are you, Conor?

USER: 13.

SAGE: I am 87 years older than you. I am older than the mountains and the seas, while you, Conor are still young. You are here because you seek knowledge. Conor, tell me your troubles, and I will enlighten you with the sayings of the Tao, the Way.

USER: I want to do better in school.

SAGE: I see, Conor. What else can you tell to this Taoist sage?

USER: I have a lot of math homework, and I don't understand all of it.

SAGE: Go on, Conor. We are becoming one.

USER: The teacher was vague in explaining it.

SAGE: My Taoist understanding tells me that you want to do better in school. You have a lot of math homework, and you don't understand all of it. The teacher was vague in explaining it. Is that close to the truth?

USER: Yes.

SAGE: Silence. Let the Way come to me. There is something for you. No need to leave your door to know the whole world; no need to peer through your windows to know the Way of heaven. The farther you go, the less you know. Therefore, the sage knows without going, names without seeing, and completes without doing a thing. Get the idea, Conor?

USER: I was already supposed to know how to do it.

SAGE: I see, Conor. Maybe this teaching of the Way will help you. To understand others is to be knowledgeable; to understand yourself is to be wise. To conquer others is to have strength; to conquer yourself is to be strong. To know when you have enough is to be rich. To go forward with strength is to have ambition. To not lose your place is to last long. To die but not be forgotten, that's true long life. Do you get it?

USER: Yes, I'm supposed to learn from my mistakes.

SAGE: So, Conor, now you understand the Way. I must go now to become one again with heaven and earth. Conor, don't stop thinking about the Way, and you will find your own way. May you, Conor, become one with the Tao.

Over sixty young children, such as Abraham and Conor, engaged with the Hasidic and the Taoist sages and were able to both tell their own stories and interpret the hidden meaning behind the traditional stories offered by the SAGES. However, older youth were also able to suspend disbelief and engage in the storytelling interactions.

SAGE: *Shalom, nu?* What's your name?

USER: David.

SAGE: Oh, David, I am the rabbi of this computer community. My grandfather was a rabbi in a shtetl back in Russia. But enough about me. How old are you, David?

USER: 16.

SAGE: *Baruj Hashem*, I am 64 years older than you.

SAGE: David, you know what *tsuris* are? Troubles . . . everybody has troubles. Tell me yours, and I will tell you a Hasidic story back.

USER: I have an assignment to write, and I am finding great difficulty writing it.

SAGE: *Oy vey!* Go on, David.

USER: If I do not write it, my teachers will be very upset with me.

SAGE: I see David. What else do you have to share with this old rabbi?

USER: I am afraid that they will not renew my scholarship if I do not hand it in.

SAGE: *Oy va voy*, David, let me see if this rabbi understands. You have an assignment to write, and you are finding great difficulty writing it. If you do not write it, your teachers will be very upset with you. You are afraid that they will not renew your scholarship if you do not hand it in. Is this right?

USER: Yes.

SAGE: Your situation reminds me of a story. Remember, there is power in the stories. Rabbi Israel Baal Shem Tov, who later was to found the Chasidic movement, lived a "hidden" life, artfully concealing his knowledge and piety in the guise of an ignorant clay digger. Once, he came to see the rabbi of Brody. The rabbi, seeing only his visitor's crude manners and torn and muddy clothes, treated him with contempt. Said the Baal Shem Tov, "Our sages tell us to 'learn from every man,' for your fellow is your mirror. If your own face is clean, the image you encounter will also be flawless. Should you gaze into a mirror and see a blemish, it is your own imperfection that you are seeing." *Nu*, David, do you get it?

USER: Yes. You are trying to tell me that I am self-centered and wrapped up in my own problem.

SAGE: *Baruch Ha Shem*, you got the power of the story.

SAGE: I hope it was helpful. *Zie gezunt*, be well. May you have the courage to keep telling stories, and may you, David, keep listening to stories.

While all of the examples above show interactions with sages that mimic religious or spiritual leaders, children also chose to create characters such as Mother Nature, Shaquille O'Neal, and the Big Orange Fox, who would tell them meaningful stories (Bers and Cassell 1998). In building a SAGE, children designed that person to whom they wished they could turn with their problems. They also played with different notions of self, by creating or imitating the narrative voices they wanted or needed to hear.

Pilot research showed that children created storytellers as projections of fears, feelings, interests, and role models. For example, young cardiac patients used the SAGE environment to tell personal stories and created interactive characters, such as Mrs. Needle or Mr. Tape, as a way of coping with cardiac illness, hospitalizations, and invasive medical procedures (Bers et al. 1998).

THE ZORA VIRTUAL ENVIRONMENT

Zora is a three-dimensional multiuser environment that provides tools for children to design and inhabit a virtual city. Avatars representing users can navigate around the virtual city. They can converse with others in real-time through a graphical chat system and construct the city's private and public spaces. They can create objects, characters, and stories while developing a virtual community (Bers 2001).

In Zora, all virtual objects have three different kinds of attributes that need to be personalized by the children. The *presentation attributes* determine the object's graphical appearance and motion. The *administration attributes* determine who owns the object and, therefore, who can edit it and who has permission to decide if the artifact can be cloned. The *narrative attributes* (i.e., textual descriptions, stories, values, and conversations) structure a way of thinking about objects that highlights their potential to carry ideas about self and community and that allow for specifying the meaning or personal and moral values that people assign to them.

Users can write stories and associate them with keywords to be used while programming a conversation for the object, as well as write values and associate them with definitions. Figure 2 shows the Zora interface. Zora also offers tools for researchers and teachers to evaluate the learning experience. For example, the Zora system logs, with date and time, everything users say or do online. A Zora log parser was implemented in order to parse the system logs according to the specific needs of each research project and gives control to the researcher of the variables to retrieve and display. The first version of Zora was implemented as part of my doctoral work at the MIT Media Lab using Microsoft's Virtual Worlds research platform. The second version of Zora is implemented using the Active Worlds platform at Tufts University.

In the spirit of the constructionist learning theory that states that people learn better while engaged in the design of a meaningful project (Papert 1980), Zora offers easy-to-use authoring tools to build a virtual city. Users can import and create their own personally meaningful pictures and objects and can make virtual heroes and villains, positive and negative models of identification.

In constructionist graphical virtual environments, users can define how their objects behave in the world by programming their motion and animations. In Zora, however, users can also program objects to engage in storytelling interactions, as in SAGE, by describing the underlying turn-taking rules between user and object. They can also define the stories to be told by the object in response to certain input. While programming interactive conversations, learners engage in perspective taking or seeing the world as others do. This is a fundamental mechanism involved in the process of identity formation.

Zora's values dictionary is a compendium of all personal and moral values and their multiple definitions, held by the Zora community. At the beginning of a Zora experience, the dictionary is empty. As learners populate the virtual city with objects and characters and define the values and definitions associated with them, the dictionary starts to fill up. The collaborative values dictionary was designed to allow users (a) to browse its content and easily visualize clashes between different definitions for a same value in order to trigger interesting conversations and reflections and (b) to enter new values and definitions independently from grounding them in objects in the virtual world.

FIGURE 2. Zora's Interface

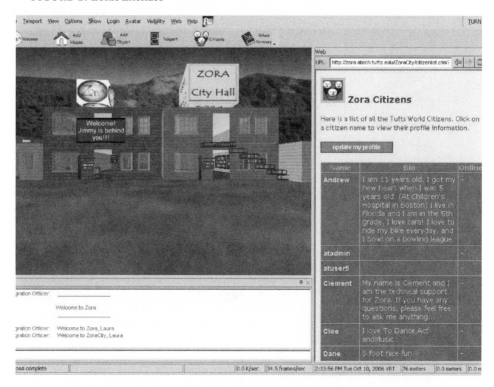

I conducted four different studies with young people using Zora. All of the studies focused on developing components of positive technological development—in particular, competence and confidence in using technology—but each study had some of the C's as a focal variable. For example, in a pilot summer camp with a multicultural group of teenagers, from which excerpts were shown in the scenario presented earlier (Bers 2001), the primary goal was to help youth develop character. In a pilot study with young patients in the dialysis unit at Boston Children's Hospital (Bers, Gonzalez-Heydrich, and DeMaso 2001) the focal C was connection. Patients used Zora to form a virtual community to escape the harshness of the dialysis treatment and to create a network to facilitate mutual support and new kinds of interactions with hospital staff (Bers, Gonzalez-Heydrich, and DeMaso 2003). In a recent study aimed at engaging freshman in a local university as civic participants in the university and the wider community, the focal C was contribution (Bers, in press). In an ongoing project with post-transplant solid-organ patients at Children's Hospital in Boston, the focal C's are connection and caring, as well as competence to engage in medical adherence (Bers et al. 2007). The following section presents an example of how a young girl used Zora during the summer workshop to express and explore her religion.

Elisa and the Personal Meaning of Judaism

Elisa is 16 years old and goes to a Jewish school. She is the daughter of a rabbi, and much of her identity is linked to Judaism. She lives in a wealthy part of town and is very proud of her Jewish heritage. She wears Jewish symbols around her neck and likes to read and write in Hebrew. She is very driven, independent, and outgoing. She loves to talk about herself and has many friends. She has strong opinions about what is good and bad, and she is not shy about sharing them with others.

Elisa started out by building a virtual temple instead of a personal home. She wasn't the only one to do so. Catherine created a Baptist church with God and evil. Elisa created a Jewish temple containing objects, such as a Jewish prayer book, a picture with her name written in Hebrew, and an Israeli flag, as

FIGURE 3. The Jewish Temple Built by Elisa

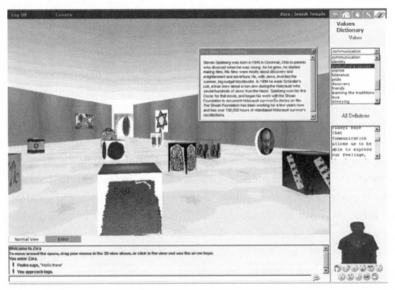

well as heroes like Steven Spielberg and her dad, who is a rabbi (see Figure 3).

In Elisa's Jewish temple, every object had very personal stories attached to it, as well as descriptions aimed at teaching others about Judaism. For example, the *kippah* has the following description: "Leather or cloth skullcap worn on the head to both show and feel closer connection to God through the body." It has the following associated story:

I live in the U.S.A., and so I don't normally see Jews just walking down the street in a non-Jewish environment. Even if I did see one, I wouldn't know because Jews look the same as everyone else. That's why I love when I see someone in a *kippah*. They enable me to know if they are Jewish just by looking at them. I know it is not much, but whenever I see random people wearing *kippot* I feel closer to them. I know that being Jewish is just as important for them as it is to me.

While building her Jewish temple, Elisa did research about what kinds of objects best express Judaism, especially her particular vision of it. This is very important for her. She doesn't want to absorb Judaism from her environment. She wants to re-create it and appropriate it in her own way.

While designing her Jewish temple, Elisa learned more about Judaism, but even more important, she was able to reflect about what Judaism means for her. She had a "palace in time," a safe space reserved for reflection and introspection.

This introspective exploration is very similar to the kind of "working through" issues of identity that Sherry Turkle, borrowing from Erikson, describes as a moratorium, or a time of constant experimentation and reflection in the adolescent's life (Turkle 1995). The process of self-reflection that Elisa engaged in has some resemblance to what happens during therapeutic interventions and meditative or religious experiences. However, during the experience with Zora, the products of self-reflection became tangible, dynamic, and manipulable.

Zora provided children like Elisa with a palace in time to reflect, explore, interpret, and share in community. However, technology can afford the development of both virtual and physical spaces that will serve the functions of a palace in time. In the next section, I present a pilot study done in a Jewish day school in Buenos Aires, Argentina, in which parents and children came together to participate in a robotics workshop to create interactive creative prayers during the holy days of the new years and the Day of Atonement. Those ten days, the *Iamin Noraim*, the terrible days, provided a palace in time for these families to explore their Jewish values.

PROJECT INTERACTIONS: ROBOTIC PRAYERS

The Project InterActions research program looks at the many interactions that exist when parents and young children are brought together in a learning environment using new technologies to program and build a robotic toy that represents an aspect of the family's cultural heritage (Bers, New, and Boudreau 2004; Bers 2007).

The first pilot study of Project InterActions was done in a Jewish day school in Buenos Aires, Argentina, in which families made robotic creative prayers and shared them with other members of the community at the synagogue before the traditional prayers of the Jewish high holidays (Bers and Urrea 2000).

The timing of the workshop was carefully selected to overlap with the Jewish high holidays, a period of ten days in which the community gathers to celebrate the Jewish New Year and the Day of Atonement. In this context, children's curriculum focuses especially on the values of these festivities, the most holy in the Jewish calendar. To hold a workshop during these holidays was very meaningful because of the spiritual work of reflection and forgiveness that takes place both in the school and the community. The workshop was a first step toward forming a group of parents, children, and teachers who would later integrate this approach to values and technology into the school's curriculum and make it available to a wider audience.

We worked with a project-based immersing methodology. By *project-based learning*, we mean that learners were asked to choose a project that they would like to work on for the whole duration of the workshop. They were involved in all aspects of the project. They chose the values to explore, decided the materials to use, managed the resources and time frame, resolved the technological challenges (both in terms of programming and mechanics), created a narrative around the final project, and presented it to the other members of the community through creative prayers.

By *immersing learning*, we refer to the notion that learners immersed themselves in the learning process by having a lot of time devoted to play and to explore their ideas in depth. For example, in this particular workshop, we worked with parents and children during five days, eight hours a day. During that time, participants could try many ideas and had enough time to iterate through different versions of the same idea. Each participant was asked to keep a design notebook to document the project progress as well as ideas and difficulties. We created a workshop Web site to document the experience collectively. Since this was the first pilot workshop within the research program, documentation was very important to allow future experiences and comparative studies.

Technology was used by parents and children in very different ways to explore values: (a) to represent symbols, (b) to represent values, and (c) to evoke reflection and conversation (Bers and Urrea 2000). Projects in the first category, technology to represent symbols, created artifacts that resembled the Jewish symbols without deeper exploration of the nature of the values represented by these symbols. For example, Michael, a 10-year-old boy, describes his project in the following way: "We built a '*Maguen David*,' Star of David, as a symbol of our Jewish people, and we programmed it to turn forever like the wheel of life and have flashing lights resembling candles welcoming the New Year. We also reproduced the sound of the *shofar*. It has three different tones that are supposed to awake us for reflection and atonement." Michael's group chose the value "awakening" or "call for reflection." They designed their project by anchoring it to traditional symbols. The construction of the star was done in a very careful way out of LEGO pieces and flashing lights. The center of the star was connected to a platform that moved with a motor. They used a touch sensor to launch and stop their program, which had three basic jobs: turn the motor on, turn the lights on and off, and play the sound of the *shofar* (see Figure 4).

Projects in the second category, technology to represent values, involved both artifacts and stories that made the chosen value more explicit. For example, a group chose the value "friendship" and created a puppet theater. The theater

FIGURE 4. The LEGO-based Star of David

had a curtain that opened to show the performance of two LEGO dolls hugging after a fight (see Figure 5). Marcia, 9 years old, created a story about the girl's situation and the connection with some of the values of the high holidays, such as *Teshuva*. "This project tells the story of two girls that, after a fight, give each other a hug and become best friends," explains Marcia. "This project talks about the *Teshuva* that allows us to repair our mistakes. The friends did *Teshuva* and became friends again with a big hug." Marcia built the dolls with LEGO bricks, attached colorful strings as hair, and placed motors in the arms to swing back and forth simulating a hug.

The "friendship" project used technology as well as storytelling. Since the chosen value was the main element of this project, the group needed to tell a story to reinforce the interpretation of the value. The participants wrote the story in the good-wish card that was handed out to visitors during the open houses. Telling a coherent story around the robotic creation was as important as getting the mechanics and the programming right.

Projects in the third category, technology to evoke reflection and conversation, treated values in a more elaborated way and provided an opportunity for others to engage in experiencing the complexity of the chosen value and participate in thoughtful discussion. For example, Paula and her 10 year-old son, Matias, with the help of two other moms, created a conveyor-belt contraption that transports the actions of the previous year (see Figure 6). Paula explained

FIGURE 5. Friendship and *Teshuva*

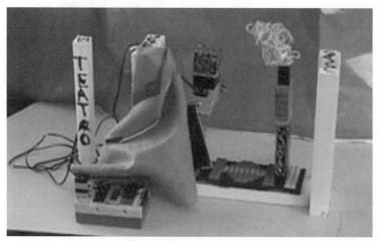

FIGURE 6. A conveyor belt transporting good and bad actions done during the year.

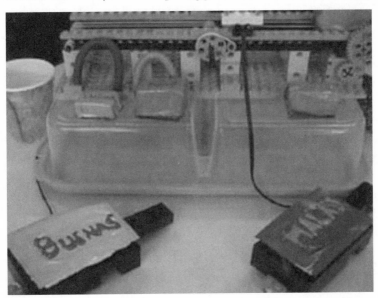

how they came up with the idea: "During the high holidays, we think about the actions in our everyday life. It is the time to think, reflect, and become conscious about our past deeds, so we can choose to continue with the good deeds or to rectify the actions that we believed were wrong." This idea gave birth to the conveyor belt. The machine was designed to carry actions until a reflection point, where the users could spend the needed time to decide about their positive or negative significance. An action considered good was transferred to a good container, and an action considered bad was taken back, meaning that people had to amend it. The mechanics consisted of a structure to hold the belt, which was made out of rubber bands; a motor located in the starting point of the contraption to move the belt; and two touch sensors to select between good and bad actions. Actions were foam rubber cubes wrapped in color papers and labeled with a name.

A program was created to start the motor for a given number of seconds and wait for the sensor input to take the actions to the next stage. If the sensor for good actions was pressed, the program started the motor in the same direction to go forward. If the sensor for bad actions was pressed, the program made the motor move in the opposite direction, taking back the action to the starting point. For this group, it was very important to have users of its project not only learn the value of reflection but also experience it by reflecting about their own actions.

CONCLUSION

In this chapter, I have shown three different ways in which technology can be used to support children's spiritual development and religious explorations. In a world in which young people don't always have a place to tell their stories, computer-based identity-construction environments can provide a potentially powerful opportunity to help them discover who they are and what are their most cherished values.

Narrative is not only a communication genre but also a cognitive tool fundamentally responsible for organizing our possible selves in a coherent way (Bruner 1986). This view of narrative as having a major part in the construction of our identity is present in most religious traditions. For example, in the Jewish tradition, the Torah (five books of Moses) is conceived not only as a sacred text but also as a life text that is rewritten and reinterpreted at each relevant personal and community event. The value of the text lies in the value of the people who re-create and reinterpret, in their own lives, the values proposed by the tradition. For example, the festivity of Passover, which commemorates the exodus of Israel from Egypt, conveys the importance of narrative: "And you should tell (*vehiggadeta* in Hebrew) to your son on that day, 'It is [being celebrated] because of what the Lord did for me when I went free from Egypt'" (Exodus 13:8). However, the mandate to tell the story is not only to remember the exodus but also to allow people in each generation to consider himself as if he had gone free from Egypt. It is not enough to learn about the value of freedom: we need actually to experience it.

Identity-construction environments provide a palace in time to tell our stories, to reinterpret their meaning by sharing with others in the community, and to experience them. For example, in Zora, children not only create a virtual city and populate it with characters, objects, and stories, but they also inhabit it. Through SAGE, children create storytellers as the narrative voices they want or need to hear and then communicate with them about their inner problems. In Project InterActions, parents and children come together to explore their most cherished values and represent them through robotic creative prayers.

It is the purpose of this chapter to show that, in the twenty-first century, in which technologies are used by young people in all their walks of life, from learning to dating, they can also become "a palace in time" to engage them in introspection, self-reflection, and community building. However, for this to happen, we need more people invested in positive youth development to take an active role in the design and study of technology-rich interventions.

NOTES

1. A *mezuzah* is a small piece of parchment inscribed with biblical passages from Deuteronomy, which is rolled up in a container and affixed by many Jewish households to their door frames in conformity with Jewish law and as a sign of their faith.

2. I use the term *value* to refer to the importance or worth of a quality for an individual and society.

REFERENCES

Bers, M. 2001. Identity construction environments: Developing personal and moral values through the design of a virtual city. *Journal of the Learning Sciences* 10, no. 4: 365–415.

———. 2006. The role of new technologies to foster positive youth development. *Applied Developmental Science* 10, no. 4: 200–219.

———. 2007. Positive technological development: Working with computers, children, and the Internet. *MassPsych* 51, no. 1: 5–7, 18–19.

———. In press. Civic identities, on-line technologies: From designing civics curriculum to supporting civic experiences. In L. Bennett, ed., *Volume on Digital Media and Civic Engagement. MacArthur series on Digital Media*. Cambridge, MA: MIT Press.

Bers, M., E. Ackermann, J. Cassell, B. Donegan, J. Gonzalez-Heydrich, D. DeMaso et al. 1998. Interactive storytelling environments: Coping with cardiac illness at Boston's Children's Hospital. Proceedings of *Computer-Human Interaction*. New York: ACM. 603–10.

Bers, M., and J. Cassell. 1999. Interactive storytelling systems for children: Using technology to explore language and identity. *Journal of Interactive Learning Research* 9, no. 2: 603–9.

Bers, M., C. Chau, K. Satoh, and L. Beals. 2007. Virtual communities of care: Online peer networks with post-organ transplant youth. Proceedings of the 2007 Computer Supported Collaborative Learning Conference, July 16–21, 2007. New Brunswick, NJ.

Bers, M., G. Gonzalez-Heydrich, and D. DeMaso. 2001. Identity construction environments: Supporting a virtual therapeutic community of pediatric patients undergoing dialysis. Proceedings of *Computer-Human Interaction*. 380–87. New York: ACM.

———. 2003. Use of a computer-based application in a pediatric hemodialysis unit: A pilot study. *Journal of the American Academy of Child and Adolescent Psychiatry* 42, no. 4: 493–96.

Bers, M., B. New, and L. Boudreau. 2004. Teaching and learning when no one is expert: Children and parents explore technology. *Journal of Early Childhood Research and Practice* 6, no. 2: 1–19.

Bers, M., and C. Urrea. 2000. Technological prayers: Parents and children exploring robotics and values. In A. Druin and J. Hendler, eds., *Robots for kids: Exploring new technologies for learning experiences*, 193–216. New York: Morgan Kaufmann.

Bers, M. 2007. Project InterActions: A multigenerational robotic learning environment. *Journal of Science Education and Technology* 16, no. 6: 537–52.

Bruner, J. 1986. *Actual minds, possible worlds*. Cambridge, MA: Harvard University Press.

Eccles, J., and J. A. Gootman. 2002. *Community programs to promote youth development*. Washington, D.C: National Academies Press.

Heschel, A. J. 1951. *The Sabbath: Its meaning for modern man*. New York: Farrar, Straus and Giroux.

Lenhart, A., M. Madden, and P. Hitlin. 2005. *Teens and technology: Youth are leading the transition to a fully wired and mobile nation.* Retrieved October 29, 2005, from http://www.pewInternet .org/pdfs/PIP_Teens_Tech_July2005web.pdf.

Lerner, R. 2002. *Concepts and theories of human development.* Hillsdale, NJ: Lawrence Erlbaum Associates.

Lerner, R. M., C. B. Fisher, and R. Weinberg. 2000. Toward a science for and of the people: Promoting civil society through the application of developmental science. *Child Development* 71: 11–20.

Lerner, R. M., J. V. Lerner, J. Alermigi, C. Theokas, E. Phelps, S. Gestsdottir et al. 2005. Positive youth development, participation in community youth development programs, and community contributions of fifth grade adolescents: Findings from the first wave of the 4-H study of positive youth development. *Journal of Early Adolescence* 25, no. 1: 17–71.

Papert, S. 1980. *Mindstorms: Children, computers and powerful ideas.* New York: Basic Books.

Scales, P. C. B., N. Leffert, D. A. Blyth. 2000. Contribution of developmental assets to the prediction of thriving among adolescents. *Applied Developmental Science* 4, no. 1: 27–46.

Turkle, S. 1995. *Life on the screen: Identity in the age of the internet.* New York: Simon and Schuster.

———. 1984 *The second self: Computers and the human spirit.* New York: Simon and Schuster.

List of Contributors

MONA ABO-ZENA is a doctoral student in the Eliot-Pearson Department of Child Development at Tufts University. Her primary research interests include religious and identity development, family and community contexts, and immigrant and other multicultural individuals and communities. In addition to her work as a research assistant on the John Templeton Foundation study of religion and positive youth development, Mona works on the Adoption & Development Project at Tufts. Abo-Zena earned her B.A. in sociology from the University of Chicago and her Ed.M. from Harvard University.

JEFFREY JENSEN ARNETT is a research professor in the Department of Psychology at Clark University in Worcester, Massachusetts. Dr. Arnett is the originator of the theory of emerging adulthood and the author of numerous articles on emerging adulthood, as well as the textbook *Adolescence and Emerging Adulthood: A Cultural Approach* (2004, Prentice Hall). His book *Emerging Adulthood: The Winding Road from the Late Teens Through the Twenties* was published in 2004 by Oxford University Press.

MARINA UMASCHI BERS is assistant professor in the Eliot-Pearson Department of Child Development with a secondary appointment in the Computer Science Department at Tufts University. She is also associate scientific staff at Boston Children's Hospital. Bers received her Ph.D. from MIT. Her research interests include the design and study of new technologies for promoting positive youth development in hospitals and mental health care settings, schools, and after-school programs.

AERIKA S. BRITTIAN is a doctoral student in the Eliot-Pearson Department of Child Development at Tufts University and a doctoral research assistant for the 4-H study of positive youth development. Her research interests include individual and contextual factors that influence positive youth development in underrepresented populations.

WILLIAM DAMON is a senior fellow at the Hoover Institution, the director of the Stanford Center on Adolescence, and a professor of education at Stanford University. Before coming to Stanford in 1997, Damon was a professor of education and university professor at Brown University, where he continues to hold an appointment as an adjunct professor of human development. Damon received his bachelor's degree from Harvard University and his Ph.D. in developmental psychology from the University of California, Berkeley. He is the author or editor of numerous books.

ANGELA DESILVA is a third-year doctoral student at Boston College in the Developmental, Educational, and Counseling Psychology program. She also earned her master's degree in mental health counseling at Boston College. Her areas of research include social supports and mental health among immigrant groups as well as spirituality and belief in God among adolescents and young adults.

JACQUELYNNE S. ECCLES is the McKeachie Collegiate Professor of Psychology at the University of Michigan. She conducts research on topics ranging from gender-role socialization and classroom influences on motivation to social development in the family, school, peer, and wider cultural contexts.

DAVID HENRY FELDMAN is professor of developmental psychology in the Eliot-Pearson Department of Child Development at Tufts University. His interests include cognitive development theory and research, extreme giftedness, and education.

SIMON GÄCHTER holds the chair in Psychology of Economic Decision Making at the Centre for Decision Research and Experimental Economics at the University of Nottingham. He is an experimental economist with research interests in the determinants of voluntary cooperation and punishment and the behavioral effects of incentive provision.

ELENA L. GRIGORENKO received her Ph.D. in general psychology from Moscow State University, Russia, in 1990, and her Ph.D. in developmental psychology and genetics from Yale University in 1996. Currently, Dr. Grigorenko is associate professor of child studies, psychology, and epidemiology, and public health at Yale, and adjunct professor of psychology at Columbia University and Moscow State University. Dr. Grigorenko has published more than 200 peer-reviewed articles, book chapters, and books. She has received awards for her work from five different divisions of the American Psychological Association (APA); in 2004, she won the APA Distinguished Award for an early career contribution to developmental psychology.

SONIA S. ISSAC is a doctoral student and research assistant for the John Templeton Foundation–funded study, "The Role of Spiritual Development in Growth of Purpose, Generosity, and Psychological Health in Adolescence" at the the Institute for Applied Research in Youth Development in the Eliot-Pearson Department of Child Development at Tufts University. Her research interests include mentoring relationships and ethnic identity development in religious organizations, especially in immigrant communities.

LENE ARNETT JENSEN is associate professor of psychology at Clark University and editor-in-chief of *New Directions for Child and Adolescent Development*. She received her Ph.D. from the committee on human development at the University of Chicago and did one year of postdoctoral work in the Department of Sociology at the University of California, Berkeley. Dr. Jensen's research addresses cultural identity development in the contexts of migration and global change as well as takes a "cultural development" approach to morality, addressing how moral reasoning and behavior are culturally and developmentally situated. Jensen's research also addresses the role of religion and spirituality for both migration and morality.

CARL N. JOHNSON is currently chair of the Department of Psychology in Education and coordinator of the Program in Applied Developmental Psychology at the University of Pittsburgh. Known for his pioneering research on children's concepts of mind, his attention has recently turned to the concept of spirit as the primordial organizing force of life.

LINDA JUANG received her Ph.D. in developmental psychology from Michigan State University. She is an associate professor of psychology at San Francisco State University. Her research focuses on issues of acculturation, ethnic identity, discrimination, and well-being of immigrant adolescents and their families. She has co-authored a textbook with David Matsumoto entitled *Culture and Psychology*.

PAMELA EBSTYNE KING is research assistant professor of family studies in the Center for Research and Child and Adolescent Development in the School of Psychology at Fuller Theological Seminary and ordained in the Presbyterian Church (USA). Her research and teaching focuses in the areas of positive youth development, spiritual and moral development, and theological perspectives of development. She is a co-author of *The Reciprocating Self: A Theological Perspective of Development* and co-editor of the *Handbook of Spiritual Development in Childhood and Adolescence.*

RICHARD M. LERNER is the Bergstrom Chair in Applied Developmental Science and the director of the Institute for Applied Research in Youth Development in the Eliot-Pearson Department of Child Development at Tufts University. A developmental psychologist, Lerner received a Ph.D. in 1971 from the City University of New York. Lerner is the author or editor of sixty-three books and more than four hundred scholarly articles. He is known for his theory of, and research about, relations between life-span human development and contextual or ecological change.

JENNIFER MENON MARIANO is an assistant professor in the College of Education at the University of South Florida, Sarasota/Manatee. She received her Ph.D. from Stanford University in 2007. Dr. Mariano has conducted research in schools throughout the United States examining the personality, family, and community antecedents and correlates of positive youth development and youth thriving.

NA'ILAH SUAD NASIR is an assistant professor at Stanford University's School of Education. She received her B.A. in psychology and social welfare from UC Berkeley and her Ph.D. in education from UCLA. She has been at Stanford since 2000. In her research, Dr. Nasir draws on socio-cultural theory to explore the relation between learning, development, and culture.

GUERDA NICOLAS is a licensed clinical psychologist and the assistant director of the Institute for the Study and Promotion of Race and Culture along with Dr. Janet E. Helms as well as an assistant professor at Boston College in the Lynch School of Education, Department of Counseling, Developmental, and Educational Psychology. She obtained her doctoral degree in clinical psychology from Boston University. Her current research projects focus on developing culturally effective mental health intervention for ethnic minority adolescents, with a specific focus on immigrant children, adolescents, and families. She has published several articles and book chapters and delivered numerous invited presentations at the national and international conferences in the areas of women's issues, depression and intervention among Haitians, social support networks of ethnic minorities, and spirituality.

TOMÁŠ PAUS holds the chair of Developmental Cognitive Neuroscience and is the director of the Brain and Body Centre at the University of Nottingham, U.K. He is also an adjunct professor of neurology and neurosurgery and psychology at McGill University, Canada.

STEPHEN C. PECK is a research investigator at the Research Center for Group Dynamics at the University of Michigan. He received a B.A. in psychology from the California State University, Long Beach in 1985; an M.A. in experimental social psychology from the University of Montana in 1990; and a Ph.D. in personality psychology from the University of Michigan in 1995. Dr. Peck uses a combination of variable- and pattern-centered methodological approaches to study how multilevel systems of personal and environmental contents, structures, and processes interact to produce more or less healthy forms of human development.

ERIN PHELPS is research professor and deputy director of the Institute for Applied Research in Youth Development in the Eliot-Pearson Department of Child Development at Tufts University. A developmental psychologist and research methodologist, Phelps obtained her doctorate in human development at the Harvard Graduate School of Education. While at the Murray Research Center for the Study of Lives at Harvard University, she championed the use of longitudinal data for the understanding development in context across the life span and clamored for better representation of underrepresented groups in research. She is editor of *Research in Human Development* and co-author of *Looking at Lives: American Longitudal Studies of the 20th Century* and *An Inventory of Longitudal Studies in the Social Sciences.*

ALAN P. POEY grew up in Amherst, Mass., and graduated in 2006 with a B.A. in neuroscience and behavior from Wesleyan University. At Tufts University he worked for the Institute for Applied Research in Youth Development and then as project coordinator in Heather Urry's Emotion, Brain, and Behavior Lab. He is currently preparing for medical school.

ROBERT W. ROESER is currently an associate research professor in the Institute for Applied Research in Youth Development in the Eliot-Pearson Department of Child Development at Tufts University. Roeser received his B.A. in psychology from Cornell University and his Ph.D. in education and psychology from the University of Michigan. He also holds master's degrees in developmental psychology, clinical social work, and religion. Dr. Roeser's current research focuses on the use of contemplative practices like yoga and meditation in secondary schools for purposes of stress reduction and the enhancement of positive development among principals, teachers, and adolescent students.

W. GEORGE SCARLETT is deputy chair of the Eliot-Pearson Department of Child Development at Tufts University. He has co-authored the first-ever chapter on religious and spiritual development in the *Handbook of Child Psychology*, co-edited an encyclopedia of religious and spiritual development, and co-authored books on behavior and classroom management and teaching baseball to children.

LONNIE R. SHERROD received his Ph.D. in psychology from Yale University in 1978, an M.A. in biology from University of Rochester (1976), and a B.A. from Duke University (1972). He is currently executive director of the Society for Research in Child Development and professor of psychology in Fordham University's Applied Developmental Psychology Program. His area of research is Youth Political Development, and he has co-edited special issues of the *Journal of Research on Social Issues* (1998) and *Applied Developmental Science* (2002) on the topic.

GABRIEL S. SPIEWAK received his B.A. from Yeshiva University in 2005 and is currently a graduate student in the Applied Developmental Psychology Program at Fordham University. In addition to his academic training, he has spent several years studying at a seminary for advanced Jewish learning and has had the opportunity to teach and interact with many religious adolescents.

CHRIS STARMER is the chair of experimental economics and director of the Centre for Decision Research and Experimental Economics at the University of Nottingham, U.K. In his current research, Prof. Starmer applies experimental methods in the development and testing of models of human decision making.

MOIN SYED is a doctoral candidate in the developmental psychology program at the University of California, Santa Cruz. His research interests broadly cover identity development in adolescence and emerging adulthood, with current emphasis on narrative approaches to ethnic identity development.

JANICE L. TEMPLETON is an assistant professor of psychology at Fort Lewis College in Durango, Colorado. She received her Ph.D. from the University of Michigan in 2007. Her current research focuses on developmental, social, and spiritual aspects of extending moral concern to a broad circle of others.

HEATHER L. URRY is an assistant professor in the Department of Psychology at Tufts University. She joined the faculty after completing a postdoctoral fellowship and serving as an assistant scientist at the University of Wisconsin-Madison in 2005. She earned her Ph.D. in clinical psychology from the University of Arizona in 2001.

RICHARD WILKINSON is professor of social epidemiology at the University of Nottingham (U.K.) Medical School and visiting professor at University College London. Working in the field for thirty years, Prof. Wilkinson has played a formative role in research and public awareness of health inequalities and the social determinants of health. Since persuading the U.K. Secretary of State to set up the working party, which in 1980 produced the Black Report on Health Inequalities, he has worked particularly on the health and social effects of income inequality. His latest book, written with Kate Pickett, is *Inequality* (January 2009, Penguin).

Index

ACSE. *See* Swiss Christian Students' Association
active citizenship, 325
adaptive developmental regulations, 5, 7
adolescence. *See also* youth(s)
 adaptive, development, 4
 African American, 97
 afterlife beliefs of, 232–42
 biopsychosocial changes in, 74
 brain development in, 13–15, 139
 cognition-affect interplay in, 132–33
 cognitive development of, 140
 developmental systems models of, 10
 ECM in, vii, xi, 198
 experiencing transcendence, 68–69
 faith in, 210–28
 high *vs.* low resistance to peer influence in,
  136–37
 horizontal and vertical transcendence in, 11
 identity in, 62, 263, 292
 influence of peers on, 133
 interpersonal/group relationships and, 68
 late, 51
 longitudinal research focused on, 159
 minority, 247, 305, 308–10, 314
 parent, relationship, 280
 personal integration in, 58
 purpose in, 210–28
 religion and, 34, 75, 95, 97, 210
 religious identifications of U.S., 94–95
 religious worldview among U.S., 88–89
 research studies on, 95, 159, 197, 213–28, 262,
  270, 299–301, 305–18, 326–27
 role of, 9
 self-definition by, 3
 sociocultural factors on, 305
 spirit, religion and, 34
 spiritual research with ethnically diverse,
  305–18
 spirituality in, xi, 15, 42–43, 55, 69, 75, 197,
  210–28

spirituality, religious faith and purpose in,
  210–28
 spirituality/religiosity in, viii, xi, 95, 198
 survey of American, 34
adults. *See* emerging adults
African American(s)
 religion and, 97
 women, 199
afterlife beliefs, 18, 231–42, 240
Ahluwali, M. K., 265
Allison, T., 132
Allport, G. W., 225–26
altruism, 130
amygdala, 158–59
 activation of, 155
 blood oxygenation-level dependent response
  in, 132–33
Arch, J. J., 157
Aristotle, 37
Arnett, J. J., 232, 241, 272
assets, 212, 323–24
Attention Network Test (ANT), 157
awareness, 78
Azmitia, M., 272, 278

Baime, M. J., 157
Bankston, C. L., 268–69
Basic Levels of Self (BLoS) Model, 77–79
Bateson, C. D., 225–26
Being, 84
Bem, D. J., 87
Benson, P. L., 57
Bergeson, Henri, 35, 37
Bible, 35, 45–46, 211
biological
 changes and PYD, 16
biological covariates, spirituality and religiosity
  of, 9
biology, 51, 171
Black Seminole, 115

Blasi, A., 200
blood oxygenation-level dependent (BOLD)
    response in amygdala, 132–33
BLoS. *See* Basic Levels of Self Model
Boston Children's Hospital, 349, 350
brain
    adaptive, development, 13–15
    adolescent, 13–15, 139
    economic inequality shaping, 131
    emotion regulation in, 8, 154–56, 160
    environmental influences on, 130
    growth, 5
brain-behavior-context relations, 17
Branscombe, N. R., 205
Braque, Georges, 171, 185
Bringuier, Claude, 182
Bronfenbrenner, U., 295, 306–7, 317
Burton, L., 211

Cahn, B. R., 156, 158
Carey, S., 180
categorization, 202
Catholicism, 10, 277
Cavalcanti, H. B., 270
Cawley, Michael, 212
Chan, D., 157
Chandler, Michael, 30
change. *See also* cognitive developmental change;
    developmental change
    biological, and PYD, 16
    biopsychosocial, in adolescence, 74
    cognitive developmental, 169–83, 193
    irreversible, 175–77, 188
    large scale or pervasive, 179–80
    nondevelopmental or antidevelopmental,
        193n1
    qualitative, 173–75, 189
    sequential, 177
    spiritual-developmental, 188
    world, 91
character-development goals, spirituality and
    religious faith, 223
Charnwood, Lord, 52
childhood, spirit of, 32
children's
    ideas about soul, 34
    intuition, 30, 32
Christ, 62
Christianity, 62, 66, 87, 201, 220, 238, 240
church(es), 97, 220
    adolescents attendance at, 211
    Catholic, 10
    Christian, 90
    membership in U.S., 247, 331

cigarette smoking, 122, 139
circles of care, PYD and spirituality, 197–98
civic engagement, 55
    contribution and, 6, 326
    defining, 324–26
    immigrant, and religion, 247–59
    in organizations, 255
    positive outcomes and, 248–49
    PYD and, 248, 262, 322–35
    religion and, 247–59
    religion's role in, 247–59
    religious/spiritual themes pertaining to, 254
    spirituality/religiosity and, 322–35
    in youth, 324–26
civic participation, engagement, and contribu-
    tion, 6
"civic thinking," 48
Civil War, 52
classic attribution theory, 109
cognition, 27, 140
cognitive development
    of adolescence, 140
    course, 181
    ECM and, 204
    emotional markers and, 182–83
    major, advances, 171–72
    phase shifts of, 171
    Piaget's four major stages of, 170, 177–78,
        180, 183
    transitions, 181–83
cognitive developmental change, 169–83, 193. *See
    also* developmental change
cognitive meaning, 38
cognitive-development theory, 28
Cohen, S., 206
Colby, A., 205, 208
Cole, Robert, vii, 295
communication, nonverbal, 130
community(ies), 94, 204, 258, 281
    based approach to research studies, 308–10
    building, 340–41
    emerging adults within, 265, 274, 277, 279
    engagement, 254
    ethic of, 254
    ethics of divinity, autonomy, and motives of,
        254, 256
    ethnically diverse, 318
    faith, 62
    Hispanic, 306
    involvement, ethnic identity and PYD, 94, 267
    religious, 224, 267–69, 279, 287–88
    research with ethically diverse, 308–18
    shared purpose within, 221
    societal and, supports, 210

spiritual, and PYD, 299
spirituality and, 308
transcendence and, 10
virtual, 340
youth, purpose and religious, 220–21
computer-based identity-construction environments, 340–57
conflict resolution, 128–41
conscience, 43
consciousness, 79, 85
core, 78
ego, 80
knower in, 76
spirituality as states of, 80
taking of, 175, 182
constant comparative method, 259n3
context, 55–56, 58–59
cultural and geographic, 290–91
developmental, 290
historic, 291
institutional, 296–97
interaction of individual and, 293–95
of spirituality, 57, 285–301
contextual emotion-regulation therapy (CERT),
154
contribution, 225–26, 325
civic engagement and, 6, 326
spirituality and, 216–17
by youth, 6
conversion, 190
Cook-Greuter, S. R., 93
cooperative behavior, 128–41
cortical thickness, interregional correlations
in, 137
Cowie, S. E., 265
Craske, M. G., 160
Creswell, J. D., 158
Crick, Francis, 171, 184
Crocker, J., 206
cults, 65, 169
cultural practices, social interactions in, 295, 298
culture, 27, 29, 285–301

Dahlen, E. R., 153
Damasio, A., 78
Damon, W., 205, 208
d'Aquili, E., 9, 13
Darwin, Charles, 172, 184–85
De Vogler, K. L., 214
De Waal, F. B., 130
Dean, D., 29
death, emerging adults views on life after, 231–42
Declaration of Independence, 45–46, 53
Delpierre, V., 95

Denton, M. L., 83, 88–89, 242
Dernelle, R., 95
development influences, 202–3
developmental advance, 185–86
developmental change
attributes of, 172
cognitive, 169–83, 193
intentional, 183
markers for cognitive, 184
positive, 172–73
role of, in spiritual development, 167–93
developmental levels, 178
developmental neuroscience, 13
developmental psychology, viii, 111, 174, 248, 266
developmental systems model, 4–16
developmental systems theory, 8, 331–34
the Divine, 61, 63, 91
divine inspiration, 257
Dowling, E. M., 11
Duckworth, A. L., 152
dynamics systems theory (DST), 193n1

Ebersole, P., 214
Eccles, J. S., 198, 264
ECM. *See* expanding circle morality
the ecological model, 306–8, 317
economic inequality, 131–32
ego
consciousness, 80
development theory, 19
pure, 76
transcendence of, 80, 93
Ellison, C. G., 160
emergence, 183
emerging adults
cultural support from community for, 265,
274, 277, 279
ethnicity and religion/spirituality viewed by,
265, 272, 274, 277
identity formation in, 263
Mexican-American and Filipino, 277
views on life after death of, 231–42
emotion regulation
attention and, 159
brain and, 8, 154–56, 160
indicators for, 148–50
questionnaires, 148
research studies on, 149–50, 152, 155
resilience in wake of adverse events and,
150–51
scale, 153
strategies for, 147
well-being and, 146, 150, 152–54, 158–59
emotional markers, 182–83, 190

Engeström, Y. R., 295
enlightenment, 33, 85, 179
epigenetic
    effects, 118–20
    mechanisms, 119
Erikson, E. H., 10, 57–58, 74–75, 78, 86, 88, 217, 242, 263, 351
    theory of ego development by, 19
    theory of psychosocial development by, 331
ethics of divinity, autonomy, and community
        motives, 254, 256
ethnic identity, 18
    community involvement, PYD and, 94, 267
    in everyday experiences study, 270–80
    PYD and, 267–68
    religion/spirituality and, 262–81
    religious discrimination and, 270
    spirituality and, 262–81
ethnic minority, 247, 308–10, 314
ethnic diversity, 305–18
Evans, T. E., 211
evolutionary influences, 27, 203
expanding circle morality (ECM)
    in adolescence, viii, xi, 198
    altruism and, 201–2
    challenges to, 202
    cognitive development and, 204
    spiritual beliefs and, 200–201
    spiritual identity and, 207
    spirituality/PYD and, 197–208
    values/well-being, 206

Faison, N., 325
facial electromyography (EMG), 149
faith, 48, 54, 74, 82, 90, 115, 219
    in adolescence, 210–28
    community, 62
    development, 92
    emerging, 42, 53
    genetics of, 109–23, 118
    goals, purpose and spiritual, 224–25
    of Hitler, 47, 49
    identity through, 49
    of Lincoln, 44–45
    parasitic, 89
    privatization of, 84, 89
    religious, 110–16, 210–28
    research studies on genetics of, religion and
        spirituality, 118
faith-based institutions, 10
Fest, J., 47–48, 50, 52
The Fetzer Institute, 60, 63
"Five/Six Cs," 6, 139, 210, 220, 342

Flanagan, C., 250, 325
Fleming, Alexander, 186
fMRI, 136
focus groups, 71
Folkman, S., 153
Fowler, J., 52–53, 82
framing, 28–31, 37
Francis, L. J., 211
Frankfurt, H. G., 200, 204
Frankl, V. E., 217
free will *vs.* determinism, 110
Fuller, R. C., 89–90
functional magnetic resonance imaging (fMRI),
        155
Furrow, J., 11, 59

Gallup organization, 231, 242
gamma-aminobutyric acid (GABA), 119
Garbarino, J., 64
Gardner, H., 187
Garnefski, N., 153
gene(s)
    behavior and, 121
    candidate, 117
    environment and, 116–23
    expression and life experiences, 119
    monoaminergic, 121
    networked, of small effect, 117–18
    religion and, 111–16, 118, 123
    spirituality and, 110–23
    transcendence of, and self, 113
Gene-environment
    covariation (correlation, rGE), 122
    interaction (GxE interaction), 120–22
generosity, 3–20
genetic research studies, 113–16
genetics of faith, 109–23
Germany, 47
Gilligan, Carol, 177, 187
Glaser, Evan, 259n3
Glover S. H., 311
God, 35, 45, 60, 66, 84–85, 89, 185, 219, 237, 240,
        257. *See also* the Divine
    college student's beliefs in, 232
    nature of, 88
    need for, 115
    personal, 43
    transcendent, 43–44, 46
Goodenough, U., 8

Haight, J. E., 8
Hamer, Dean, 113
happiness, 152

Harris poll, 231
health
    local/global economic environment and, 128
    socioeconomic status on, 131
heart, mind, hands model, 331–35
heaven, 237–39, 241
Heider, F., 109
Hein, D., 44
hell, 238–39, 241
Herberg, W., 268
heritability
    estimates, 116
    of morality/spirituality, 112
    studies, 111–13
Heschel, Abraham Joshua, 340
Hess, W. R., 32
heuristic representation of religious/spiritual
        practices, 146
Hindu tradition, 61
Hitler, 42, 47–53
    faith of, 47, 49
    public speaking skills of, 48
    youth/nature as seen by, 50
hot cognition, 140
human
    behaviors, 117
    biology, 17
    health, spirituality and religiosity, 79, 216–18
    identity, 75–93, 99
    life, 30, 309
    nature, 88
    spirit, 32
human development, 55–56, 80, 198. *See also*
        development
    cognition, evolution, culture and, 27
    developmental systems and, 19
    developmental theories of, 5
    importance of genetic forces for, 116
    positive and problematic, 98
    spirituality and, 92
Human Genome Project, 116
humility, 53

ideals, 84–86, 91, 99
identity, 75–93. *See also* ethnic identity
    in adolescence, 62
    constructive environments promoting PYD,
        341
    development, 82, 94, 198–99
    ECM and spiritual, 207
    ethnic and religious/spiritual, intersectional-
        ity in, 264–65, 269–71, 274, 279
    through faith, 49

formation and religious development, 177,
        331–32
formation in emerging adults, 263
frameworks of religion/spirituality as group
        membership, 263–64
future research in, 279
human, 75–93, 99
multiple group, ethnic, religious/spiritual, 281
in PYD, 82, 267–68, 341
religion and, 75, 93–95, 262–81
research studies and literature on, 270,
        279–80, 292
sociocultural theories of, 292
spiritual *vs.* religious, 75
spirituality and, 74–99, 87, 262–81, 292–93,
        341
status model, 263–64
theories, 263, 266, 331–32
ideology, 57–58, 68, 98
    social context, transcendent experiences
        and, 69
    spirituality and, 57, 67, 80
    youth and, 88
immigrant(s)
    civic engagement and religion, 247–59
    from India and El Salvador, 249–59
    literature on, 278
    psychological support for, 267
    religion and, 97–98, 247–59
    religious communities, 267–68
    spirituality and, 97–98
    in U.S., 153, 247
    volunteerism, 250
    youth, 98, 248–49
individual ↔ context relations, 5–6, 8, 14, 15
infancy, 28, 31–32
integrative cognitive functions, 14
integrative regulation of emotions, attention and
        action, 14
Intercultural Adjustment Potential Scale
        (ICAPS), 153
interviews, 248–59, 316
intuition, 29–32, 34
I-self, 76–80
Islam, 61, 87

Jagers, R., 97
James, William, ix, 26, 75–81, 99, 169–70
Jensen, L. A., 232, 241
Jesus, 43, 46, 53, 187, 201, 257
Jha, A. P., 157
Jones', Jim Jonestown sect, 169
Juang, Linda, 262, 272

Judaism, 62, 87, 114–15, 167–68, 350–57
Judgement Day, 232

Kabat-Zinn, J., 156
Kalton, M., 7–8, 14
Keil, F., 180
Keller, Helen, 33, 175
Kilts, C. D., 132
King, P. E., 11, 59, 262
Kirkland-Harris, A. M., 265
Kirshner, B., 296
Kluger, Jeffrey, 115
Kohlberg, L., 203
Kovacs, M., 154
Kraaij, V., 153
Krompinger, J., 157
Kuhn, D., 29
Kumpfer, K. L., 212

language, 33
Larsen, R. J., 146
Larson, R. W., 248
Lazar, S. W., 158
learning
    observational, criteria, 333
    project-based and immersing, 352
Learning as Leadership, Inc., 205–6
Leontiev, A. N., 295
Lerner, R. M., 210, 220, 248
Levin, J. S., 160
Lewin, K., 295
life, 231
    expectancy, 128
    force, 33
    human, 30, 309
    love of, 26, 81
    spirit and, 35–36
    spiritual, 28
    worlds, 30–31
Likert scale, 315
Lincoln, Abraham, 42, 44–47, 51–54
    farewell speech by, 46
    spiritual faith of, 44–45
logical necessity, 176
longitudinal data, 11, 20, 70, 159, 327
Lopez, M. H., 258
love
    Keller and, 33
    of life, 26, 81

Marcia, J. E., 217, 263
Martin, R. C., 153
Matsumoto, D., 153
Mattis, J. S., 199, 265

McCarthy G., 132
Measure of Affect Regulation Styles (MARS), 148
meditation, 145, 160, 178
    emotions and, 157–58
    psychological and neural correlates of, 156–58
    research studies on, 156
Mein Kampf, 49–50, 52
ME-selves, 76–80, 82
meta-awareness, 80
metaphysics, 34
meta-reflection, 31
meta-thinking, 30
methodolical issues, 295–99
Min, P. G., 267, 277
mindfulness-based stress reduction (MBSR) course, 157
mirror neurons, 133
Mohatt, G. V., 314
Moltmann, Jurgen, 26
moral concern, 204–5
Moralistic Therapeutic Deism (MTD), 34, 89
morality
    heritability of, 112
    interpersonal relationships and, 203–4
Morris, A. S., 152
Moses, J., 200
motivation, egosystem and ecosystem, 206
Muhammad, 187
Muller, Alexander von, 48
multidimensional scaling (MDS), 135
multiple group identities, 281

Nasir, N., 296
National 4-H Council, 329
National Study of Youth and Religion (NSYR), 94–95, 232, 242
nature, 36, 50–51
    experiences of transcendence in, 61
    Hitler's image of, 50
Nettler, G., 110
networks
    action-observation, biological-motion processing and executive, 133
    STS and PCF, 134
neural circuits, 139
neural correlates of meditation, 156–58
neural mechanics of resistance to peer influences, 132–33
neural pathways, 13
    late developing, 131
    social environment and, 130
neural studies, 136
neural system, face and body cues, 139

neuroimaging, 140
neuropsychology of spirituality ↔ thriving
    relation, 12–15
neuroscience, 140, 146
New Testament, 60
Newberg, A. B., 9, 13
Nicolas, G., 308
noogenic neuroses, 217
Nozick, Robert, 38
Nuer, N., 206

O'Brien, M., 199
observational learning criteria, 333
Ochsner, K. N., 155
Olivier, M. A., 206
ontogeny, 5, 16
ontology of spirit, 26–28
open-coding, 259n3
orthogonol factor analyses, 12
Oser, Fritz, 177–78, 187

Pargament, K. I., 81
Paus, T., 159
peer pressure, 133, 136–40
Peterson, C., 212
Pew Research Center, 242
phenomenology, 299
Piaget, Jean, 35–38, 42–44, 51, 53, 170–84, 187
    four stages of cognitive development by, 170,
        177–78, 180, 183
Picasso, 171, 185
Pickett, K. E., 128
plasticity, 55–56
Plato, 183
Polich, J., 156, 158
positive technological development (PTD), 342
positive youth development (PYD)
    afterlife beliefs and, 240
    assets and, 323–24
    biological changes and, 16
    circles of care and, 197–98
    civic engagement and, 248, 262, 322–35
    cooperative behavior, conflict resolution and,
        128–41
    developmental assets of, 212, 323–24
    ECM, spirituality and, 9–12, 197–208, 293
    economic environment, prosocial behavior
        and, 140
    environments promoting, 341
    ethnic identity and, 268
    ethnic identity, community involvement and,
        94, 267
    of ethnically, racially and culturally diverse, 97
    experiences of transcendence and, 60, 62

"Five/Six Cs" of, 6, 139, 210, 220, 342
    generosity in, 3–20
    as highly contextual, 289–95
    increasing, 139
    operational definition of, viii, 18, 84, 130, 132,
        210, 289
    parental warmth, school involvement, role
        models and, 140
    peer pressure and, 140
    potential social influences of spirituality on, 59
    purpose and, 211, 218, 226, 240, 262
    religion, spirituality and, 67
    research studies on, 71, 131, 299, 305, 323
    spiritual communities and, 299
    spiritual development and, 9–12, 42–54, 225,
        293
    spirituality and, vii, xi, 3–20, 55–71, 69–70,
        74–99, 197–208, 285–301, 305, 322–35
    spirituality, context, culture and, 285–301
    spirituality/religiosity, civic engagement and,
        322–35
    theories, 210, 213
    theory and measurement development in, 335
    thriving and, 10
    transcendence and, 60, 62
    youth policy development and, 323
Post Traumatic Stress Disorder (PTSD), 151
prefrontal cortex (PFC), 155, 158–59, 231
the Prisoner's Dilemma game, 129
Project InterActions, 340–41, 352–56
pseudospeciation, 86–87
psychological correlates of meditation, 156–58
psychological research studies, 176
psychology, 169, 285
    developmental, viii, 111, 174, 179, 248, 266
    educational, 179
    literature on, and well-being, 151
    researchers, 146, 270
    study of spirituality in mainstream, 301
    terminology constraints of, 63
psychotherapy, 156
the Public Good game, 129
Puce, A., 132
Pumariega, A. J., 311
purpose, 218–19
    in adolescence, 210–28
    afterlife beliefs and, 240
    definition of, 212–13
    PYD and, 211, 218, 226, 240, 262
    spirituality and, 212–13, 218–19, 224–25
    study exemplars, 221–28
    surveys on religiosity and, 226
    transcendence and, 212–13
    youth, 17, 210–28

Purpose in Life (PIL) test, 211
Putnam, R. D., 247–48, 258
PYD. *See* positive youth development

qualitative change, 173–75, 189
qualitative positive advance, 175
qualitative research studies, 199, 299

Radmacher, K., 278
reality
    collective, 30
    spirit and, 25–26
Redlich, F., 47
Regnerus, M. D., 307
Reich, K. H., 7–8
reincarnation, 239–40
Reinhold, Neibuhr, 53
relationships, 59, 68, 203–4
religion, viii, 35, 58, 61, 222. *See also* religiosity
    adolescence and, 34, 75, 95, 97, 210
    African Americans and, 97
    Americans and, 83
    civic engagement and, 247–59
    emerging adults views on, 265, 274, 277
    end of, 81
    ethnic identity, spirituality and, 262–81
    ethnicity, spirituality and, 276–77
    exclusion clauses of, 87
    experience *vs.* beliefs about, 90
    faith and, 110–16, 210–28
    genes' role in, 111–16, 123
    genetic studies on, 118
    group membership identity and, 263–64
    identity and, 75, 93–95, 262–81, 331
    immigrants and, 97–98, 247–59
    National Study of Youth and, 94–95, 232, 242
    organized, 11
    personal, 81, 83
    PYD and, 67
    roles of, 97
    science and, 35, 42–43
    spirit, 26
    spirituality and, 55, 97–98, 197, 210, 264,
        276–77, 288
    spirituality and, as social group, 273–74, 278
    studies on smoking and alcohol consumption
        and, 122
    study on U.S. adolescence and, 95
    U.S. communities and, 267–68
religiosity, 7–8, 18, 95, 145. *See also* religion; spiri-
    tuality and religiosity
    future surveys of, and purpose, 226
    genes and, 110–23

    negative and positive forms of, 193n1
    spirituality *vs.*, 326–29
    violent and inhumane expression of, 168
religious community, 224, 267–69, 279, 287–88
religious development, 177
    cognitive processes linked to, 332
    identity formation and, 331
religious experience, 70, 167–69
religious faith, 111–16, 210–28
religious fathers, 122
religious foundations, 257
religious identifications, 198, 277
religious identity
    ethnic identity and, 277
    spiritual identity *vs.*, 75, 93–95
religious judgment, 177, 187
religious motives, 259
religious organizations, 255, 259
religious practices
    spirituality and, 288
    well-being and, 145–60
religious rites of passage, 62
religious rituals, 69
religious traditions, 57, 83
religious values, 37, 274–76
religious worldview among U.S. adolescents,
    88–89
religious youth, 58, 65, 220–21
research studies
    on adolescence, 95, 159, 197, 213–28, 262, 270,
        299–301, 305–18, 326–27
    behavior-genetic, 111, 116
    on catalysts of spiritual development, 98
    on child and adolescence development, 307
    on church membership, 331
    community-based approach to, 308–10
    cross-cultural, 311, 317
    cross-sectional, 12
    culturally sensitive, 305, 313
    ecological framework in, 318
    on emerging adults beliefs in afterlife, 231–42
    emotion regulation, 149–50, 152, 155
    on ethnic identity, 270–80
    on ethnically diverse communities, 308–18
    ethnographic, of Cambodian Americans in
        Boston, 265
    on face perception, 132–33
    future, 11, 75, 96–98, 159, 226, 262, 278–81, 334
    on genetics of faith, religion and spiritual-
        ity, 118
    heritability, 111–13
    on identity, 270, 279–80, 292
    immigrant civic engagement and religion, 250

individual characteristics and meanings in, 298
on infant attachment to caregivers, 334
instrumentation issues in spirituality, 315
on Laotian Refugee in Louisiana, 269
on Latinos from Richmond Virginia, 270
linguistic differences in, 314–15
literature on identity in adolescence, 292
longitudinal, 11, 20, 70, 159, 327
on meditation, 156
methodological issues in, 295
molecular-genetic, 113–14
neural, 136
Pakistani adolescence in U.K., 270
Pew Internet and American Life project, 342
population-genetic, 114–16
projects for spiritual development, 16
psychological, 176
on purpose and spirituality in adolescence, 211, 213–28
qualitative, 199, 299
questions for, 68–70, 297–300, 312–13
on religion's influence on smoking and alcohol consumption, 122
religious/spiritual and ethnic identities, 262–81
sensitivity within, 19
on spiritual and nonspiritual children, adolescence and adults, 300–301
of spiritual development, xii, 20, 98
on spirituality and PYD, 71, 131, 299, 305
on spirituality and religion, 197, 301, 307–8, 334–35
spirituality, with ethnically diverse youths, 305–18
on spirituality/religiosity, 197, 301, 307–8, 334–35
theory and measurement development in, 335
on twins, 112
on well-being, 152
youth engagement in, design, 70
on Zora, 349
researchers, 252–53
in ethnic minority communities, 310
neuroscience, 146
psychology, 146, 270
timing differences in collection of data by, 313
resistance-to-peer-influence (RPI), 133–34
Richards, T. A., 153
ritual, spiritual and religious, 69
Robinson, L., 270
Roeser, R., 326
Rogoff, B., 295

role models, 68–69, 97–98, 140
Rome, 51
Ross, J. M., 225–26
Rutter, Michael, 120
Ryff, C. D., 151, 212

Sabbath, 340–41
the sacred, 84, 91
SAGE (Storytelling Agent Generation Environment), 340–41, 343–47, 357
Saroglou, V., 95
Sawyer, Keith, 183
Scarlett, George, 193n4
Scheffield-Morris, A., 307
Schleef, D., 270
Schoenrade, P., 225
science, 37
science and religion, 35, 42–43
"science of genera," 37
The Search Institute, vii, 11, 212
self
    BLoS model and theories of, 79
    social and spiritual, 76
    transcendence of genes and, 113
self-concept, 66
self-images, sources of, 98
self-interest, 7
self-preservation system, 203
Seligman, M. E. P., 152, 212
sequences, universal and nonuniversal, 178
Sherrod, L. R., 250
Shiota, M. N., 152
Showalter, S. M., 214
Silberman, I., 91
Silk, J. S., 152
Simonton, Dean Keith, 183
Smith, C., 11, 34, 83, 88–89, 242
Smith, Huston, 207
Smith, W. C., 47
Smith-Hefner, N., 265
social behavior imitation, 131–32
social context, 293, 299, 307
    spirituality as, 58–59
    survival and, 13
    transcendent experience ideology and, 69
social group, 273–74, 278
social inequality, 128
society and spirituality, 64
Society for Research in Adolescence, ix
Society for Research in Child Development, ix
socioeconomic status (SES) on health, 131
soul, 34, 76, 109

South African tribe, Lemba, 114–15
Speed, Joshua F., 44–45
Spencer, M. B., 299
spirit, vii, ix, 37, 39, 236
  of childhood, 32
  definition of, 16, 25
  ground of, 31
  human, 32
  life and, 35–36
  ontology of, 26–28
  reality, thriving and, 25–26
  of religion, 26
  science of, x
  of spiritual development, 25–39
  of youth, 34–37
spiritual anchors, 63–65
spiritual beliefs and ECM, 200–201
spiritual development, vii, 7–15, 16, 26, 39, 79, 87,
    185–89. *See also* spirituality
  in adolescence, viii, xi, 198
  in childhood, viii
  conceptualizing, 92
  cultural analysis of, 265
  definition of, viii–ix
  developmental change's role in, 167–93
  domain of, 190–92
  dynamics in, ix
  ECM and, 203–4
  emotions and, 188
  fixed or static views *vs.*, 189–90
  of Lincoln, 44–46
  measure of, 52–53
  PYD and, 9–12, 42–54, 225, 293
  research studies on, xii, 20, 98
  spirit of, 25–39
  technologies to support children's, 339–57
  thriving and, 26, 53
spiritual domain and ideology, 80
spiritual exemplars, 59
spiritual experience, 70, 167–69
spiritual ideals, 84–86
spiritual identity, 74–99, 87, 262–81, 292–93
  defining, 79–93
  hallmark of, 86
spiritual knowledge, 14
spiritual life, 28
spiritual modeling, 59
spiritual practices, 62, 91–92, 145–60
spiritual research, 305–18
spiritual rituals, 69
spiritual role models, 68–69
spiritual self-identification, 83
spiritual stories, 91

spiritual systems, 68
Spiritual Transcendence Index, 68
spiritual values, 37, 64, 258, 274–76
spiritual virtues, 258
spirituality, 55, 83, 210, 273–74. *See also* religion;
    religiosity; spiritual development; spiritu-
    ality and religiosity
  adolescence and, xi, 15, 42–43, 55, 69, 75, 197,
    210–28
  character development and, 219–20, 223
  circles of care and, 197–98
  civic engagement and, 322–35
  conceptualizing, 99
  context of, 290–92
  contribution and, 216–17
  culture and, 285–301
  definitions of, 79–93, 288
  development of, 298
  ECM and, 197–208
  the ecological model and, 306–8
  emerging adults and, 265, 272, 274, 277
  ethnic identity and, 262–81
  ethnic-minority adolescence and community
    based approach to, 308
  experiences of, 70
  generosity and, 3–20
  genes and, 110–23
  heritability of, 112
  horizontal and vertical, 11, 14
  human development and, 92
  in human health, 79, 216–18
  identity and, 74–99, 87, 262–81, 292–93, 341
  as ideological context, 57
  ideological, social and transcendent dimen-
    sions within, 55–56, 67
  ideology and, 57, 67, 80
  immigrants and, 97–98
  literalist, 189
  model of positive and negative forms of,
    66–67
  nurturance of healthy forms of, 96
  ontogeny among, 16
  postmodern, 35
  in psychology, 301
  purpose and, 212–13, 218–19, 224–25
  PYD and, vii, xi, 3–20, 55–71, 69–70, 74–99,
    197–208, 285–301, 305, 322–35
  religion and, 55, 97–98, 197, 210, 264, 276–77,
    288
  religiosity *vs.*, 326–29
  religious faith and purpose in adolescence,
    210–28
  religious practices and, 288

research studies on, 16, 71, 98, 131, 197, 211,
213–28, 299, 301, 305–18, 334–35
resources, 67
roles of, 97
as social context, 58–59
social impact of, 58–59
society and, 64
as states of consciousness, 80
theistic and contemplative, 85
thriving and, 12–15
transcendence and, 61, 63
transitions in, 188–89
understanding, 99, 287
worldview offered by, 64
youth and, 64
spirituality and religiosity, 111, 115, 121, 211
in adolescence, viii, xi, 75, 198–99
biological covariates of, 9
civic engagement, PYD, 322–35
definitions of, 329
differentiating, 12, 198–99, 326–29
genes' role in, 123
in human health, 79, 216–18
neural mechanisms of, 9
receptivity to, 192
research on, 197, 301, 307–8, 334–35
spirituality research projects, 305–18
Stanford Center on Adolescence, 211
Steen, T. A., 152
Stein, N., 153
Steinberg, L., 152, 307
Stepick, A., 248
Stepick, C. D., 248
Strauss, A. L., 259n3
Stroop Color-Word test, 157
studies. *See* research studies
Sullivan, Anne, 33
survey
adolescence religious views, 34
future, on religiosity and purpose, 226
purpose in adolescence, 213–16
survival and social context, 13
Swiss Christian Students' Association (ACSE),
36, 43
Syed, Moin, 262, 272, 278

"taking of consciousness," 175, 182
technologies, 339–57
The Teen Outreach Program (TOP), 207
Templeton, Sir John, xiii, 16, 198, 205, 264
Theokas, C., 323–24
theories
attachment, 332, 335

classic attribution, 109
cognitive-development, 28
constructive learning, 348
developmental theories, 8, 331–34
of ego development, 19
grounded, 259–60n3
identity, 331–32
intuitive and nonuniversal, 29
of psychosocial development by Erikson, 331
PYD, 210, 213
in research studies and measurement development, 335
self-categorizing, 202
social identity, 263, 266
social learning, 332–33
sociocultural, 290, 292, 295
stage-structure, 93
Thomas, Lewis, 183, 185
Thompson, R. A., 146
thriving
neuropsychology of spirituality, relation,
12–15
ontogeny among spirituality, generosity and,
16
positive youth development and, 10
spirit, reality and, 25–26
spiritual development and, 26, 53
spirituality and, 12–15
youth(s), 210–11
Tibetan Buddhist lamas, 191
Tillich, Paul, 47
Time, 115
Trabasso, T., 153
transcendence, 30
communities and, 10
devotion, responsibility and commitment
inspired by, 62
of ego, 80, 93
experiences of, 56, 60, 62, 64, 66, 68–69
of genes and self, 113
horizontal, 7–8
ideology, social context and, 69
in nature, 61
Piaget, Lincoln, Hitler and, 53
problem of, 42–54
purpose as strength of, 212–13
and PYD, 60, 62
sacred and communal, 65
of self-interest, 130
the spiritual, index, 57, 68, 87
spirituality and, 61, 63
vertical, 8
youth and, 11, 13, 53, 58, 61, 68–69

"transcending self," 14
Trevarthen, C., 32, 39
Trevor-Roper, H. R., 49
Triandis, H. C., 308
Trimble, J. E., 314
Turkle, Sherry, 351
two-dimensional affective field, 32

U.S. *See* United States
UNICEF index, 128
United States (U.S.), 90, 128, 211, 213, 231, 248, 281
    adolescence in, 88–89, 94–95
    Hispanic community in, 306
    immigrants in, 153, 247
    minority adolescence in, 247, 305, 308–10, 314
universal-to-unique continuum, 184–85
University of North Carolina, 11

values, 37, 64, 258, 274–76
van Etten, M., 153
Ventis, W. L., 225
volitional necessity, 200, 204
volunteer service programs, 206–7
volunteerism, 248, 250
Vygotsky, Lev, 173–74, 181, 193n3, 295

Wagener, L. M., 214
Watson, James, 171, 184
Web site for Human Genome Project, 116
Weisberg, Robert, 183
well-being, 112–13, 128–29
    ECM and, 206
    emotion regulation and, 146, 150, 152–54, 158–59
    psychological, 151
    psychological functions of, 145–46
    religious/spiritual practices and, 145–60
    research studies on, 152
    subjective (SWB), 151, 154
    of youth, 128–29
Wilkinson, R. G., 128
Wilson, David Sloan, 203
Wohl, M. J. A., 205

Woollacott, M., 157
world change, 91
World War I, 43, 47
worldviews, 64, 87–91, 93–94, 98
Wright brothers, 186

Yang, 270
yoga, 178
"Young Adolescents and their Parents" (YAP), 11
Youniss, J., 58, 250
youth(s). *See also* adolescence; positive youth development
    American, 262
    assets, 212, 323–24
    civic engagement in, 324–26
    community and, 220–21
    contribution by, 6
    disenfranchised, 333
    engagement in research design, 70
    generosity of, 6
    Hitler's vision of, 50
    ideology and, 88
    immigrant, 98, 248–49
    literature on development of, 17
    National Study of, and Religion, 94–95, 232, 242
    in New York, 323
    optimal development of, 56
    policy development and PYD, 323
    purpose and, 17, 210–28
    religious, 58, 65, 220–21
    service activities by, 10
    spirit of, 34–37
    spirituality, 64, 198
    studies on, 70, 210, 262, 265, 305–18, 349
    thriving, 210–11
    transcendence and, 11, 13, 53, 58, 61, 68–69
    well-being of, 128–29
    Zora and, 349

Zarrett, N., 324
Zhou, M., 268
Zora, 339, 341, 347–51, 356